AF744461

ADVANCED MAGNETIC MATERIALS

Edited by **Leszek Malkinski**

Advanced Magnetic Materials
Edited by Leszek Malkinski

Published by InTech
Janeza Trdine 9, 51000 Rijeka, Croatia

Copyright © 2012 InTech
All chapters are distributed under the Creative Commons Attribution 3.0 license, which allows users to download, copy and build upon published articles even for commercial purposes, as long as the author and publisher are properly credited, which ensures maximum dissemination and a wider impact of our publications. After this work has been published by InTech, authors have the right to republish it, in whole or part, in any publication of which they are the author, and to make other personal use of the work. Any republication, referencing or personal use of the work must explicitly identify the original source.

As for readers, this license allows users to download, copy and build upon published chapters even for commercial purposes, as long as the author and publisher are properly credited, which ensures maximum dissemination and a wider impact of our publications.

Notice
Statements and opinions expressed in the chapters are these of the individual contributors and not necessarily those of the editors or publisher. No responsibility is accepted for the accuracy of information contained in the published chapters. The publisher assumes no responsibility for any damage or injury to persons or property arising out of the use of any materials, instructions, methods or ideas contained in the book.

Publishing Process Manager Romina Skomersic

Technical Editor Teodora Smiljanic

Cover Designer InTech Design Team

Additional hard copies can be obtained from orders@intechopen.com

Advanced Magnetic Materials, Edited by Leszek Malkinski
p. cm.
ISBN 978-953-51-0637-1

We are IntechOpen, the world's leading publisher of Open Access books

Built by scientists, for scientists

4,000+
Open access books available

116,000+
International authors and editors

120M+
Downloads

151
Countries delivered to

Our authors are among the
Top 1%
most cited scientists

12.2%
Contributors from top 500 universities

CLARIVATE ANALYTICS
BOOK CITATION INDEX
INDEXED

WEB OF SCIENCE™

Selection of our books indexed in the Book Citation Index
in Web of Science™ Core Collection (BKCI)

Interested in publishing with us?
Contact book.department@intechopen.com

Numbers displayed above are based on latest data collected.
For more information visit www.intechopen.com

For EU product safety concerns:
IN TECH d.o.o., Prolaz Marije Krucifikse Kozulić 3, 51000 Rijeka, Croatia,
info@intechopen.com or visit our website at intechopen.com.

Meet the editor

Dr Leszek Malkinski received his Ph.D. degree in Physics in 1991 from the Institute of Physics of the Polish Academy of Sciences for his study of magnetoelastic effects in amorphous alloys. He gained his experience in magnetic materials working as a visiting scholar in the leading magnetic laboratories in Poland, Austria, Germany, Hungary, Italy, Switzerland and the U.S.A. Since 2002 he has been with the Department of Physics and the Advanced Materials Research Institute of the University of New Orleans, U.S.A where he currently holds a position of Associate Professor of Physics and Materials Science. He is an editor of "Nomagnetism" and a co-author of more than 100 peer-reviewed articles on magnetoelastictiy, exchange bias, spintronics, multiferroics and spin dynamics.

Contents

Preface

Recent progress in information technology, wireless communication, biotechnology and microelectronics requires advanced technologies and new magnetic materials to meet demands of modern devices. This collection of eight chapters provides an up-to-date review of recent trends and developments in technology, characterization methods, theory and applications of modern magnetic materials with original, never published contributions from the renowned scientists in the field of magnetism. The book is addressed to a diverse group of readers which include students, engineers and researchers in the fields of physics, chemistry, bioengineering, electronics and materials science, who wish to enrich their knowledge about advanced magnetics. Depending on the discipline represented by the readers they are invited to read entire book or select chapters of particular interest. In order to help with the choice of appropriate chapters below I summarize their content:

Chapter 1. The first chapter reports on fabrication method of amorphous magnetic nanowires with the glass coating using rapid solidification of the melt. The original measurement methods and magnetic properties of the nanowires with the diameters ranging from 90 nm to 13 μm are presented. In particular, the article targets the effect of fabrication conditions and post-fabrication treatment on mobility and velocity of domain walls in the nanowires, which can be used in future racetrack memories, magnetic domain wall logic devices, domain wall diodes and oscillators, and other devices.

Chapter 2. Future microwave devices for wireless communications require extended frequency range. This chapter treats about new developments in technology and application of a new microwave material - barium hexagonal ferrite films with extremely high anisotropy, large saturation magnetization and low losses in the microwave range. The author demonstrates original results regarding prototype notch filters and phase shifters operating at millimeter range wavelengths which employ the hexagonal ferrite films.

Chapter 3. Spin electronics (or spintronics) is an emerging field of science which takes advantage of magnetic moments to build nonvolatile random access memories and other digital devices. In order to compete and eventually replace semiconductor-based electronics, the spintronic devices must use materials with nearly 100% spin polarization. The magnetite is one of the best candidates for the spintronic applications. However, its performance in existing devices is drastically reduced by the atomic structure at the surface which differs from that of the bulk. This chapter presents studies of surface reconstruction of the magnetite and describes methods for increasing spin efficiency in spintronic devices by preserving the Fe_3O_4 structure of the surface.

Chapter 4. Multiferroic composites consisting of ferromagnetic and ferroelectric materials provide a unique way of converting magnetic field into electric field. This process involves stresses at the interface between these materials and takes advantage of the piezoelectricity of the ferroelectric and the magnetostriction of the ferromagnetic phase. This chapter describes original methods of synthesis of multiferroic core-shell nanoparticles and indicates entirely new biomedical applications of such nanoparticles. Using external alternating magnetic fields it is possible to produce local electric fields near multiferroic nanoparticles which can control opening and closing of voltage-gated ion channels in mammalian cells. Ion channels are involved in generation and propagation of action potentials in nerves and their malfunction can lead to multiple diseases such as cystic fibrosis, diabetes, cardiac arrhythmias, neurological disorders or hypertension. The proposed mechanism has also potential for a new method of cancer treatment.

Chapter 5. One of challenges in the design of portable electronic devices is the optimization of on-chip inductors for the power conversion. This optimization takes into account multiple factors, such as size, a choice of fabrications method, frequency range of operation, energy losses and cost of the device. This chapter provides detailed discussion about theory, design, fabrication methods and measurements of essential parameters characterizing electroplated micro-inductors for DC-DC conversion operating with the switching frequency up to 100 MHz.

Chapter 6. The focus of this chapter is on the relation between magnetic properties and disorder in the Fe-Al system. Depending on the composition and the microstructure, different magnetic and structural orders can exist in this system including a spin-glass order. Detailed experimental studies based on magnetometry, the Mössbauer effect and X-ray diffractometry, as well as theoretical models of this system show that the atomic disorder, which can be controlled by a mechanical treatment or an annealing, leads to significant increase in the lattice parameters and the magnetization compared to those in the ordered structures. The contribution of disorder to the magnetism of these alloys depends on the Fe content of the alloy and is the largest close to the equi-atomic FeAl alloy, but in $Fe_{75}Al_{25}$ alloy it is similar to the one given by the volume change.

Chapter 7. Direct characterization of the magnetic material parameters based on measurements of magnetic devices with complicated geometry is a complex problem. This chapter proposes the state-of-the-art methodology to extract materials characteristics from the measurements of electromagnetic devices. This original algorithm is robust because all uncertainties present in electromagnetic devices are taken into account in a stochastic framework. The algorithm is validated by applying it for the identification of the magnetic properties of the material in an electromagnetic device - an electromagnetic core inductor.

Chapter 8. The last chapter tackles the problem of modeling of magnetic properties. Macroscopic samples and devices consist of a large number of magnetic domains which interact and evolve during magnetization process. Therefore, modeling of magnetic hysteresis is one of the most difficult tasks in magnetism. This chapter provides a detailed description of a new model for modeling of magnetization processes in magnetic materials. It is based on a stochastic Preisach approach and uses hyperbolic analytical approximation of the Everett integral. The theory offers modeling of coercive force, remanent magnetization, hysteretic losses, Barkhausen effect, eddy current losses, exchange bias and other essential properties of the magnetic materials. It is applicable to a large variety of magnetic materials.

Research on magnetic materials is a dynamically expanding field of science and this book demonstrates some highlights rather than complete overview of this vast discipline. I hope that reading this book will spark interest of the readers and encourage them to study magnetism of advanced materials. It will also bring better understanding of challenging technologies which become a part of our life, whenever we use computers, cellphones and other modern devices.

Dr. Leszek Malkinski
Associate Professor of Physics and Materials Science,
Department of Physics and the Advanced Materials Research Institute,
University of New Orleans, New Orleans, Louisiana,
USA

1

Rapidly Solidified Magnetic Nanowires and Submicron Wires

Tibor-Adrian Óvári, Nicoleta Lupu and Horia Chiriac
National Institute of Research and Development for Technical Physics Iaşi
Romania

1. Introduction

Magnetically soft amorphous glass-coated microwires are suitable for numerous sensor applications. Their typical dimensions - metallic nucleus diameter of 1 to 50 μm and glass coating thickness of 1 to 30 μm - make them promising candidates for high frequency applications, especially given their sensitive giant magneto-impedance (GMI) response in the MHz and GHz ranges (Torrejón et al, 2009). The magnetic properties of amorphous microwires are determined by composition, which gives the sign and magnitude of their magnetostriction, as well as by dimensions - metallic nucleus diameter, glass coating thickness, and their ratio - which are extremely relevant for the level of internal stresses induced during preparation. The magneto-mechanical coupling between internal stresses and magnetostriction is mainly responsible for the distribution of anisotropy axes and domain structure formation. Microwires generally display a core-shell domain structure in their metallic nucleus, with orthogonal easy axes, e.g. axial in the core and circumferential or radial in the shell, as schematically shown in Fig. 1.

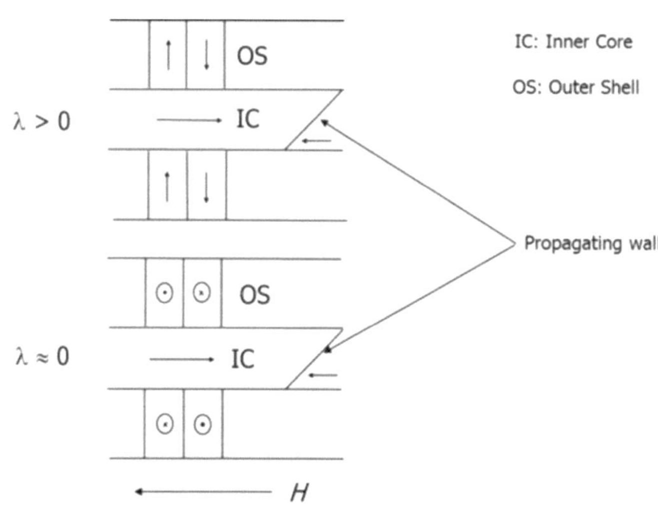

Fig. 1. Typical core-shell domain structures in amorphous glass-coated microwires with positive ($\lambda > 0$) and nearly zero magnetostriction ($\lambda \approx 0$), respectively.

An axially magnetized core, usually encountered in amorphous microwires with large and positive magnetostriction, but also in nearly zero magnetostrictive ones if their nucleus diameter is larger than 20 μm (Chiriac et al., 2007a), leads to the appearance of the large Barkhausen effect (LBE), that is a single step reversal of the magnetization in the core when the sample is subjected to a small axial magnetic field. LBE takes place through the propagation of a pre-existent 180° domain wall from one microwire end to the other, as illustrated in Fig. 1.

Ferromagnetic nanowires are aimed for novel spintronic applications such as racetrack memory, magnetic domain wall logic devices, domain wall diodes and oscillators, and devices based on field or spin-current torque driven domain wall motion (Allwood et al., 2005; Finocchio et al., 2010; Lee et al., 2007; Parkin et al., 2008). These applications require nanowires with characteristics that can be accurately controlled and tailored, and with large domain wall velocities, since the device speed depends on domain wall velocity. At present, spintronic applications which require magnetic nanowires are based on planar nanowires prepared by expensive lithographic methods (Moriya et al., 2010).

Recently, the large values of domain wall velocity reported in amorphous glass-coated microwires have offered new prospects for the use of these much cheaper rapidly solidified materials in spintronic applications, subject to a significant reduction in their diameter (Chiriac et al., 2009a). The amorphous nanowires are composite materials consisting of a metallic nucleus embedded in a glass coating prepared in a single stage process, the glass-coated melt spinning, at sample lengths of the order of 10^4 m (Chiriac & Óvári, 1996). In order to overcome the experimental difficulties related to the fabrication of such ultra-thin wires and to drastically reduce the typical transverse dimensions of microwires (1 to 50 μm for the metallic nucleus diameter), the apparatus used for the preparation of the rapidly solidified nanowires has been significantly modified. These efforts have led to the successful preparation and characterization of rapidly solidified submicron wires with the metallic nucleus diameter of 800 nm, reported less than 2 years ago (Chiriac et al., 2010). Figure 2 (a) shows the SEM images of a submicron amorphous wire with the nucleus diameter of 800 nm, whilst Fig. 2 (b) illustrates the optical microscopy image of the submicron amorphous wire in comparison with two typical amorphous microwires with the nucleus diameters of 4.7 and 1.8 μm, respectively. These results have opened up the opportunity to develop nanosized rapidly solidified amorphous magnetic materials for applications based on the domain wall motion.

This first success has been shortly followed by the preparation and characterization of amorphous glass-coated submicron wires with metallic nucleus diameters down to 350 nm (Chiriac et al., 2011a), in which domain wall velocity measurements have also shown very promising results (Óvári et al., 2011).

The well-known methods employed in the experimental studies have been extensively modified in order to allow one to perform complex measurements on such thin wires, especially due to the high sensitivity required to measure a single rapidly solidified ultra-thin wire (Corodeanu et al., 2011a).

Following the same path, we have been able to produce rapidly solidified amorphous nanowires through an improved technique. The diameters of the as-quenched nanowires were ranging from 90 to 180 nm (Chiriac et al., 2011b). These new materials are useful for

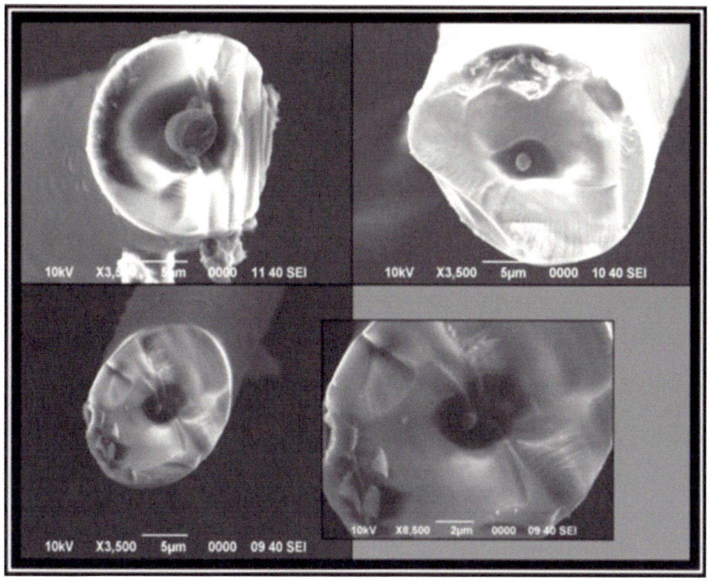

Fig. 2a. SEM images of a submicron amorphous wire with the nucleus diameter of 800 nm.

Fig. 2b. Optical microscopy images of the submicron amorphous wire in comparison with two typical amorphous microwires with the nucleus diameters of 4.7 and 1.8 μm, respectively.

applications in both domain wall logic type devices and in novel, miniature sensors. The accurate control of the domain wall motion could be performed without irreversible modifications of the wire geometry, as recently pointed out (Vázquez et al., 2012). Nevertheless, there are several issues to be addressed before these new materials can reach their full practical potential: their integration in electronic circuits, the use of lithographic methods to prepare the miniature coils required to inject and trap domain walls, the clarification of the role of glass coating and whether or not it should be kept, removed or just partially removed - and in which stages of the device development, issues related to the manipulation of wires with such small diameters, etc.

Figure 3 shows two SEM micrographs of a glass-coated $Fe_{77.5}Si_{7.5}B_{15}$ amorphous magnetic nanowire with the metallic nucleus diameter of 90 nm and the glass coating of 5.5 μm, taken at different magnifications.

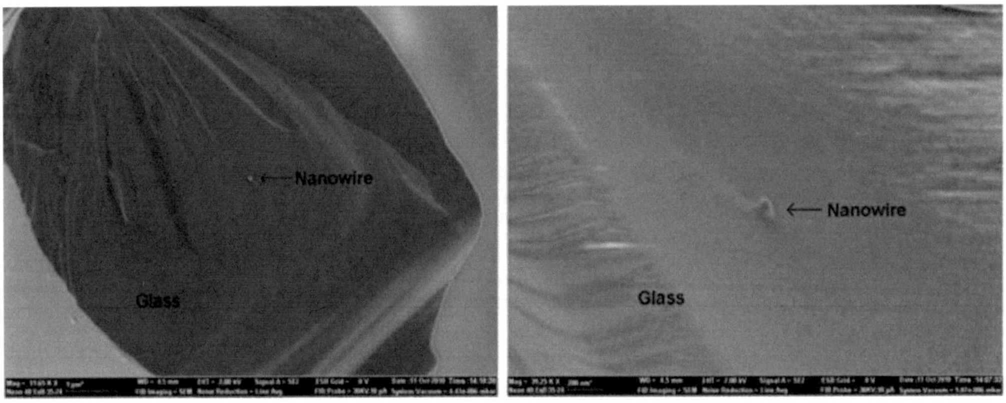

Fig. 3. SEM micrographs at two different magnifications of a rapidly solidified amorphous nanowire with positive magnetostriction having the metallic nucleus diameter of 90 nm and a glass coating thickness of 5.5 μm.

A new method for measuring the domain wall velocity in a single, ultrathin ferromagnetic amorphous wire with the diameter down to 100 nm has been developed in order to measure such novel nanowires (Corodeanu et al., 2011b). The method has been developed in order to increase the sensitivity in studying the domain wall propagation in bistable magnetic wires in a wide range of field amplitudes, with much larger values of the applied field as compared to those employed when studying the wall propagation in typical amorphous microwires. The newly developed method is especially important now, when large effort is devoted to the development of domain wall logic devices based on ultrathin magnetic wires and nanowires.

Besides the spintronic applications, the investigation of rapidly solidified amorphous submicron wires and nanowires is aimed towards the understanding of the changes in the magnetic domain structure, which makes the bistable behavior possible, and in the switching field, at submicron level and at nanoscale.

2. Experimental techniques for the characterization of rapidly solidified amorphous nanowires and submicron wires. Domain wall velocity measurements

2.1 Magnetic characterization

Given the ultra-small diameters of rapidly solidified submicron wires and nanowires (metallic nucleus diameters between several tens of nanometers and hundreds of nanometers), the use of the classical characterization techniques employed for typical microwires with diameters between 1 and 50 μm (Butta et al., 2009; Kulik et al., 1993) in order to measure their basic magnetic properties, e.g. to determine their magnetic hysteresis loops, is not viable due to the low sensitivity and signal-to-noise ratio (SNR). Therefore, in order to investigate the magnetic properties of a single ultrathin magnetic wire, a reliable measuring system has been developed (Corodeanu et al., 2011a). The new procedure has been employed to measure a single ultrathin magnetic wire, i.e. a submicron wire or a nanowire, using a digital integration technique. The new experimental set-up has been developed in order to increase the sensitivity and to extract from the noisy signal a reliable low frequency hysteresis loop for a single submicron wire or nanowire.

The main components of the measuring system used in the experiments are: the magnetizing solenoid, the system of pick-up coils, a low-noise preamplifier, a function generator, and a data acquisition board. A schematic of the system is shown in Fig. 4.

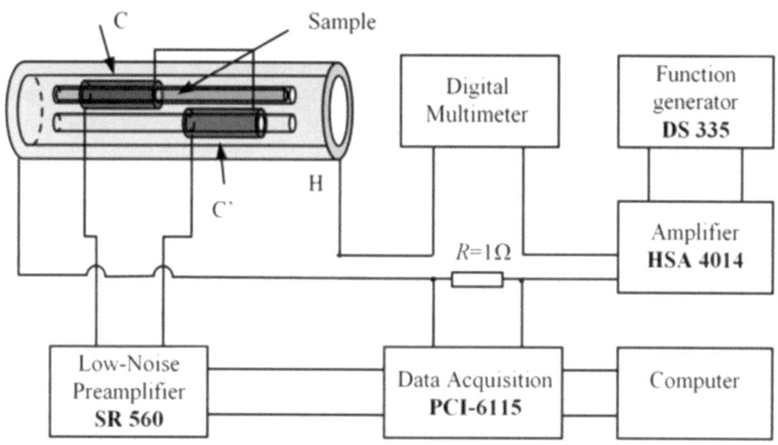

Fig. 4. Schematic of the experimental system employed for the magnetic characterization of rapidly solidified amorphous nanowires and submicron wires.

The magnetizing solenoid is powered by a Stanford Research DS 335 function generator through a high power bipolar amplifier HSA 4014, being capable of generating magnetic fields up to 30,000 A/m. Two pick-up coils connected in series-opposition are used in order to avoid any induced voltage in the absence of the sample. Each pick-up coil is 1 cm long and has 1,570 turns, wound with enameled 0.07 mm copper wire on a ceramic tube with an outer diameter of 1.8 mm and an inner diameter of 1 mm. A 1 Ω resistor (R) is used to provide a voltage proportional to the applied magnetic field. The voltage induced in the pick-up coil is amplified up to 50,000 times using a Stanford Research SR560 low-noise preamplifier in order to obtain a measurable value of the induced voltage and a high SNR. The voltage drop on the resistor R and the amplified induced voltage from the pick-up coil system are digitized using a National Instruments PCI-6115 four channels simultaneous data acquisition board. The acquisition of the signals was done using a sampling frequency between 800 kHz and 10 MHz (with 5,000 to 62,500 points/loop at 160 Hz). The acquired signals have been processed using LabVIEW based software.

Two methods have been employed to measure the hysteresis loops. For the first one, it was necessary to make an average over a large number of acquired signals, while for the second one only two recordings of the signal were required (with and without the sample), followed by digital processing to trace the hysteresis loop.

For the first method, in order to extract the useful signal from the noisy one, two sets of data have been acquired. First, the signal from the pick-up coil system with no sample in it has been acquired; this 'zero signal' contains information about any possible miss-compensation of the pick-up coils either due to the imperfect winding of the pick-up coils or of the magnetizing solenoid. It is necessary to mention at this point that the field generated by the magnetizing solenoid cannot be perfectly uniform. The non-uniformity of the field together with the imperfections in the winding of the pick-up coils will affect the shape of the small induced signal. This effect of the measuring system on the induced signal has to be removed in order to obtain the clearest possible signal from the sample. Subsequently, the induced signal has been digitally integrated, and the integrated signal was averaged over a large number of measurements. In this way, an apparent hysteresis loop of the system with no sample was recorded. An external trigger has been used to avoid any phase mismatch when averaging.

Averaging was done over the integrated signals rather than the induced signals. This was due to the very small width of the peaks (of the order of 10 μs) in the induced signals, as well as due to the fact that they were not always in the exact same place (magnetization reversal does not always occur at the same value of the field). Averaging such signals would cancel the sample signal together with the noise.

The next step was to insert the sample, using a glass capillary, inside one of the pick-up coils, taking care to avoid any displacements of the coils within the system. The induced signals have been measured again. The corresponding signals were integrated and the integrated signal was averaged as in the previous case in order to obtain the apparent loop of the system with the sample.

The intrinsic hysteresis loop of the sample is obtained in this method by subtracting the apparent loop without the sample from the apparent loop with the sample. A low pass filter has been employed for further noise reduction, taking care not to alter the shape of the sample peaks.

The magnetization process of these materials results in a square hysteresis loop since the magnetization reversal takes place in a single step. Therefore, the induced signal displays a peak, and should be null in rest (zero induced signal since magnetization does not change). Based on this, a second, faster method is proposed to obtain a less noisy hysteresis loop.

In this second method, the signals induced with and without sample are acquired only once. The signal without sample is subtracted from the signal with sample. The SNR is still too small to obtain a good integration. Therefore, a window method has been employed (Butta et al., 2009). Considering that everything outside the peak area of the signal must be zero, two windows were used to select the peak area from the sample signal while the rest of the noisy signal was numerically forced to become null. Using digital integration of the windowed signal and an accurate selection of the peaks, the hysteresis loop of the sample is obtained.

First, a wider range which includes the peak area is selected. The hysteresis loop which corresponds to this selection displays a noisy jump in magnetization, as if the sample magnetization values would be larger and then smaller than the actual values. Therefore, the selection range is progressively reduced in several steps, with integration being performed at each step, in order to reduce the noise as much as possible in the region of the magnetization jump. Special care was taken to avoid cutting the peaks. The selection accuracy does not affect the measured value of coercivity and only slightly influences the magnetization (less than 5% for the thinnest sample which has been used in the experiments).

Figure 5 shows the hysteresis loops of the same nanowire obtained through both methods.

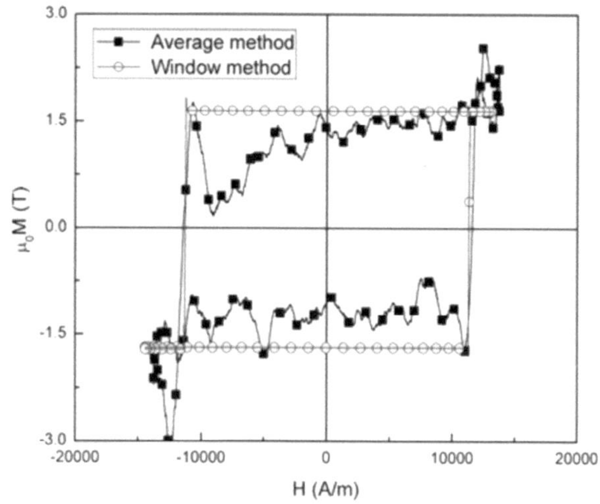

Fig. 5. Hysteresis loop of a 133 nm $Fe_{77.5}Si_{7.5}B_{15}$ nanowire covered by 6 μm of glass measured using both methods (averaged and window).

Thus, a reliable method for the precise magnetic measurement of ultrathin wire shaped samples, e.g. single nanowire, has been developed. The combination of the two methods proposed for hysteresis loop measurements leads to an accurate characterization of materials such as submicron wires and nanowires with diameters down to 100 nm: the first method provides information about the profile of the hysteresis loop and magnetic behavior of the sample (bistable or not), while the second one removes almost all the noise resulting in a valid noise-free loop.

2.2 Domain wall velocity measurements

Magnetic bistability is one of the key characteristics of amorphous glass-coated submicron wires and nanowires which make them important for applications. The magnetic bistable behavior represents the one-step reversal of the magnetization along such samples at a certain value of the applied magnetic field, value which is called switching field (Komova et al., 2008). The actual reversal consists in the displacement of a 180° domain wall along the entire length of the sample. The characteristics of the wall propagation, especially its velocity, are essential for the properties of the domain wall logic devices which could be developed. Therefore, it is extremely important to measure the domain wall velocity and its field dependence with high accuracy, in order to determine the wall mobility and to correctly predict characteristics such as operating speed of the future devices.

Therefore, the development of a new method for measuring the domain wall velocity in a single magnetic wire with dimensions ranging from those of a typical microwire (1 - 50 μm) to those of a submicron wire (hundreds of nm) and further down to a nanowire (100 nm) was required. Such a method was also necessary due to the increment of the field range in which the wall velocity needs to be measured. The new experimental set-up was developed in order to increase the sensitivity and to study the domain wall mobility and damping mechanisms in bistable magnetic wires in a wide range of the applied field amplitude.

The main problem addressed with the proposed method of measuring the domain wall velocity is related to two important factors which change drastically as the wire diameter decreases from the range of microns to that of submicrons and further down to nanometers: the wall velocity values are very large and the propagation fields become extremely large. Due to these two reasons, the existing measuring methods are inefficient in providing accurate values for the domain wall velocity, mainly due to the nucleation of additional domain walls which propagate among the pick-up coils, rendering the whole measurement incorrect.

The measurement of the domain wall velocity is based on the classical method developed by Sixtus and Tonks (Sixtus & Tonks, 1932). The original method has been improved by various authors, e.g. (Hudak et al., 2009), in order to study the wall propagation in different types of materials under various circumstances. A schematic diagram of the new experimental set-up proposed in this work is shown in figure 6. The experimental set-up consists of a long solenoid (37 cm long, 2 cm in diameter, 2335 turns with a field to current constant of 6214 Am^{-1}/A) powered by a Stanford Research Systems DS 335 function generator through a high power bipolar amplifier HSA 4014, and two compensated systems of four pick-up coils placed within the solenoid. Each pick-up coil system consists of four identical coils Cx and Cx' (x = 1, 2, 3, 4) connected in series-opposition in order to obtain a compensated system able to provide only the sample signal and almost zero signal in the absence of the sample. The compensation is especially important for measurements performed on submicron wires and nanowires due to the small sample induced signal relative to the field induced one in the case of non-compensated systems.

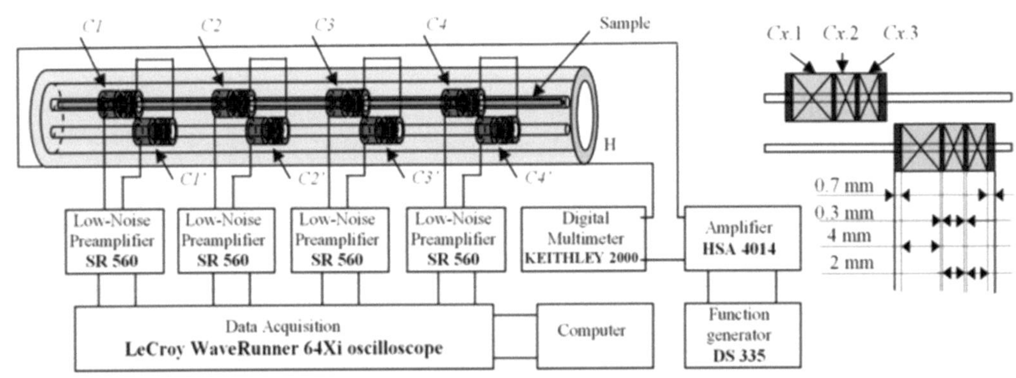

Fig. 6. Schematic of the system of four pairs of compensated pick-up coils.

Each pick-up coil is composed of three windings: a 4 mm long one with 1800 turns (Cx.1) and two 2 mm long ones with 800 turns (Cx.2 and Cx.3) placed close to the first one on the right side, all of them wound with enameled 0.07 mm copper wire on a ceramic tube with the outer diameter of 1.8 mm and the inner one of 1 mm. Cx.1, Cx.2 and Cx.3 are connected in series in the given order, with Cx.2 being wound in the opposite direction as compared to Cx.1 and Cx.3 in order to create a clear separation between the peaks induced in Cx.1 and

Cx.3. In order to reduce the self-oscillation of the system, an appropriate kilo-ohm resistor has been connected in parallel with each winding (4 kΩ for *Cx*.1 and 2 kΩ for *Cx*.2 and *Cx*.3, respectively). The pick-up coils are placed at certain distances among them (1 to 4 cm - depending on the wire characteristics).

There have been four pick-up coils used to measure the domain wall velocity in order to detect if any additional domain walls are nucleated in the measuring space and to obtain this way an accurate value of the wall velocity and not just an apparent one. The compensated system of pick-up coils can be employed to measure the wall velocity for bistable wires with diameters from tens of micrometers down to 100 nanometers (the smallest wire diameter tested) and for large values of the magnetizing field. For small wire diameters (below a few μm) the signals from the pick-up coils were amplified up to 50,000 times using four Stanford Research Systems SR560 low noise preamplifiers in order to obtain a measurable value of the induced voltage and a good signal to noise ratio.

The amplified signals were digitized using a four-channel LeCroy WaveRunner 64Xi oscilloscope, each of the four pick-up coils being connected to an input channel of the oscilloscope (*Cx* to input *Ix* , where $x = 1, 2, 3, 4$). An external trigger has been used in order to synchronize the acquired sample signal with the driving field. The current passing through the magnetizing solenoid (sinusoidal with a frequency of 160 Hz) was measured using a Keithley 2000 multimeter.

The acquired signals were processed using LabVIEW based software.

The inductive method is the most employed and straightforward technique used to measure the domain wall velocity in bistable microwires (Chiriac et al., 2009b; Garcia-Miquel et al., 2000; Ipatov et al., 2009). Various measuring configurations with two, three or four measuring points on the wire length were previously reported, each of them being a step forward for an enhanced and more precise measurement of the domain wall velocity in this type of wires.

The system with only two pick-up coils is not the most adequate for wall velocity measurements, since in the case of the wire, additional domain walls can nucleate at both ends of the wire and even at different points on the wire length when the driving field is large enough. Therefore, the signal picked up from one of the two coils is not precisely determined to be the result of the same domain wall as the signal picked up by the other coil. An apparent higher velocity than the real one can be recorded in this case.

Other measuring systems consist of four pick-up coils distributed on the wire length at a certain distance among them (Chiriac et al., 2008). This configuration provides information about the direction of the propagating wall. It also allows one to measure three values of the wall velocity and, if all of them are equal, then it is clear that there is a single wall propagating within the wire and the recorded wall velocity is the real one and not an apparent one. The main disadvantage of such a system is the impossibility to exactly identify the direction of the domain wall displacement through each coil.

This shortcoming has been solved by a recently proposed configuration which includes four pairs of pick-up coils which allow one to identify the direction of the wall propagation when it passes through each pair (Chiriac et al., 2009b). The main disadvantages of this set-up appear when the velocity is measured at large values of the applied field and/or when the measured wire has such a small diameter that the sample induced signal is much smaller than the signal induced by the external field. In these cases, it is practically impossible to distinguish the sample-generated peaks in the recorded signal.

To overcome this matter, the following solution has been developed: two identical strips with four pairs of pick-up coils have been made according to the description given in (Chiriac et al., 2009b), with 2000 turns for the large coil and 800 turns for the smaller one, in order to increase the sensitivity. Each pair of coils from the first strip has been connected in series-opposition with a pair of coils from the second strip in order to cancel the signal induced by the applied field. The recorded signal had two peaks. However, this solution cannot be employed in the case of the ultrathin wires such as submicron wires and nanowires, because in this case the increment of the number of turns for each coil (made to increase sensitivity) leads to the impossibility to separate the two peaks. To overcome this new problem and to detect the direction of the domain wall displacement through the pick-up coil system, the measuring system has been improved with pick-up coils having three windings.

Figure 7 shows the compensated signals from each winding ($V_{Cx.1}$, $V_{Cx.2}$, $V_{Cx.3}$ – Figs. 7 a, b, c), the composed signal for all three windings $V_{Cx.1+2+3}=(V_{Cx.1})+(V_{Cx.2})+(V_{Cx.3})$, (Fig. 7 d), and the composed signal for the two windings wound in the same direction $V_{Cx.1+3}=(V_{Cx.1})+(V_{Cx.3})$, (Fig. 7 e), as they result from the separately acquired signals (x = 1, 2, 3, 4 – the number of the composite pick-up coil) for an $Fe_{77.5}Si_{7.5}B_{15}$ glass-coated microwire with the metallic nucleus diameter of 30 μm and the glass coating thickness of 25 μm. Signals have been acquired for two amplitudes of the applied field: 2 kA/m and 20 kA/m. The largest signal is given by *Cx*.1 and therefore its maximum is used as the marker for velocity calculations. The signal given by *Cx*.3 is smaller and is used to determine the direction of the wall movement in the pick-up coil system. For small values of the applied field and small

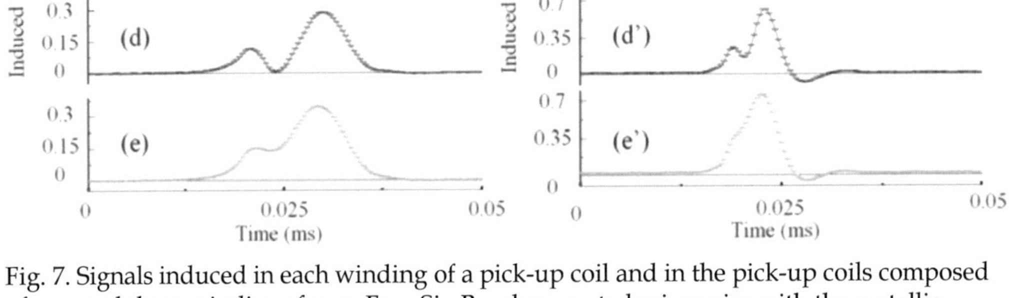

Fig. 7. Signals induced in each winding of a pick-up coil and in the pick-up coils composed of two and three windings for an $Fe_{77.5}Si_{7.5}B_{15}$ glass-coated microwire with the metallic nucleus diameter of 30 μm and the glass coating of 25 μm for two values of the applied field amplitude (2 kA/m and 20 kA/m): a, a') signal in *Cx*.1; b, b') signal in *Cx*.2; c, c') signal in *Cx*.3; d, d') signal in the pick-up coil made of three windings; e, e') signal in the pick-up coil with two windings.

wall velocity values the system with pairs of two windings is enough to accurately detect the direction of the wall movement (Fig. 7 e).

The difficulties appear when the applied field and wall velocity increase and a net distinction between the positive peaks is no longer possible (Fig. 7 e'). By adding a new winding (*Cx*.2) in series-opposition with *Cx*.1 and *Cx*.3 (physically placed between these two) a net difference between the positive peaks appears, and the direction of wall displacement can be accurately determined even for larger applied fields and a corresponding wider range of wall velocity values.

The functionality of the system with pick-up coils composed of three windings has been tested on $Fe_{77.5}Si_{7.5}B_{15}$ amorphous glass-coated wires with metallic nucleus diameters from 30 μm down to 100 nm. The amplitude of the applied field was ranging between a few A/m and 2.5 kA/m. For small values of the metallic core diameter (below 1-2 μm) four low noise preamplifiers (Stanford Research Systems SR560) have been used in order to obtain a measurable voltage from each compound pick-up coil.

Figure 8 shows the peaks generated by the displacement of the domain wall in the sequence C4 → C3 → C2 → C1, for an $Fe_{77.5}Si_{7.5}B_{15}$ glass-coated submicron amorphous wire with the metallic nucleus diameter of 500 nm and the glass coating thickness of 6.5 μm. The direction of the domain wall movement through each coil is determined from the order of the high and low amplitude peaks. The direction is from left to right (C1 → C2 → C3 → C4) when the high amplitude peak appears ahead of the low amplitude one, and from right to left (C4 → C3 → C2 → C1) when the high amplitude peak appears after the low amplitude one (see figures 7 and 8). Taking into account the succession of the amplitude peaks, one can observe that a single domain wall is propagating through the wire from right to left and the recorded wall velocity is therefore valid in this case, being 935 m/s at 3650 A/m for the tested submicron wire sample.

The distance between two neighboring pick-up coils was 40 mm. In some cases supplementary domain walls can be nucleated in the measuring space (Garcia-Miquel et al., 2000; Hudak et al., 2009), usually when a large field is applied and/or some defects are present in the wire structure. Therefore, for very high amplitudes of the applied field, the distance between two adjacent pick-up coils is reduced in order to measure the wall velocity, sometimes even down to 10 mm. This flexibility allows one to have a single domain wall moving through the measuring space for very high fields and for any sample diameter from 50 μm down to 300 nm.

However, for wires with the metallic nucleus diameter of 300 nm (see figure 9) and below, it is extremely difficult to discern the secondary peak (the smaller one) from the noise, since the amplitude of this peak is at the same level as the noise. Even so, measurements can be performed and the velocity of the domain wall is valid if the peaks are in the right order and all three measured values are equal. For the thinnest wire tested - the amorphous nanowire with the metallic nucleus diameter of 100 nm - the measurement of the wall velocity is even more difficult, as it displays a very large switching field (of about 11 kA/m), domain wall velocities above 1 km/s, and the distance which ensures that only propagation of a single domain wall takes place is very small, i.e. less than 3 cm.

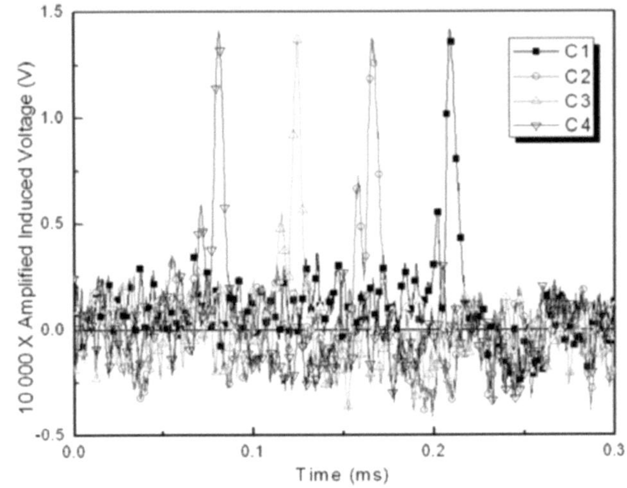

Fig. 8. Signal induced by the propagating wall in the sequence C4 → C3 → C2 → C1 in an $Fe_{77.5}Si_{7.5}B_{15}$ submicron amorphous wire with the metallic nucleus diameter of 500 nm. Applied field: 3,650 A/m.

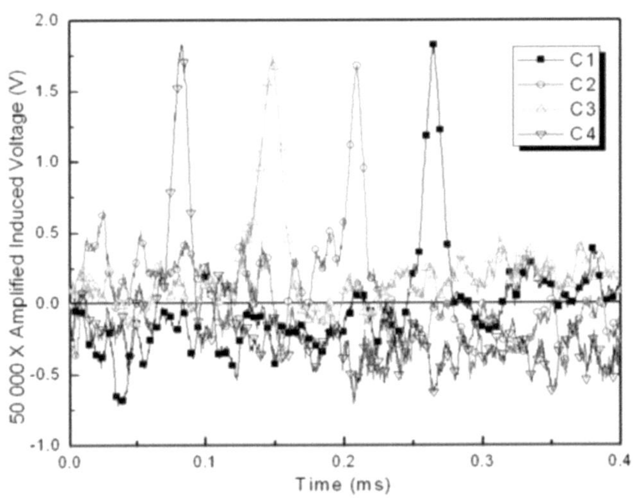

Fig. 9. Signal induced by the propagating wall in the sequence C4 → C3 → C2 → C1 in an $Fe_{77.5}Si_{7.5}B_{15}$ submicron amorphous wire with the metallic nucleus diameter of 300 nm and 6.5 μm glass thickness at 7,200 A/m.

Under these circumstances, the use of pick-up coils composed of three windings is no longer efficient. As a result, for wires with nucleus diameters below 300 nm, which display shorter propagation distances for the single wall, another specific system has been developed. The new specific system has four compensated 6 mm long pick-up coils with 1500 turns placed at 0.5 mm next to each other one, and wound with enameled 0.07 mm copper wire on a ceramic tube with an outer diameter of 1.8 mm and the inner diameter of 1 mm. The distance between two adjacent coil centers (6.5 mm) and the time interval between the two corresponding peaks has been used to calculate the domain wall velocity. The resulting peaks and schematic view of the pick-up coil system are presented in figure 10.

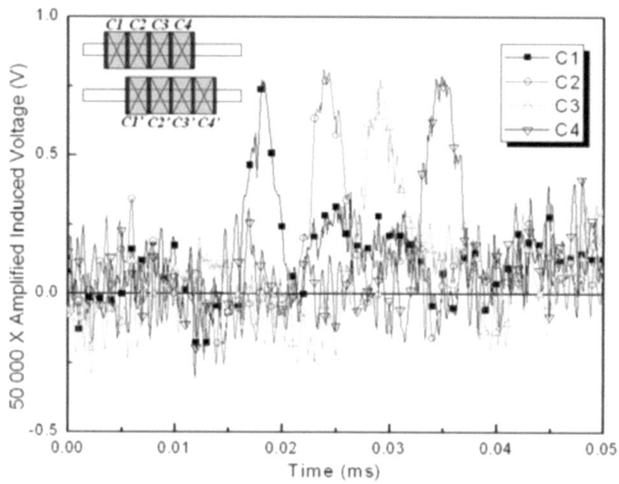

Fig. 10. Specific system of pick-up coils and signal induced by the wall in the sequence C1 → C2 → C3 → C4 for an $Fe_{77.5}Si_{7.5}B_{15}$ nanowire with 100 nm nucleus diameter and 6 μm glass coating. Applied field: 1.5 kA/m.

Precision of the measured value of the wall velocity in this case is also ensured by the right order of the recorded peaks (C1 → C2 → C3 → C4) and by the close values of all three recorded velocities ($V1$ = 1182 m/s, $V2$ = 1226 m/s, and $V3$ = 1140 m/s => $V \approx$ 1182 m/s). In the case of ultrathin wires some variations of the recorded velocity values always appear mainly due to the very low signal to noise ratio. The position of the maximum point in the peak, used for velocity calculation, is strongly affected by the noise.

A LabView application has been developed in order to reduce the measuring time and obtain a large number of points on the domain wall velocity versus field curves. A window, in which positive peaks are detected, is created and a zoom on this window is made to have all four peaks in view, in order to have always a visual control of the shape and order of the peaks. The software detects the position of the highest point from each trace and calculates three velocity values corresponding to the domain wall passing from C1 to C2, from C2 to C3, and from C3 to C4, respectively. The software records the velocities and relative peak positions and returns error messages if the succession of the peaks is not correct (C1 → C2 → C3 → C4 or C4 → C3 → C2 → C1) and if the recorded velocities differ by more than a certain predefined percent.

Figure 11 illustrates a comparative plot of the domain wall velocity vs. applied field for $Fe_{77.5}Si_{7.5}B_{15}$ amorphous microwires, submicron wires and nanowires, i.e. wires with different diameters of the metallic nucleus from the thickest (microwires with a 30 μm metallic nucleus) down to the thinnest (100 nm nanowires). The observed non-monotonic dependence of domain wall velocity on wire diameter is in agreement with previously

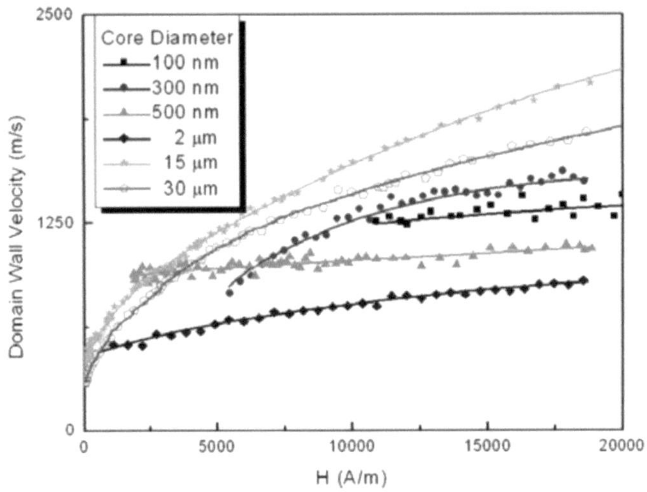

Fig. 11. Domain wall velocity in $Fe_{77.5}Si_{7.5}B_{15}$ amorphous glass-coated wires having various diameters of the metallic nucleus (microwires, submicron wires, and nanowires).

reported result for submicron amorphous wires (Óvári et al, 2011). Such comprehensive result would not have been possible without the development of the wall velocity measuring set-ups presented above.

The development of this new method was mainly requested by the inaccuracy of the current wall velocity measuring methods when the investigated samples are extremely thin, e.g. nanowires, which require very large fields to propagate a domain wall, fields which can nucleate additional domain walls and thus result in incorrect wall velocity values. Another reason which required the development of the novel method was the increment in sensitivity in order to measure domain wall velocity in wires with diameters down to 100 nm, in which the signal to noise ratio is very small. The proposed system is able to measure domain wall velocities between 50 and 2400 m/s for samples in which the magnetic flux is as low as 1.27×10^{-14} Wb. The availability of this new method is timely and of great importance at present, when much work is undertaken in order to develop novel domain wall logic devices which employ magnetic nanowires.

3. Magnetic behavior of rapidly solidified amorphous nanowires and submicron wires

3.1 800 nm submicron wires

The bulk and surface magnetic behavior of submicron amorphous wires have been investigated in order to compare them to the well known magnetic behavior of amorphous microwires and to monitor the changes induced as the threshold toward submicron dimensions is crossed.

The bulk magnetic behavior of the submicron wires has been studied by means of inductive hysteresis loops, obtained using a fluxmetric method. Due to the small value of the induced voltage in case of submicron wires, the signal was amplified using a Stanford Research Systems SR560 low-noise voltage preamplifier, and subsequently fed into the integrating fluxmeter.

The surface magnetic behavior has been investigated by magneto-optical Kerr effect (MOKE) in longitudinal configuration, using a NanoMOKE II magnetometer, produced by Durham Magneto Optics Ltd. In this case, the rotation of the plane of polarization was proportional to the magnetization component parallel to the plane of incidence. A polarized light of He-Ne laser (λ = 635 nm) was reflected from the wire to the detector. The diameter of the light beam was 2 μm and the penetration depth of the laser light is 9 nm. The plane of incidence was parallel to the wire axis. The following surface MOKE hysteresis loops have been measured: axial magnetization (M_Z) vs. axial field (H_Z), M_Z vs. perpendicular field ($H_\perp$), and M_Z vs. helical field ($H_{\theta Z}$).

Ferromagnetic resonance (FMR) measurements have been performed in order to study the magnetic anisotropy from the surface region of submicron wires and to correlate the results with the MOKE results. The FMR spectra were determined with an X-band spectrometer using the modulation technique. The DC magnetic field was modulated with an alternating field having a frequency of 1 kHz and the amplitude of 10 Oe. The working frequencies were 8.5, 9.5, and 10.5 GHz, respectively.

Figure 12 shows the axial inductive hysteresis loop of a submicron wire measured at 50 Hz. One observes that the submicron amorphous wire displays an unusual magnetic behavior, unlike microwires with the same composition and metallic nucleus diameters below 20 μm, which typically display an almost anhysteretic axial loop (Zhukov et al., 2003).

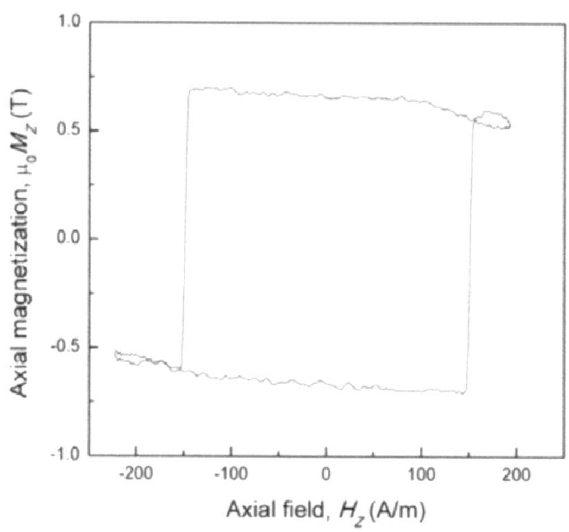

Fig. 12. Axial inductive hysteresis loop of a $(Co_{0.94}Fe_{0.06})_{72.5}Si_{12.5}B_{15}$ submicron amorphous glass-coated wire with the metallic nucleus diameter of 800 nm and the glass coating thickness of 6 μm.

On the contrary, the submicron wire is bistable even at such small diameter of the metallic nucleus, which shows shape anisotropy becomes more important than magnetoelastic anisotropy, which is prevailing in microwires with larger metallic nucleus diameters (several microns up to 20 μm). The value of the axial bulk switching field is 149 A/m, which is quite small for such low dimension.

Figure 13 illustrates the M_Z vs. H_Z MOKE surface hysteresis loop of the submicron amorphous wire. One observes that axial bistability is maintained even in the 9 nm deep

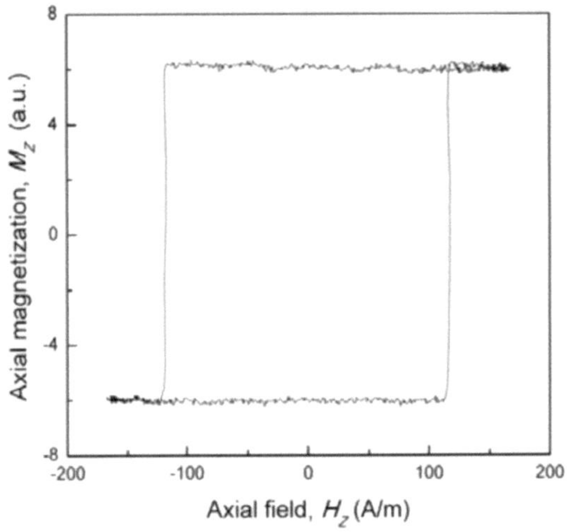

Fig. 13. Axial magnetization vs. axial field MOKE hysteresis loop for the $(Co_{0.94}Fe_{0.06})_{72.5}Si_{12.5}B_{15}$ submicron amorphous glass-coated wire.

surface region, and the surface axial switching field of 118 A/m is in the same range as the bulk value. This behavior is quite surprising, given the expected large values of the circumferential compressive stresses induced in the surface region during the preparation of the submicron wire. Nevertheless, it supports the above statement about the increased importance of shape anisotropy, and it shows that the submicron wire displays an axial component of magnetization in the surface region, as opposed to regular microwires with similar composition in which the outer shell is mostly circumferential.

Figure 14 shows the M_Z vs. $H_\perp$ MOKE surface loop of the submicron wire. The jump in magnetization is still observed, however the perpendicular switching field of 1600 A/m is one order of magnitude larger than the bulk axial switching field. This was expected, as $H_\perp$ does not act on the axial component of the magnetization M_Z, but only locally on the circumferential component M_θ, so a quite large field is required to switch the resultant surface magnetization $\boldsymbol{M}$ by acting only locally on one of its components, i.e. M_θ. The result of magnetization switching is monitored through the other component - M_Z.

Figure 15 illustrates the M_Z vs. $H_{\theta Z}$ MOKE surface loop for the same submicron amorphous wire. The magnetization jump is observed at a smaller value of the helical switching field, of 89 A/m. Results from figures 13 through 15 support the existence of helical magnetic anisotropy in the surface region of the submicron wire, rather than either solely axial or circumferential anisotropy. The small value of the helical switching field confirms the competition between magnetoelastic anisotropy and shape anisotropy and indicates a smaller overall anisotropy toward the surface, which is also in agreement with the decrease of the surface axial switching field (118 A/m) as compared to the value of the bulk axial switching field (149 A/m).

FMR has been employed to further investigate the surface anisotropy of the submicron amorphous wire. The values of the FMR line width at various frequencies for the submicron amorphous wire are listed in Table 1.

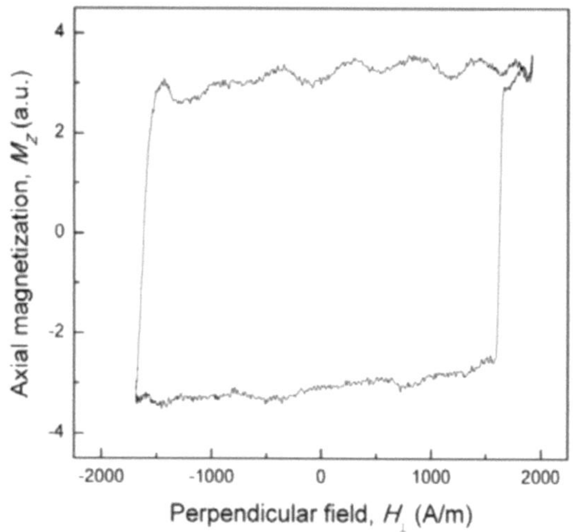

Fig. 14. Axial magnetization vs. perpendicular field MOKE hysteresis loop for the $(Co_{0.94}Fe_{0.06})_{72.5}Si_{12.5}B_{15}$ submicron amorphous glass-coated wire.

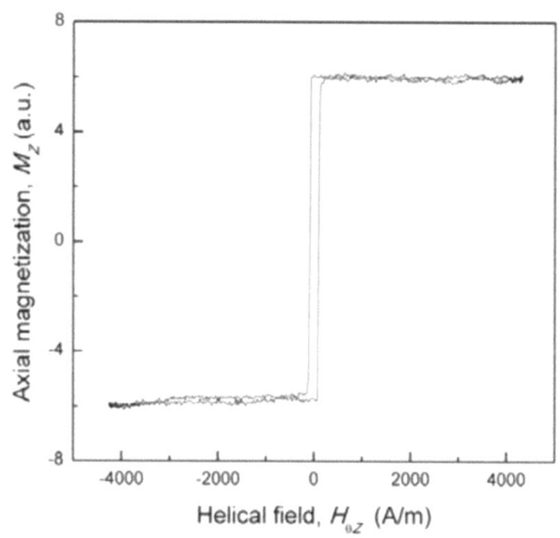

Fig. 15. Axial magnetization vs. helical field MOKE hysteresis loop for the $(Co_{0.94}Fe_{0.06})_{72.5}Si_{12.5}B_{15}$ submicron amorphous glass-coated wire.

Frequency (GHz)	Line width (kA/m) Submicron wire	Line width (kA/m) 6.5 μm microwire
8.5	5.65	11.94
9.5	6.29	13.53
10.5	6.37	14.57

Table 1. FMR line width values at various frequencies for the $(Co_{0.94}Fe_{0.06})_{72.5}Si_{12.5}B_{15}$ submicron amorphous glass-coated wire and for a microwire with the same composition having a 6.5 μm nucleus diameter and a 9.5 μm glass coating thickness.

The FMR line width at 8.5 GHz is less than half the value for a microwire with a 6.5 μm nucleus diameter, given for comparison. This shows that surface anisotropy is better emphasized in submicron wires as compared to microwires with typical dimensions, in agreement with larger stresses induced during preparation of these thinner samples. An increase of the resonance frequency to 9.5 and further to 10.5 GHz results in the increase of the line width, which shows a higher degree of anisotropy spread toward the surface and supports the above mentioned smaller overall anisotropy in the surface region.

Thus, the well known magnetic behavior of a typical nearly zero magnetostrictive microwire changes when the metallic nucleus diameter enters the submicron range. Shape anisotropy becomes dominant. Nearly zero magnetostrictive submicron wires are fully bistable, bistability being maintained even in a very thin 9 nm surface layer.

3.2 Submicron wires with diameters between 350 and 800 nm

A more significant effect of the reduction in the metallic nucleus diameter on the magnetic behavior of rapidly solidified $(Co_{0.94}Fe_{0.06})_{72.5}Si_{12.5}B_{15}$ nearly zero magnetostrictive and $Fe_{77.5}Si_{7.5}B_{15}$ positive magnetostrictive submicron wires has been investigated in a comparative manner. The investigated submicron wire samples display metallic nucleus diameters ranging from 350 to 800 nm. Their overall axial magnetization process has been studied by measuring the bulk inductive hysteresis loops using a fluxmetric method. A special attention has been also paid to the surface magnetic behavior of these thinner samples, which has been studied by means of MOKE and FMR.

Figure 16 shows the bulk axial hysteresis loops of a $(Co_{0.94}Fe_{0.06})_{72.5}Si_{12.5}B_{15}$ sample and an $Fe_{77.5}Si_{7.5}B_{15}$ sample with close values of the metallic nucleus diameter. The nearly zero magnetostrictive wire has a metallic nucleus diameter of 510 nm and the glass coating of 6.5 μm, whilst the positive magnetostrictive sample has the metallic nucleus diameter of 530 nm and a glass coating of 9.7 μm.

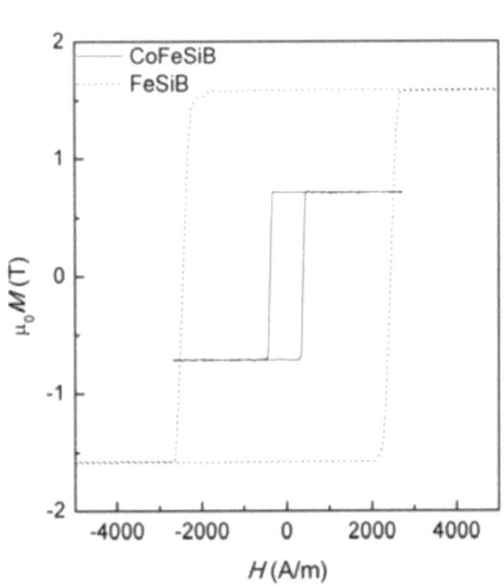

Fig. 16. Axial inductive hysteresis loops for a $(Co_{0.94}Fe_{0.06})_{72.5}Si_{12.5}B_{15}$ submicron amorphous wire with the metallic nucleus diameter of 510 nm and the glass coating of 6.5 μm and for an $Fe_{77.5}Si_{7.5}B_{15}$ sample with the metallic nucleus diameter of 530 nm and the glass coating of 9.7 μm.

The most important observation is that both types of submicron wires are indeed bistable, as proven by their rectangular hysteresis loops, which makes them suitable for spintronic applications. As concerns the particular aspects of these rectangular loops, besides the known differences in their saturation magnetization, one also observes the large difference between the coercivity values. The Co-based nearly zero magnetostrictive sample displays a coercivity of 350 A/m, whilst the Fe-based positive magnetostrictive submicron wire has a much larger coercivity of 2450 A/m. The correlation between dimensions, internal stresses and coercivity has been extensively studied in the case of the larger amorphous glass-coated microwires (Chiriac & Óvári, 1996). Such considerations also apply to the submicron wires, which are similarly composite wires prepared using the same technique. Therefore, coercivity is expected to decreases if the metallic nucleus diameter increases and/or the glass coating thickness decreases. Indeed, for a positive magnetostrictive submicron wire with the nucleus diameter of 670 nm and the glass coating of 6.5 μm, coercivity reaches down to 2000 A/m. This value shows that the coercivity of positive magnetostrictive submicron wires is strongly influenced by the magnetoelastic coupling between internal stresses and magnetostriction. However, the differences between the coercivity values in Fig. 16 cannot be entirely attributed to the different strength of the magnetoelastic coupling in positive and nearly zero magnetostrictive samples. A contribution is also given by the different nature of the uniaxial anisotropy: magnetoelastic in case of the positive magnetostrictive sample and shape anisotropy for the nearly zero magnetostrictive wire.

Coercivity increases to 4235 A/m for the thinnest positive magnetostrictive submicron wire (nucleus of 350 nm and coating of 6.5 μm). This is an expected consequence of the reduction in the metallic nucleus diameter, but it is difficult to estimate the role of shape anisotropy in this case.

Figure 17 illustrates the MOKE surface axial hysteresis loop of the 350 nm $Fe_{77.5}Si_{7.5}B_{15}$ submicron wire. The coercivity of the surface loop has the same value as the coercivity of the bulk loop. In case of the 510 nm nearly zero magnetostrictive submicron wire, the MOKE surface loop also shows the same coercivity value as the bulk one. These results show that the analyzed samples are bistable in their entire volume.

However, there is no information on whether or not the anisotropy is perfectly uniaxial in the near-surface region. FMR has been employed in order to study this particular aspect of the anisotropy distribution.

Figure 18 shows the FMR spectra of the $Fe_{77.5}Si_{7.5}B_{15}$ submicron amorphous wire with a metallic nucleus diameter of 350 nm. One observes that, irrespective of frequency, the derivative resonance spectra display a single resonance field. These results indicate that in such ultrathin submicron wires one can expect a uniaxial anisotropy which may show the presence of a single domain structure instead of the well known core-shell magnetic structure found in microwires (Chiriac & Óvári, 1996).

For comparison, figure 19 shows the FMR spectra of a typical $Fe_{77.5}Si_{7.5}B_{15}$ amorphous microwire with the metallic nucleus diameter of 22 μm and the glass coating thickness of 20 μm. One observes that the resonance peaks are split in this case, showing a complex anisotropy, which may be related to the typical outer shell and to the interdomain wall between the outer shell and the inner core encountered in microwires of this size (Chiriac et al., 2007b).

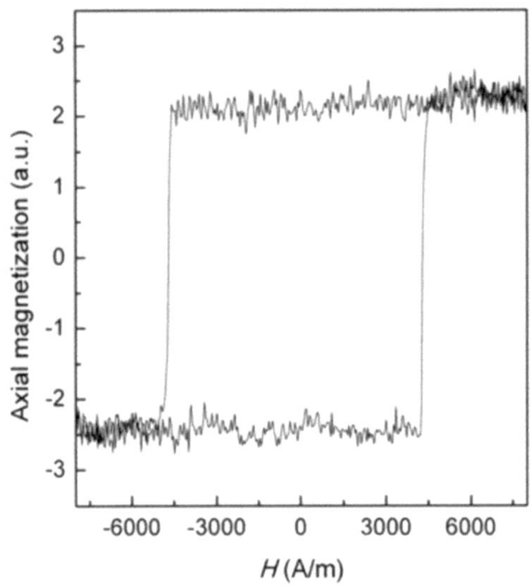

Fig. 17. MOKE surface axial hysteresis loop of an $Fe_{77.5}Si_{7.5}B_{15}$ submicron wire with the metallic nucleus diameter of 350 nm and the glass coating thickness of 6.5 μm.

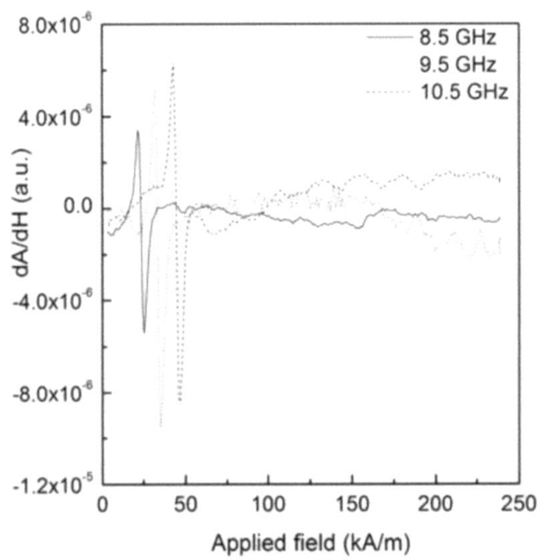

Fig. 18. FMR spectra of the $Fe_{77.5}Si_{7.5}B_{15}$ submicron amorphous wire with a metallic nucleus diameter of 350 nm and the glass coating thickness of 6.5 μm.

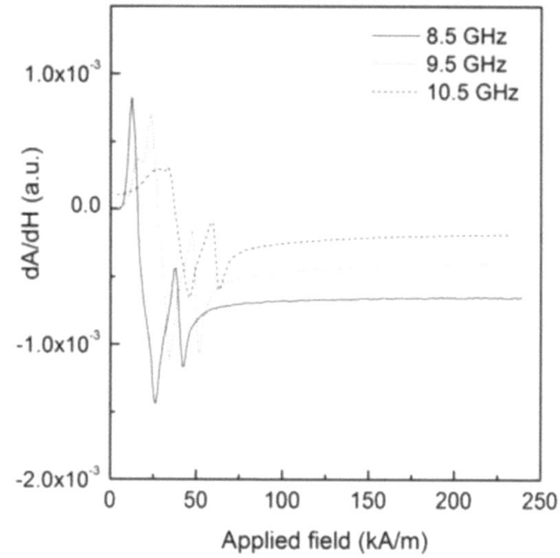

Fig. 19. FMR spectra of an $Fe_{77.5}Si_{7.5}B_{15}$ amorphous microwire with the metallic nucleus diameter of 22 μm and the glass coating thickness of 20 μm.

The most important result is that split resonance peaks have been found in this study even for $Fe_{77.5}Si_{7.5}B_{15}$ submicron wires with metallic nucleus diameters down to 500 nm. Figure 20 shows the FMR spectra of the $Fe_{77.5}Si_{7.5}B_{15}$ submicron wire with the metallic nucleus diameter of 530 nm and the glass coating thickness of 9.7 μm.

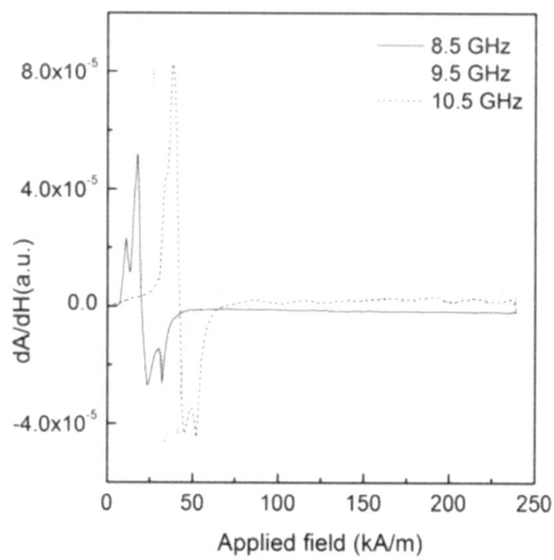

Fig. 20. FMR spectra of the $Fe_{77.5}Si_{7.5}B_{15}$ submicron wire with the metallic nucleus diameter of 530 nm and the glass coating thickness of 9.7 μm.

This shows that, even though the wire is fully bistable in its entire metallic nucleus, in the near-surface region there is a complex anisotropy which is different from the expected pure axial one. This is presumed to be a remnant of the radial anisotropy encountered in amorphous microwires with the same composition (Chiriac & Óvári, 1996). It still produces effects which are detected by means of FMR due to the very large internal stresses produced by the existence of a large glass coating compared to the diameter of the metallic core. However, in the range from 500 nm to 350 nm an essential change occurs: shape anisotropy becomes much more important than the magnetoelastic one. Therefore, the 350 nm submicron wire is not only fully bistable, but it also displays a uniaxial anisotropy associated with the expected single domain magnetic structure.

In the case of nearly zero magnetostrictive submicron wires, the FMR spectra display only one maximum for nucleus diameters of either 800 nm or 500 nm. FMR does not emphasize the helical anisotropy observed by means of MOKE (Chiriac et al., 2010). This is an indication that the region with helical anisotropy is extremely thin, given that the penetration depth of the laser light is only 9 nm, and one can state that it is more a magnetization ripple located at the very surface of the metallic nucleus, rather than a well defined region with helical anisotropy. Such statement is in agreement with the axial magnetic bistability determined by shape anisotropy in the $(Co_{0.94}Fe_{0.06})_{72.5}Si_{12.5}B_{15}$ samples.

Thus, the crucial role played by shape anisotropy in the magnetic behavior of these ultrathin magnetic wires has been emphasized once more. Shape anisotropy is the main factor that determines the bistability of nearly zero magnetostrictive submicron wires, irrespective of the diameter of their metallic nucleus. This is due to their much smaller magnetoelastic term, which originates in their small negative magnetostriction. As a result, they display a single domain magnetic structure with an ultrathin magnetization ripple at the surface. On the other hand, in positive magnetostrictive samples, magnetoelastic anisotropy still plays an important role in wires with nucleus diameters from 500 nm and up, as shown by the FMR spectra. In thinner samples, shape anisotropy becomes dominant and therefore they display a single domain magnetic structure.

3.3 Rapidly solidified amorphous nanowires

Figure 21 shows the bulk hysteresis loops for two rapidly solidified glass-coated amorphous nanowire samples. One observes the significant difference between the two switching fields - 420 A/m as compared to 7400 A/m.

Figure 22 illustrates the MOKE surface hysteresis loop for the same samples. Both loops show that the nanowires are bistable in their entire volume. This is an indication that rapidly quenched amorphous magnetic nanowires display a single-domain structure, as opposed to the classical core-shell structure (Takajo et al., 1993; Vázquez, 2001) of the thicker amorphous microwires (metallic nucleus diameters over 1 μm) and of the submicron amorphous wires with diameters between 500 and 900 nm.

Therefore, at nanoscale, irrespective of composition, sample dimensions no longer allow the formation of a complex core-shell magnetic domain structure as a result of magnetoelastic energy minimization (Chiriac et al., 1995; Velázquez et al., 1996). Hence, despite the larger values of the internal stresses in these ultra-thin rapidly solidified materials, which coupled with the large positive magnetostriction of the $Fe_{77.5}Si_{7.5}B_{15}$ alloy lead to a large magnetoelastic term, the shape anisotropy is preponderant at nanoscale. The large internal stresses induced by both rapid solidification of metal and the difference between the thermal expansion coefficients of metal and glass (Chiriac et al., 1995) give rise to quite large values of the switching field, as shown in Figures 20 and 21.

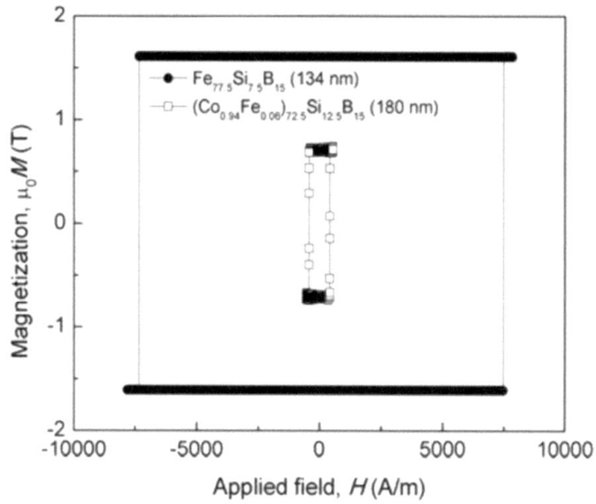

Fig. 21. Axial inductive hysteresis loops of two rapidly solidified amorphous nanowires: one with positive magnetostriction having the metallic nucleus diameter of 134 nm and the glass coating thickness of 6 μm and the other one with nearly zero magnetostriction having the metallic nucleus diameter of 180 nm and the glass coating thickness of 5.6 μm.

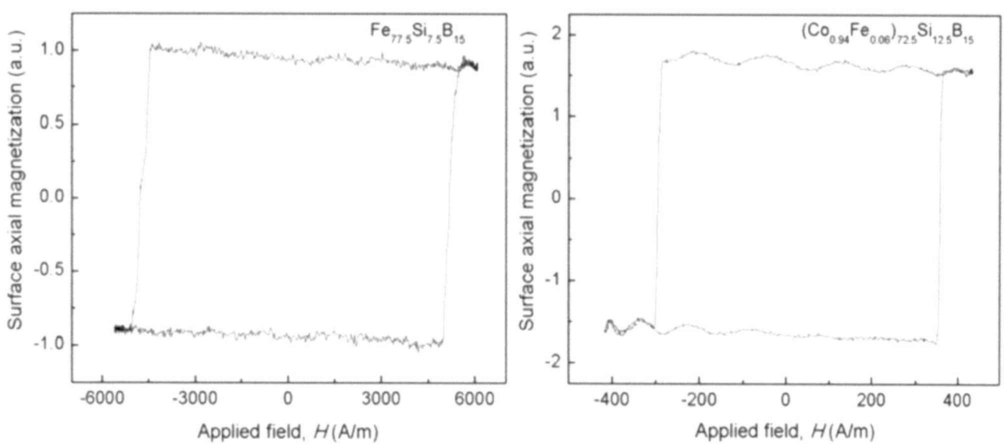

Fig. 22. Axial magnetization vs. axial field MOKE hysteresis loop for the rapidly solidified amorphous nanowire with positive magnetostriction (left) and for the one with nearly zero magnetostriction (right).

The direct relation between the value of the switching field and the glass to metal ratio is well known (Chiriac et al., 1997). In case of the $(Co_{0.94}Fe_{0.06})_{72.5}Si_{12.5}B_{15}$ samples magnetostriction is much smaller leading to smaller values of the switching field. Thus, there is a wide range of possibilities to adjust the value of the switching field by changing the composition or by partially or fully removing the glass coating. First glass removal experiments have been successful, with the glass coating being thinned down to 10 nm.

The difference between the values of the switching field for the bulk and surface loops originate in the surface defects of the metallic nucleus as well as in the demagnetizing effect, which are expected to cause a slight magnetization ripple at the surface, i.e. small local deviations of the magnetization from the axial direction. Such deviations are more easily emphasized in MOKE measurements on nearly zero magnetostrictive samples, when the field is applied transversally to the nanowire.

In order to substantiate the existence of the single domain structure in the rapidly solidified glass-coated amorphous magnetic nanowires, further investigations by means of FMR have been performed on positive magnetostrictive nanowires. The aim was to investigate the effect of the large magnetoelastic term at the wire surface, where the outer shell should exist.

Figure 23 shows the derivative microwave absorption spectrum of the $Fe_{77.5}Si_{7.5}B_{15}$ amorphous nanowire (left). For comparison, the spectrum of a rapidly solidified 800 nm submicron amorphous wire with the same composition is given (right).

The FMR spectrum of the submicron wire displays split resonance peaks, which reflect a complex anisotropy, which may indicate the presence of the complex core-shell domain structure. On the other hand, the FMR spectrum of the rapidly solidified amorphous nanowire does not display such a split, showing a single anisotropy direction at all frequencies, which supports the existence of a single domain structure.

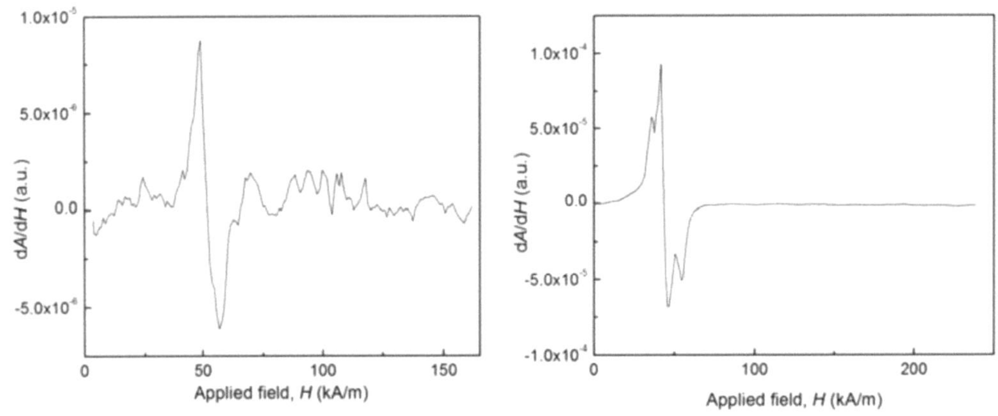

Fig. 23. Microwave absorption spectrum of the 134 nm $Fe_{77.5}Si_{7.5}B_{15}$ amorphous nanowire (left) and of an 800 nm $Fe_{77.5}Si_{7.5}B_{15}$ submicron amorphous wire (right) at 10.5 GHz.

Thus, both FMR and hysteresis loop measurements (inductive and MOKE) point to the existence of a single domain structure in rapidly solidified amorphous glass-coated nanowires. The large magnetoelastic term cannot exceed the shape anisotropy in positive magnetostrictive amorphous nanowires. This is even less likely to happen in the nearly zero magnetostrictive nanowires, given their much smaller magnetoelastic term.

The uniaxial magnetization in the whole volume of the amorphous nanowire is favored by shape anisotropy. This means that the nanowires are too thin to allow the formation of a more complex magnetic domain structure. Therefore, such nanowires are the perfect candidates for spintronic applications based on the domain wall propagation along the entire length of the sample.

4. Domain wall velocity in rapidly solidified amorphous nanowires and submicron wires

The study of the domain wall velocity in bistable submicron amorphous wires with positive and nearly zero magnetostriction prepared by rapid solidification from the melt is closely linked to the purpose for which submicron wires have been prepared, i.e. to understand the characteristics of domain wall propagation in the thinnest possible wires made by rapid solidification, in order to propose new materials for spintronic applications. The effect of wire dimensions on wall velocity is studied in conjunction with their magnetic behavior. The role of magnetostriction in these ultrathin wires is also analyzed. The employed experimental set-ups are those described in detail in section 2.2.

Figure 24 shows the dependence of wall velocity on applied field for $Fe_{77.5}Si_{7.5}B_{15}$ positive magnetostrictive amorphous submicron wires with different metallic nucleus diameters and the same glass coating thickness of 15 μm. Figure 25 illustrates the axial hysteresis loops for the same samples.

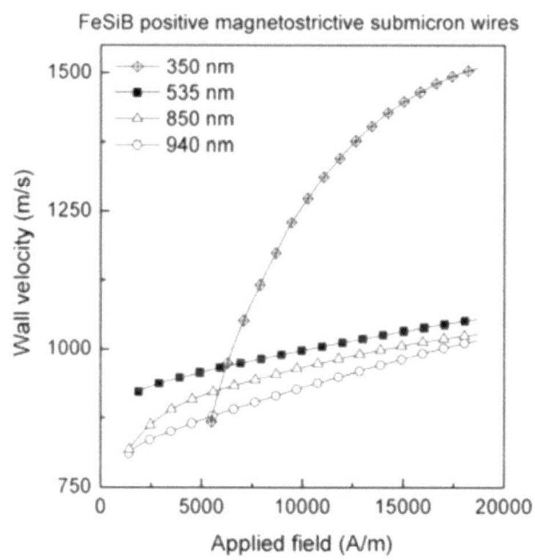

Fig. 24. Wall velocity vs. applied field for $Fe_{77.5}Si_{7.5}B_{15}$ submicron amorphous wires with positive magnetostriction, having different metallic nucleus diameters and a similar glass coating thickness (15 μm).

A correlation between the domain wall velocity and the value of the switching field is observed, i.e. the larger the switching field, the larger the wall velocity, which indicates a relation between the magnitude of the uniaxial anisotropy and the domain wall velocity. However, something is different for the sample with the 350 nm nucleus diameter: the slope of its wall velocity vs. applied field curve, i.e. the wall mobility, is significantly larger than the slope of the curves which correspond to the other samples. This shows that something is different about the uniaxial anisotropy of this sample. If in case of the thicker samples one would expect a closely related cause of the uniaxial anisotropy with the case of typical amorphous microwires, i.e. the magnetoelastic coupling between the large axial internal stresses and the positive magnetostriction, in case of the thinnest one, the role of magnetoelastic anisotropy is taken over by the shape anisotropy, as mentioned in the previous sections.

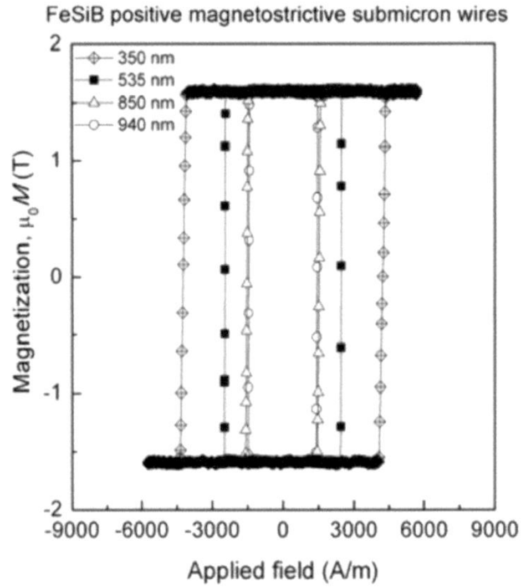

Fig. 25. Axial hysteresis loops for $Fe_{77.5}Si_{7.5}B_{15}$ submicron amorphous wires with positive magnetostriction, having different metallic nucleus diameters and a similar glass coating thickness (15 μm).

Thus, the effect of internal stresses is diminished as the metallic nucleus diameter decreases below a certain threshold, and the applied field becomes much more efficient in moving the domain wall, which results in increased mobility and velocity values. Therefore, the increased contribution of shape anisotropy results in larger wall velocity and mobility values in the case of positive magnetostrictive submicron wires.

The largest wall velocity value is close to 1500 m/s, close to the largest values reported in microwires with the same composition and with typical dimensions in the range 1-50 μm.

Figure 26 shows the dependence of wall velocity on applied field for two $(Co_{0.94}Fe_{0.06})_{72.5}Si_{12.5}B_{15}$ nearly zero magnetostrictive submicron wire samples with different metallic nucleus diameters and the same glass coating thickness of 13 μm. Figure 27 illustrates the corresponding axial hysteresis loops. The correlation between wall velocity and uniaxial anisotropy, via switching field, is also observed in the case of submicron wires with nearly zero magnetostriction. However, in this case only shape anisotropy contributes to larger wall velocity values, as opposed to the case of positive magnetostrictive microwires, in which magnetoelastic anisotropy also has some contribution, at least down to a certain threshold. Given the negative sign of magnetostriction in the nearly zero magnetostrictive samples, the magnetoelastic anisotropy would lead to transverse uniaxial anisotropy instead of an axial one. Therefore, it is clear that the magnetic bistability of these samples originates in an axial anisotropy determined by shape anisotropy only.

The maximum velocity values measured in nearly zero magnetostrictive submicron wires are slightly larger at about 1600 m/s than those measured in positive magnetostrictive ones. Nevertheless, these large velocities are obtained at much smaller values of the applied field in comparison with the case of positive magnetostrictive samples, i.e. 200 to 600 A/m as compared to 0.7 to 18 kA/m. Again, wall mobility is larger in thinner submicron wires, similar to the case of positive magnetostrictive samples, which substantiates the essential

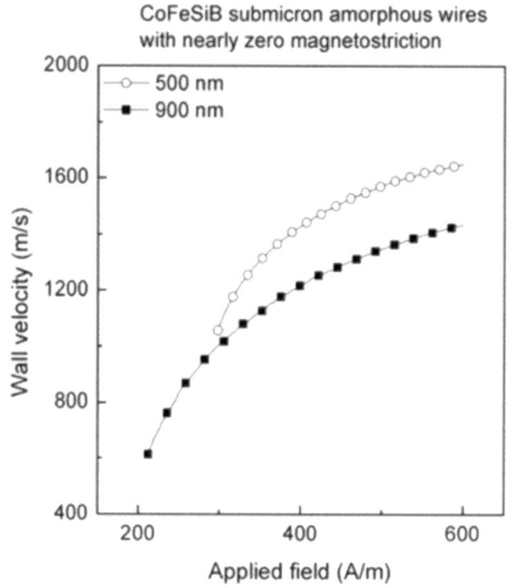

Fig. 26. Wall velocity vs. applied field for two $(Co_{0.94}Fe_{0.06})_{72.5}Si_{12.5}B_{15}$ submicron amorphous wires with nearly zero magnetostriction, having different metallic nucleus diameters and the same glass coating thickness (13 μm).

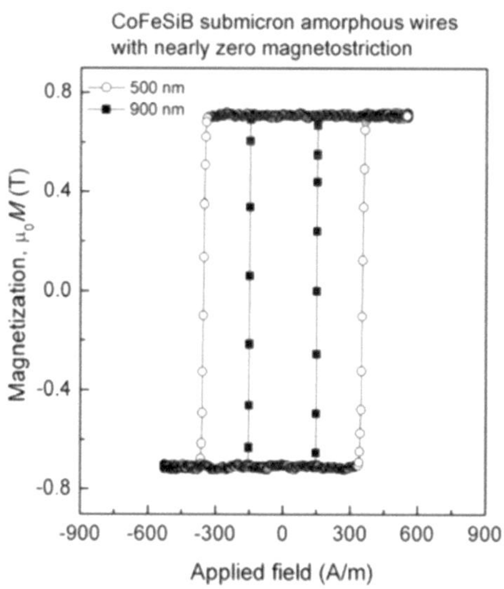

Fig. 27. Axial hysteresis loops for two $(Co_{0.94}Fe_{0.06})_{72.5}Si_{12.5}B_{15}$ submicron amorphous wires with nearly zero magnetostriction, having different metallic nucleus diameters and the same glass coating thickness (13 μm).

role played by shape anisotropy in both types of submicron wires. The wall velocity values are comparable to those measured in planar NiFe nanowires (Atkinson et al., 2003), although the mobility values are significantly larger in the case of nearly zero magnetostrictive submicron wires. Thus, both the wire dimensions and the magnetostriction are important as concerns the domain wall velocity and mobility values in rapidly solidified amorphous submicron wires. Wire dimensions influence the shape anisotropy, whilst magnetostriction affects the magnetoelastic anisotropy. Both anisotropy types play an important role in submicron wires with positive magnetostriction. In negative magnetostrictive ones, only shape anisotropy plays an essential role.

Figure 28 illustrates the field dependence of the domain wall velocity in the 134 nm rapidly solidified nanowire with positive magnetostriction (left) and in the 180 nm one with nearly zero magnetostriction (right).

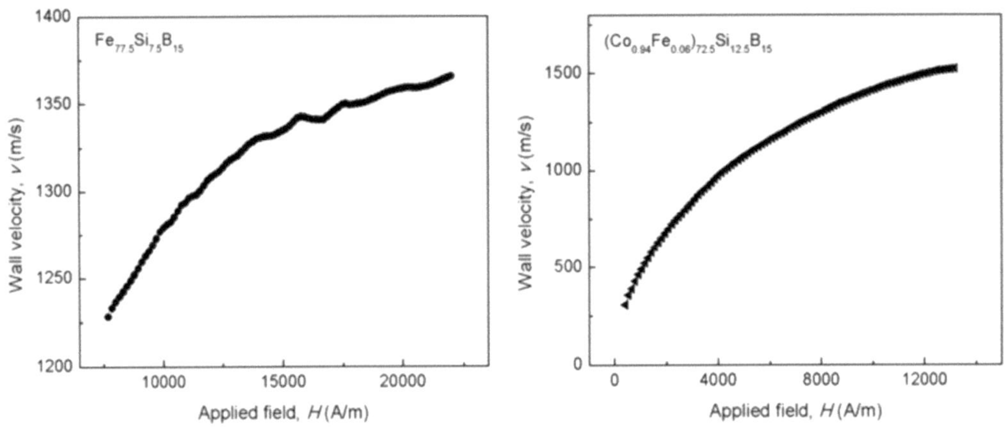

Fig. 28. Wall velocity vs. applied field for a rapidly solidified $Fe_{77.5}Si_{7.5}B_{15}$ amorphous nanowire with positive magnetostriction, having the metallic nucleus diameter of 134 nm and the glass coating thickness of 6 μm (left) and for a $(Co_{0.94}Fe_{0.06})_{72.5}Si_{12.5}B_{15}$ nearly zero magnetostrictive one with the metallic nucleus diameter of 180 nm and the glass coating thickness of 5.6 μm (right).

For the positive magnetostrictive sample the maximum wall velocity, reached at an applied field larger than 20 kA/m, is above 1360 m/s. Nevertheless, even at an applied field just above the value of the switching field the wall velocity is larger than 1200 m/s. These values are also comparable to those reported in planar NiFe nanowires (Atkinson et al., 2003) and they are expected to improve after glass removal. Although the mobility of the wall is rather small (velocity does not increase much with the applied field), this aspect is also expected to significantly improve after glass removal as a result of stress relief and decrease of the switching field. Wall velocity values are larger, over 1500 m/s, in the case of nearly zero magnetostrictive samples, and they are attained at much smaller values of the applied field. The wall mobility in these nanowires is also much larger. This shows the importance of composition for spintronic applications. These results are important as concerns the future application of rapidly solidified nanowires and submicron wires in spintronic devices.

5. Conclusions

Rapidly solidified magnetic nanowires and submicron wires have low production costs and their properties can be accurately tailored through a variety of parameters, known from their larger precursors - the amorphous microwires: the diameter of the metallic nucleus, the glass coating thickness, their ratio, and the composition, which decides the sign and magnitude of the magnetostriction constant. These tailoring parameters are adjustable through the preparation process. Post-production processing, such as various types of annealing (furnace, Joule heating, field-annealing, stress-annealing) as well as the post-production partial or full removal of the glass coating can be also used to tailor the magnetic properties. Tailoring parameters facilitate the fine tuning of nanowire and submicron wire properties. Another advantage of the rapidly solidified amorphous nanowires and submicron wires is that they can be prepared at sample lengths which basically exceed all the current requirements of applications based on nanowire samples.

The preparation of these materials has been successful since it was initially based on well known materials, which have been extensively studied at the larger micro scale. We have been able to prepare them at a much smaller scale, the nano and submicron scale, aiming to preserve their specific characteristics and properties within certain limits.

Future work will focus on tailoring the wall propagation characteristics, such as velocity and mobility. The technical solutions used for the preparation of amorphous magnetic nanowires should be extended, to permit the preparation of a much larger range of compositions. Such development would lead to novel applications, e.g. medical, various sensors, controlled motion of particles. Magnetic nanowires are also suitable for shielding applications at very high frequencies.

6. Acknowledgment

Work supported by the Romanian Ministry of Education, Research, Youth and Sports through the NUCLEU Program (Contracts No. 09-43 N, 01 02, 02 04, and PN 09-43 01 01) and through the PARTENERIATE Program under Contract No. 82-096/2008 (NADEX).

7. References

Allwood, D.A.; Xiong, G.; Faulkner, C.C.; Atkinson, D.; Petit, D. & Cowburn, R.P. (2005). Magnetic Domain-Wall Logic. *Science,* Vol.309, No.5741, (September 2005), pp. 1688-1692, ISSN 0036-8075

Atkinson, D.; Allwood, D.A.; Faulkner, C.C.; Xiong, G.; Cooke, M.D. & Cowburn, R.P. (2003). Magnetic Domain Wall Dynamics in a Permalloy Nanowire. *IEEE Transactions on Magnetics,* Vol.39, No.5, (September 2003), pp. 2663-2665, ISSN 0018-9464

Butta, M.; Infante, G.; Ripka, P.; Badini-Confalonieri, G.A. & Vázquez, M. (2009). M-H Loop Tracer Based on Digital Signal Processing for Low Frequency Characterization of Extremely Thin Magnetic Wires. *Review of Scientific Instruments,* Vol.80, No.8, (August 2009), 083906, ISSN 0034-6748

Chiriac, H.; Óvári, T.-A. & Pop, G. (1995). Internal Stress Distribution in Glass-Covered Amorphous Magnetic Wires. *Physical Review B,* Vol.52, No.14, (October 1995), pp. 10104-10113, ISSN 1098-0121

Chiriac, H. & Óvári, T.-A. (1996). Amorphous Glass-Covered Magnetic Wires: Preparation, Properties, Applications. *Progress in Materials Science,* Vol.40, No.5, (December 1996), pp. 333-407, ISSN 0079-6425

Chiriac, H.; Óvári, T.-A.; Pop, G. & Barariu, F. (1997). Effect of Glass Removal on the Magnetic Behavior of FeSiB Glass-Covered Wires. *IEEE Transactions on Magnetics,* Vol.33, No.1, (January 1997), pp. 782-787, ISSN 0018-9464

Chiriac, H.; Corodeanu, S.; Ţibu, M. & Óvári, T.-A. (2007a). Size Triggered Change in the Magnetization Mechanism of Nearly Zero Magnetostrictive Amorphous Glass-Coated Microwires. *Journal of Applied Physics,* Vol.101, No.9, (May 2007), 09N116, ISSN 0021-8979

Chiriac, H.; Óvári, T.-A.; Corodeanu, S. & Ababei, G. (2007b). Interdomain Wall in Amorphous Glass-Coated Microwires. *Physical Review B,* Vol.76, No.21, (December 2007), 214433, ISSN 1098-0121

Chiriac, H.; Óvári, T.-A. & Ţibu, M. (2008). Domain Wall Propagation in Nearly Zero Magnetostrictive Amorphous Microwires. *IEEE Transactions on Magnetics,* Vol.44, No.11, (November 2008), pp. 3931-3933, ISSN 0018-9464

Chiriac, H.; Óvári, T.-A. & Ţibu, M. (2009a). Effect of Surface Domain Structure on Wall Mobility in Amorphous Microwires. *Journal of Applied Physics,* Vol.105, No.7, (April 2009), 07A310, ISSN 0021-8979

Chiriac, H.; Ţibu, M. & Óvári, T.-A. (2009b). Domain Wall Propagation in Nanocrystalline Glass-Coated Microwires. *IEEE Transactions on Magnetics,* Vol.45, No.10, (October 2009), pp. 4286-4289, ISSN 0018-9464

Chiriac, H. ; Corodeanu, S. ; Lostun, M. ; Ababei, G. & Óvári, T.-A. (2010). Magnetic Behavior of Rapidly Quenched Submicron Amorphous Wires. *Journal of Applied Physics,* Vol.107, No.9, (May 2010), 09A301, ISSN 0021-8979

Chiriac, H.; Lostun, M.; Ababei, G.; & Óvári, T.-A. (2011a). Comparative Study of the Magnetic Properties of Positive and Nearly Zero Magnetostrictive Submicron Amorphous Wires. *Journal of Applied Physics,* Vol.109, No.7, (April 2011), 07B501, ISSN 0021-8979

Chiriac, H.; Corodeanu, S.; Lostun, M.; Stoian, G.; Ababei, G. & Óvári, T.-A. (2011b). Rapidly Solidified Amorphous Nanowires. *Journal of Applied Physics,* Vol.109, No.6, (March 2011), 063902, ISSN 0021-8979

Corodeanu, S.; Chiriac, H.; Lupu, N. & Óvári, T.-A. (2011a). Magnetic Characterization of Submicron Wires and Nanowires Using Digital Integration Techniques. *IEEE Transactions on Magnetics,* vol.47, No.10, (October 2011), pp. 3513-3515, ISSN 0018-9464

Corodeanu, S.; Chiriac, H. & Óvári, T.-A. (2011b). Accurate Measurement of Domain Wall Velocity in Amorphous Microwires, Submicron Wires, and Nanowires. *Review of Scientific Instruments,* Vol.82, No.9, (September 2011), 094701, ISSN 0034-6748

Finocchio, G.; Maugeri, N.; Torres, L. & Azzerboni, B. (2010). Domain Wall Dynamics Driven by a Localized Injection of a Spin-Polarized Current. *IEEE Transactions on Magnetics,* Vol.46, No.6, (June 2010), pp. 1523-1526, ISSN 0018-9464

Garcia-Miquel, H.; Chen, D.-X. & Vázquez, M. (2000). Domain Wall Propagation in Bistable Amorphous Wires. *Journal of Magnetism and Magnetic Materials,* Vol.212, Nos.1-2, (March 2000), pp. 101-106, ISSN 0304-8853

Hudak, J.; Blazek, J.; Cverha, A.; Gonda, P. & Varga, R. (2009). Improved Sixtus-Tonks Method for Sensing the Domain Wall Propagation Direction. *Sensors and Actuators A: Physical,* Vol.156, No.2, (December 2009), pp. 292-295, ISSN 0924-4247

Ipatov, M.; Zhukova, V.; Zvezdin, A.K. & Zhukov, A. (2009). Mechanisms of the Ultrafast Magnetization Switching in Bistable Amorphous Microwires. *Journal of Applied Physics,* Vol.106, No.10, (November 2009), 103902, ISSN 0021-8979

Komova, E.; Varga, M.; Varga, R.; Vojtanik, P.; Bednarcik, J.; Kovac, J.; Provencio, M. & Vázquez, M. (2008). Nanocrystalline Glass-Coated FeNiMoB Microwires. *Applied Physics Letters,* Vol.93, No.6, (August 2008), 062502, ISSN 0003-6951

Kulik, T.; Savage, H.T. & Hernando, A. (1993). A High-Performance Hysteresis Loop Tracer. *Journal of Applied Physics,* Vol.73, No.10, (May 1993), pp. 6855-6857, ISSN 0021-8979

Lee, J.Y.; Lee, K.S. & Lee, S.K. (2007). Remarkable Enhancement of Domain-Wall Velocity in Magnetic Nanostripes. *Applied Physics Letters,* Vol.91, No.12, (September 2007), 122513, ISSN 0003-6951

Moriya, R.; Hayashi, M.; Thomas, L.; Rettner, C. & Parkin, S.S.P. (2010). Dependence of Field Driven Domain Wall Velocity on Cross-Sectional Area in $Ni_{65}Fe_{20}Co_{15}$ Nanowires. *Applied Physics Letters,* Vol.97, No.14, (October 2010), 142506, ISSN 0003-6951

Óvári, T.-A.; Corodeanu, S. & Chiriac, H. (2011). Domain Wall Velocity in Submicron Amorphous Wires. *Journal of Applied Physics,* Vol.109, No.7, (April 2011), 07D502, ISSN 0021-8979

Parkin, S.S.P.; Hayashi, M. & Thomas, L. (2008). Magnetic Domain-Wall Racetrack Memory. *Science,* Vol.320, No.5873, (April 2008), pp. 190-194, ISSN 0036-8075

Sixtus, K.J. & Tonks, L. (1932). Propagation of Large Barkhausen Discontinuities. II. *Physical Review,* Vol.42, No.3, (November 1932), pp. 419-435

Takajo, M.; Yamasaki, J. & Humphrey, F.B. (1993). Domain Observations of Fe and Co Based Amorphous Wires. *IEEE Transactions on Magnetics,* Vol.29, No.6, (November 1993), pp. 3484-3486, ISSN 0018-9464

Torrejón, J.; Vázquez, M. & Panina, L.V. (2009). Asymmetric Magnetoimpedance in Self-Biased Layered CoFe/CoNi Microwires. *Journal of Applied Physics,* Vol.105, No.3, (February 2009), 033911, ISSN 0021-8979

Vázquez, M. (2001). Soft Magnetic Wires. *Physica B: Condensed Matter,* Vol.299, Nos.3-4, (June 2001), pp. 302-313, ISSN 0921-4526

Vázquez, M.; Basheed, G.A.; Infante, G. & Del Real, R.P. (2012). Trapping and Injecting Single Domain Walls in Magnetic Wire by Local Fields. *Physical Review Letters,* Vol.108, No.3, (January 2012), 037201, ISSN 0031-9007

Velázquez, J.; Vázquez, M. & Zhukov, A.P. (1996). Magnetoelastic Anisotropy Distribution in Glass-Coated Microwires. *Journal of Materials Research,* Vol.11, No.10, (October 1996), pp. 2499-2505, ISSN 0884-2914

Zhukov, A.; Zhukova, V.; Blanco, J.M.; Cobeño, A.F.; Vázquez, M. & González, J. (2003). Magnetostriction in Glass-Coated Magnetic Microwires. *Journal of Magnetism and Magnetic Materials,* Vols.258-259, (March 2003), pp. 151-157, ISSN 0304-8853

2

M-Type Barium Hexagonal Ferrite Films

Mingzhong Wu
Department of Physics, Colorado State University, Fort Collins, USA

1. Introduction

Magnetic garnet materials such as yttrium iron garnet (YIG) have been widely used as active components in many microwave devices.[1,2,3] These devices include resonators, filters, circulators, isolators, and phase shifters. They have had a major impact on the advancement of microwave technology. The underlying physical effects in microwave magnetic devices include ferromagnetic resonance (FMR), magnetostatic wave (MSW) propagation, Faraday rotation, and field displacement. Whatever the basis for a given device, the operation frequency is determined essentially by the FMR frequency of the garnet material. The magnetic garnets are low-magnetization, low-magnetocrystalline-anisotropy materials and, therefore, typically have a low FMR frequency in the GHz range. This imposes an upper limit on the practical operation frequency of compact YIG-based devices in the 10-18 GHz frequency range.

Presently, there is a critical need for millimeter (mm) wave devices which operate in the frequency range from about 30 GHz to 100 GHz.[4,5,6] This need is critical for three reasons. (1) Millimeter waves are recognized as a broadband frequency resource that can offer various wireless access applications. (2) The need for broadband telecommunication capabilities will mandate the use of mm-wave frequencies in next-generation satellite systems. (3) Electromagnetic radiation at mm-wave frequencies can penetrate clouds, fog, and many kinds of smoke, all of which are generally opaque to visible or infrared light.

In principle, one can extend the operation frequency of current microwave magnetic devices to the mm-wave frequency range through the use of high external magnetic bias fields. In practical terms, however, the use of high external fields is usually impractical because of the increased device size and weight, as well as incompatibility with monolithic integrated circuit technology.

One important strategy for the above-described frequency extension is to use M-type barium hexagonal ferrite $BaFe_{12}O_{19}$ (BaM) films as a replacement for those magnetic garnets. BaM films can have a very high magnetocrystalline anisotropy field. This high internal field can facilitate ferromagnetic resonance and hence device operation at mm-wave frequencies. The films can also have high remanent magnetization that can allow for device operation in absence of external magnetic fields, namely, self-biased operation, and frequency tuning using very low external fields.

To this end, significant efforts have been made in recent years that range from material preparations to structure and property characterizations and also to device applications.

Emphasis has been placed on the optimization of deposition processes for low-loss, self-biased BaM thin films,[7,8,9,10] the deposition of BaM thin films on "non-conventional" substrates, such as semiconductor substrates [11,12,13,14] and metallic substrates,[15] the fabrication of BaM thick films on semiconductor substrates,[16,17] the demonstration of BaM-based planar mm-wave devices, [18,19,20,21,22,23,10,24] the development of BaM-based ferromagnetic/ferroelectric heterostructures, [25,26,27,28,29] and the study of multiferroic effects in single-phase BaM materials.[30] A variety of different techniques have been used to fabricate BaM film materials. These include pulsed laser deposition (PLD), [7,8,9,10,11,12,28,31] liquid phase epitaxy (LPE), [32,33,34,35] RF magnetron sputtering,[36,37,19,38,39,40] molecular beam epitaxy (MBE), [14] metallo-organic decomposition (MOD), [15] chemical vapor deposition (CVD),[41] and screen printing.[16,17] The device demonstration includes both numerical [20,21,22] and experimental efforts. [18,21,22,23,10,24] The devices demonstrated include phase shifters, [21] filters, [22,23,10,24] circulators, [18] and isolators.[19]

This chapter reviews the main advances made in the field of BaM materials and devices over the past five years. Section 2 gives a brief introduction to hexagonal ferrites first and then describes in detail the structure and properties of BaM materials. This section serves to provide a background for the discussions in the following sections. Section 3 reviews the advances made in the development of BaM film materials. Section 3.1 describes the deposition of low-loss, high-remanent-magnetization BaM thin films on sapphire substrates by PLD techniques.[10] Section 3.2 discusses the deposition of BaM thin films on metallic substrates by the MOD method.[15] Section 3.3 reviews the deposition of BaM thin films on semiconductor substrates by PLD and MBE techniques.[13,14] Section 3.4 describes the fabrication of BaM thick films on semiconductor substrates by screen printing.[16,17] Section 4 reviews the demonstration of BaM thin film-based mm-wave notch filters[10,24] and phase shifters.[21] Finally, Section 5 discusses future work in the field of BaM materials and devices.

2. Structure and properties of M-type barium hexagonal ferrites (BaM)

2.1 Building blocks of hexagonal ferrites

In many solids, the atoms look like attracting hard spheres and are packed as closely as possible.[42,43] Figure 1 shows a close-packed layer of identical spheres which occupy positions A. This layer is formed by placing each sphere in contact with six others in a plane. A second and identical layer of spheres can be placed on top of this layer and occupy positions B. Each sphere in the second layer is in contact with three spheres in the first layer. A third layer of spheres may be added in two ways: they can occupy either positions A or positions C. In principle, there are an infinite number of ways of stacking the close-packed layers. Two very common stacking sequences are "ABAB…" and "ABCABC…". The first one gives a hexagonal close-packed (hcp) structure. The second one gives a structure known as face-centred cubic (fcc).

Hexagonal ferrites consist of close-packed layers of oxygen ions O^{2-}.[44,45] In certain layers, some oxygen ions are replaced by barium ions Ba^{2+}, which are approximately of the same size as oxygen ions. These close-packed layers form six fundamental blocks, S, S*, R, R*, T, and T*, among which the S*, R*, and T* blocks can be obtained simply through the rotation of the S, R, and T blocks, respectively, by 180° about the *c* axis. The different stacking of the fundamental blocks builds up materials with different structures and physical properties. Table I lists the chemical compositions and building blocks of five types of hexagonal ferrites. As indicated in Table I, M-type hexagonal ferrites are built from the stacking of S, R,

S*, and R* blocks. The structures of S and R blocks are described below. One can refer to Refs. [44] and [45] for the structure of T blocks.

Type	Formula	Build-up
M	$BaFe_{12}O_{19}$	SRS*R*
W	$BaMe_2Fe_{16}O_{27}$	SSRS*S*R*
Y	$Ba_2Me_2Fe_{12}O_{22}$	STST
Z	$Ba_3Me_2Fe_{24}O_{41}$	RSTSR*S*T*S*
U	$Ba_4Me_2Fe_{36}O_{60}$	RSR*S*TS*

Table I. Compositions and building blocks of five types of hexagonal ferrites[44,45,46]

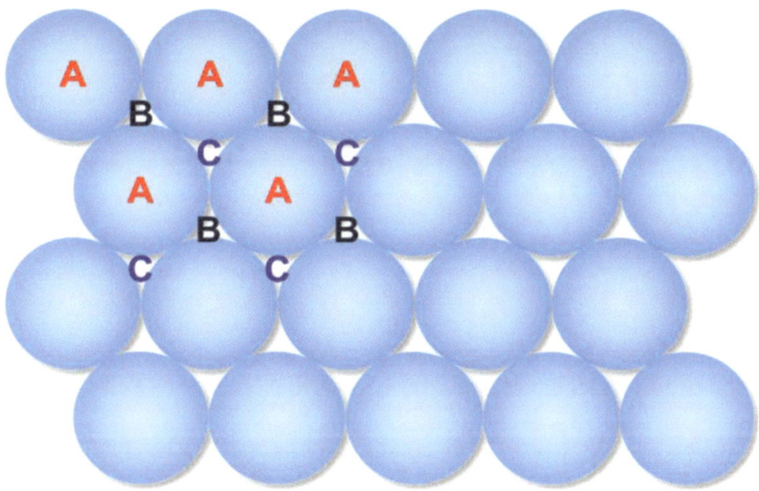

Fig. 1. A close-packed layer of spheres occupying positions A. A second and identical layer of spheres can be placed on top of this layer and occupy positions B or C.

Figure 2 shows the structure of an S block. Figure 2(b) shows a structure with oxygen layers only, and Figure 2(a) shows the top oxygen layer when viewed from above. For a better presentation, the ratio of the oxygen ion diameter to the oxygen-oxygen distance is set to be much smaller than it actually is. Figure 2(a) shows a 60° rhombus consisted of close-packed oxygen ions; and the structure in Fig. 2(b) clearly shows that an S block is built from the stacking of close-packed oxygen layers in an "ABCABC…" sequence. It is important to note that an S block consists of only two oxygen layers, although three layers are shown in Fig. 2(b). For example, one can consider an S block consisting of the top and middle oxygen layers, with the bottom layer belonging to the underneath block. Therefore, one can see that each S block contains eight oxygen ions, with four from each layer.

Figure 2(c) shows a structure with both oxygen ions and cations. The small solid circles show the cations on tetrahedral sites, while the small open circles show the cations on octahedral sites. Between the top and middle oxygen layers, there are one cation at an octahedral site and five cations at tetrahedral sites. The octahedral site is formed by three

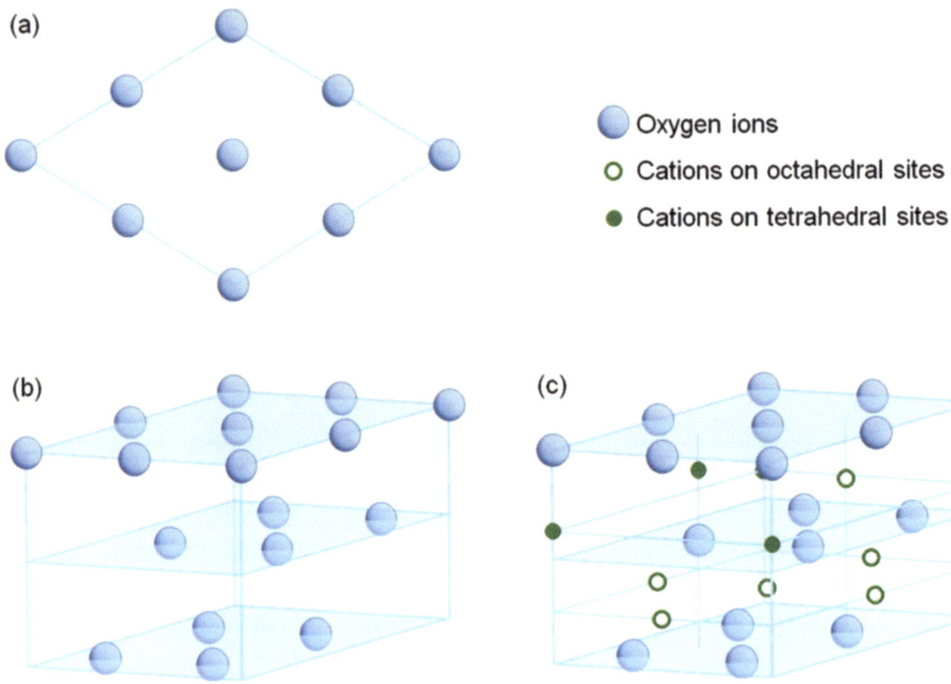

Fig. 2. Structure of S block. (a) Top oxygen layer viewed from above. (b) S block structure with oxygen layers only. (c) S block structure with both oxygen ions and cations.

oxygen ions in the top layer and three in the middle layer and, therefore, lies halfway between the top and middle oxygen layers. Among five tetrahedral cations, one occupies the tetrahedral site formed by three oxygen ions in the top layer and another in the middle layer; the other four occupy the corners of a 60° rhombus, with each shared by four 60° rhombuses. Overall, between the top and middle layers, there are one cation at an octahedral site and two cations at tetrahedral sites.

Between the middle and bottom oxygen layers, there are five cations at the octahedral sites, all of which are within a 60° rhombus halfway between the oxygen layers. Among these five cations, four are at the middle points of four rhombus sides, with each shared by two rhombuses; and the other is at the center of the rhombus and is not shared by any other rhombuses. Overall, there are three octahedral cations between the middle and bottom oxygen layers.

In total, each S block contains eight oxygen ions in close-packed plans, four cations at octahedral sites, and two cations at tetrahedral sites. If the cations are iron ions, the block contains two formula units of Fe_3O_4.

Three points should be noted about S blocks. First, the hexagonal structure of the S block is clearly shown in Fig. 2. The vertical axis of the structure is referred to as the *c* axis. Second, the S block is often referred to as a spinel block. This is because the oxygen ions and cations are so distributed that they form precisely the cubic spinel arrangement with the [111] axis vertical. Third, among the three types of fundamental blocks which make up hexagonal ferrites, the S block is the smallest one and is the only one containing no barium ions.

Figure 3 shows the structure of an R block. Figure 3(b) shows a structure with oxygen and barium ions only. Figure 3(a) shows the top layer of the structure when viewed from above. Figures 3(a) and 3(b) clearly show that an R block consists of three close-packed oxygen layers, with one oxygen ion in the middle layer replaced by a barium ion Ba^{2+}. These layers are stacking on each other in an "ABAB..." sequence. The top, middle, and bottom layers contain four, three, and four oxygen ions, respectively. Overall, each R block contains eleven oxygen ions and one barium ion.

Figure 3(c) shows a structure with all ions. The small solid circles show four cations which occupy the trigonal sites in the middle layer. As each trigonal site is shared by four 60° rhombuses, only one of these four cations belongs to the structure unit shown. The small open circles show five cations at octahedral sites. Among the five sites, one lies halfway between the top and middle oxygen layers, one lies halfway between the middle and bottom oxygen layers, and three are underneath the bottom oxygen layer.

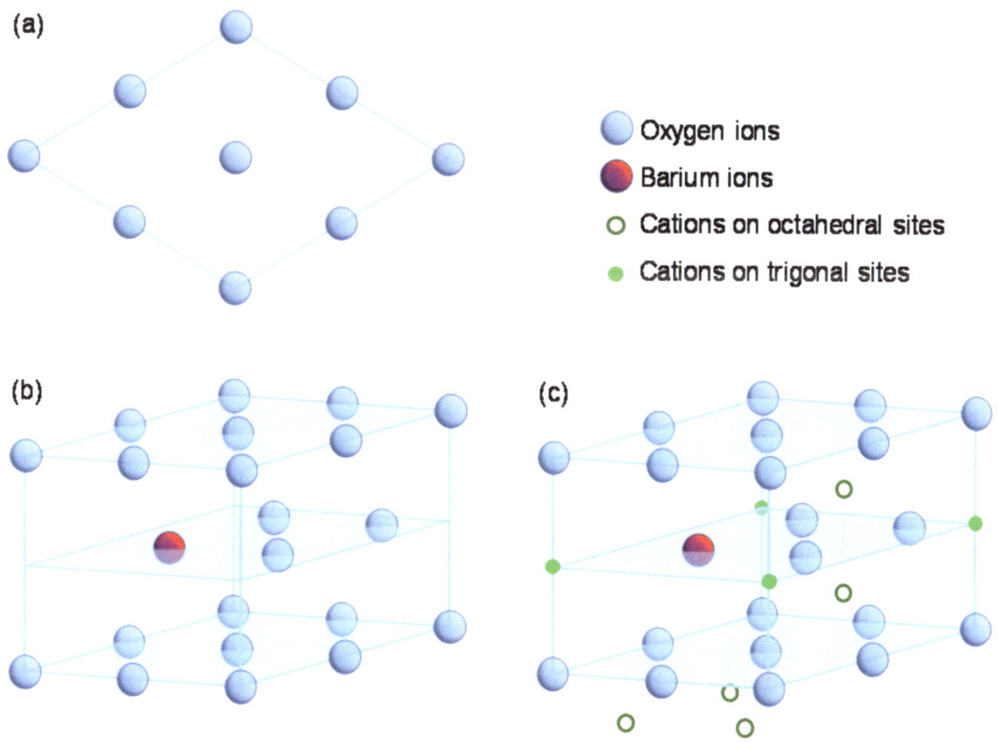

Fig. 3. Structure of R block. (a) The top oxygen layer viewed from above. (b) R block structure with oxygen and barium ions only. (c) R block structure with all ions.

In total, each R block contains eleven oxygen ions, one barium ion, one cation in a trigonal site, and five cations in octahedral sites. If those cations are iron ions, one can denote an R block by $BaFe_6O_{11}$. Note that the R block has an hcp structure thanks to the "ABAB..." stacking sequence; the *c* axis of the structure is along the vertical axis.

It should be noted that both $2Fe_3O_4$ (S block) and $BaFe_6O_{11}$ (R block) are not electrically neutral if the iron ions are trivalent. This, however, is not a problem, because hexagonal ferrites are always made up of more than one type of block. For example, the combination of one S block and one R block yields $BaFe_{12}O_{19}$ which is indeed electrically neutral when the iron ions are Fe^{3+}.

2.2 Structure and static magnetic properties of BaM materials

The M-type barium hexagonal ferrite, often called BaM, has a chemical formula of $BaFe_{12}O_{19}$, with all of the iron ions being trivalent. The crystal structure of BaM is the same as that of the mineral magnetoplumbite. Each elementary cell is formed by the stacking of S, R, S*, and R* blocks and, therefore, contains ten layers of close-packed oxygen ions. Among these ten layers, two layers contain barium ions, as shown by the middle layer in the structure in Fig. 3(b). The two layers within the S (or S*) block, the layer right above the block, and the layer right underneath the block are stacked in an "ABCABC..." sequence. The three layers within the R (or R*) block, the layer right above the block, and the layer right underneath the block are stacked in an "ABAB..." sequence.

The distribution of the iron ions in the BaM lattice sites and the orientation of their magnetic moments are summarized in Table II. A detailed description of these sites is given in Section 2.1. Magnetically, ferrite materials have majority and minority sublattices. Within each sublattice, the magnetic moments are parallel to each other. The moments in two sublattices, however, are opposite to each other. The difference between the total moments of two magnetic sublattices determines the saturation magnetization of the material. In Table II, an upward-directed arrow indicates a contribution to the majority magnetic sublattice, while a downward-directed arrow indicates a contribution to the minority magnetic sublattice.

Block	Formula	Tetrahedral	Octahedral	Trigonal	Net
S	$2Fe_3O_4$	2↓	4↑	-	2↑
R	$BaFe_6O_{11}$	-	3↑, 2↓	1↑	2↑
S*	$2Fe_3O_4$	2↓	4↑	-	2↑
R*	$BaFe_6O_{11}$	-	3↑, 2↓	1↑	2↑

Table II. Distributions of Fe^{3+} ions in a unit cell of BaM materials[44,45]

The data in Table II indicates that a full unit cell of BaM materials contains two formula units of $BaFe_{12}O_{19}$; the net magnetic moment in each unit cell is equal to the moment of eight Fe^{3+} ions. The magnetic moment of each Fe^{3+} ion is usually taken as $5\mu_B$. As a result, each unit cell of BaM is expected to have a net magnetic moment of $40\mu_B$. One can define the vertical axis of the building blocks as the *c* axis of the unit cell and one of the sides of the 60° oxygen rhombus (see Fig. 2(a)) as the *a* axis. In these terms, the lengths of the *c* and *a* axes of a BaM unit cell are about 23.2 Å and 5.89 Å, respectively. With these parameters, one can estimate the saturation induction $4\pi M_s$ of BaM as about 6680 G. This value is close to the $4\pi M_s$ value measured at low temperatures.[45] At room temperature, BaM bulk crystals usually have a $4\pi M_s$ value of about 4700 G;[47,48] while BaM thin films usually show a slightly smaller value.

BaM materials have uniaxial magneto-crystalline anisotropy, with the easy axis along the *c* axis of the hexagonal structure. The effective anisotropy field H_a is about 17 kOe.[45] This field is three orders of magnitude higher than that in YIG materials. It is this strong built-in field that facilitates ferromagnetic resonances in BaM materials at mm-wave frequencies with no need of large external bias fields and, thereby, makes the research field of BaM materials and devices very attractive and promising.

Table III gives the room-temperature structure and physical properties of BaM crystals as well as the Curie temperature of BaM materials.

Parameter	Value	Reference
Lattice constant c	23.2 Å	[45]
Lattice constant a	5.89 Å	[45]
X-ray density	5.28 g/cm^3	[44]
Curie temperature T_c	725 K	[45]
Exchange energy parameter A_{ex}	6.4×10^{-7} erg/cm	[7]
Saturation induction $4\pi M_s$	4700 G	[47], [48]
Effective anisotropy field H_a	17 kOe	[45]
Damping constant α	7×10^{-4}	[7]

Table III. Properties of M-type barium hexagonal ferrites

2.3 Ferromagnetic resonances in BaM materials

One typically makes use of ferromagnetic resonance (FMR) techniques to characterize microwave and mm-wave losses in ferrite materials. Figure 4 shows the FMR effect. Figure 4(a) gives a schematic presentation of the FMR operation, where the magnetization M absorbs energy from the microwave magnetic field h and maintains a fixed angle of precession around the static magnetic field H. The FMR effect manifests itself in a peak response in the measurement of the microwave power absorption in the material as a function of the static magnetic field, as shown in Fig. 4(b). The full width at the half maximum of this so-called FMR absorption curve is usually taken as the FMR linewidth ΔH_{FMR}. The origin of the FMR linewidth differs significantly in different materials. In ferrites, typical relaxation processes that contribute to ΔH_{FMR} include magnon-phonon scattering, two-magnon scattering, charge transfer relaxation, and processes associated with slowly relaxing impurity and rapidly relaxing impurity.[49,50] The identification and quantization of each process demand extensive measurements and numerical analyses. For this reason, one typically uses ΔH_{FMR} as a measure of the overall loss of the material. Very often, FMR measurements are carried out with field modulation and lock-in detection techniques, and the actual FMR data consist of the derivative of the power absorption curve, as shown in Fig. 4(c). In this case, one measures the peak-to-peak FMR linewidth ΔH_{pp}. If the FMR profile is Lorentzian in shape, one can convert ΔH_{pp} into ΔH_{FMR} simply by multiplying ΔH_{pp} by $\sqrt{3}$. One can also conduct FMR measurements by keeping the field constant and sweeping the frequency. The detail on the conversion between the field and frequency linewidths is given in Section 4.1.

One usually measures FMR responses in BaM films with the application of an external magnetic field H along the film easy axes. For a BaM film with the c axis out of the plane, the magnetic torque equation yields an FMR frequency of

$$f_{\mathrm{FMR}} = |\gamma|(H + H_a - 4\pi M_s) \tag{1}$$

where $|\gamma| = 2.8$ **MHz/Oe** is the absolute value of the gyromagnetic ratio. For a film with the *c* axis in the plane, the FMR frequency is given by

$$f_{\mathrm{FMR}} = |\gamma|\sqrt{(H + H_a)(H + H_a + 4\pi M_s)} \tag{2}$$

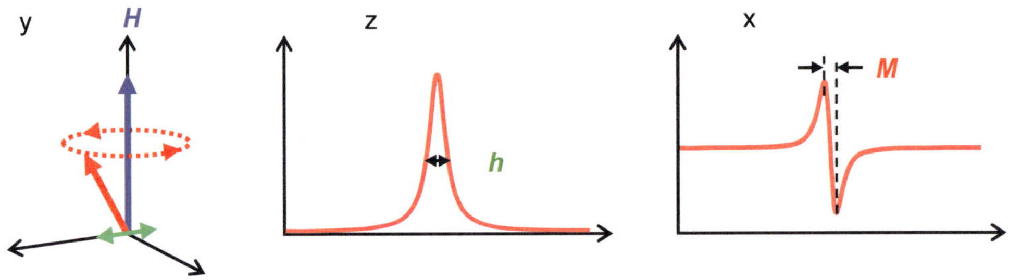

Fig. 4. Schematic presentation of ferromagnetic resonance (FMR)

The films prepared with different processes can show significantly different FMR linewidth values. The lowest linewidth so far was reported by Song *et al.* for a BaM film grown by PLD techniques.[7] This film had an out-of-plane *c*-axis orientation and was 0.85 μm thick. The film showed a 60.3 GHz FMR linewidth ΔH_{FMR} of 27 Oe, which matched the value of single-crystal BaM bulk materials. The frequency-dependent FMR measurements yielded a linear response with a slope of 0.5 Oe/GHz, which corresponded to an effective damping constant α of about 7×10^{-4}.

3. Development of M-type barium hexagonal ferrite films

This section reviews the recent advances made in the development of BaM film materials. The section consists of four subsections, each on a separate effort. The four efforts are (1) the development of BaM thin films that have both low losses and high remanent magnetizations, (2) the deposition of BaM thin films on metallic electrodes, (3) the growth of BaM thin films on semiconductor substrates, and (4) the development of BaM thick films on semiconductor substrates. The motivations and implications of each effort are described in each subsection.

3.1 Development of low-loss, self-biased BaM thin films

In terms of device applications, BaM films with narrow FMR linewidths (ΔH_{FMR}) and high remanent magnetizations (M_r) are very desirable. The narrow linewidth is critical for the realization of low insertion losses for certain devices, while the high remanent magnetization facilitates the operation of devices in absence of external magnetic bias fields, namely, self-biased operation. As mentioned above, Song *et al.* succeeded in the PLD growth of BaM thin films that showed an FMR linewidth as narrow as single-crystal BaM bulks.[7] These films, however, showed a remanent magnetization much smaller than the saturation magnetization (M_s). The main reason for this small remanent magnetization lies on the out-of-plane *c*-axis orientation. Such *c*-axis orientation gives rise to a near unity demagnetizing factor along the film normal direction and a corresponding small M_r value.

Yoon *et al.* were able to use PLD techniques to grow in-plane *c*-axis oriented BaM films with higher remanent magnetizations, at an M_r/M_s ratio of about 0.94.[8,9] Those "in-plane" films, however, had very broad FMR peaks, with a 50-60 GHz peak-to-peak FMR linewidth ΔH_{pp} of about 1150 Oe or larger. Note that the high M_r/M_s ratios in those films derived from the near zero demagnetizing factor along the in-plane *c* axis. Song *et al.* reported "in-plane" BaM films with slightly lower M_r/M_s ratios at about 0.84, but with much narrower peak-to-peak linewidths of about 250 Oe.[35] Those films, however, were made through a hybrid process that involved both PLD and LPE methods along with post-deposition surface flux cleaning.

In 2010, Song *et al.* reported the development of in-plane *c*-axis oriented BaM thin films that had both small ΔH_{FMR} and high M_r.[10] The films were grown on an *a*-plane sapphire substrate by basic PLD techniques. The high quality was realized through several changes in the substrate temperature during the deposition, along with the optimization of other PLD control parameters. The sequential changes in substrate temperature resulted in a series of BaM layers with slightly different structure properties. This quasi-multi-layered configuration served to release interfacial strain and thereby realize high-quality films. The films showed a M_r/M_s ratio that is higher than any previous BaM films, an FMR linewidth that is a factor of four smaller than those of previous PLD films, and an effective anisotropy field that closely matches the value of BaM bulk crystals.

Figure 5 gives the x-ray diffraction (XRD) data for one of those BaM films. Figure 5(a) shows the intensity vs. angle profile, and Figure 5(b) shows an XRD rocking curve. The profile in Fig. 5(a) shows three strong peaks. The central peak comes from the sapphire, while the other two are from the *m* planes of the BaM film. There are no peaks for other planes of the BaM film or the sapphire substrate, or for other phases. The rocking curve for the M(200) peak in Fig. 5(b) shows a "full width at half maximum" (FWHM) of about 0.85°. This value is very small and indicates a very small deviation of the *c*-axis orientation over the film. Note that the FWHM value is about 10% lower than that reported in Ref. [35]. These results clearly indicate that the film has a *c*-axis that is in the plane of the film and is highly oriented.

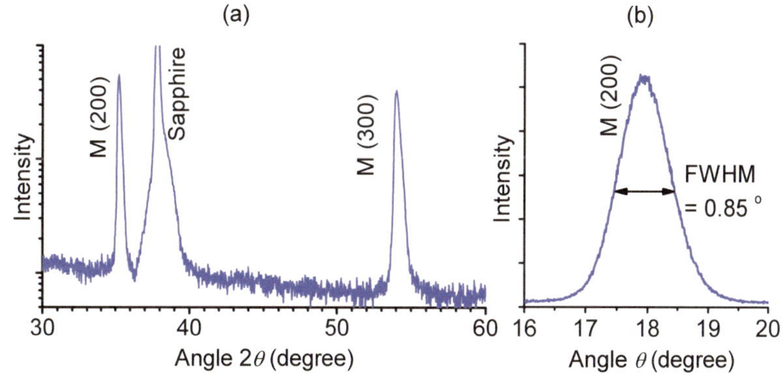

Fig. 5. (a) X-ray diffraction spectrum of a BaM film. (b) X-ray diffraction rocking curve of the same film. (After Ref. [10])

Figure 6 shows two scanning electron microscopy (SEM) images. The one in Fig. 6(a) is for the film surface, while the one in Fig. 6(b) is for the cross section of the film. The image in Fig. 6(a) shows a reasonably smooth surface and no notable holes. The image also shows many fine lines, as indicated by the black arrows. These lines correspond to fine parallel cracks along the direction perpendicular to the *c* axis. The parallelism of these cracks gives a rough measure of the good orientation of the *c* axis. The image in Fig. 6(b) shows that the film thickness is uniform at 2.52 μm and that there are no cracks at the film-substrate interface.

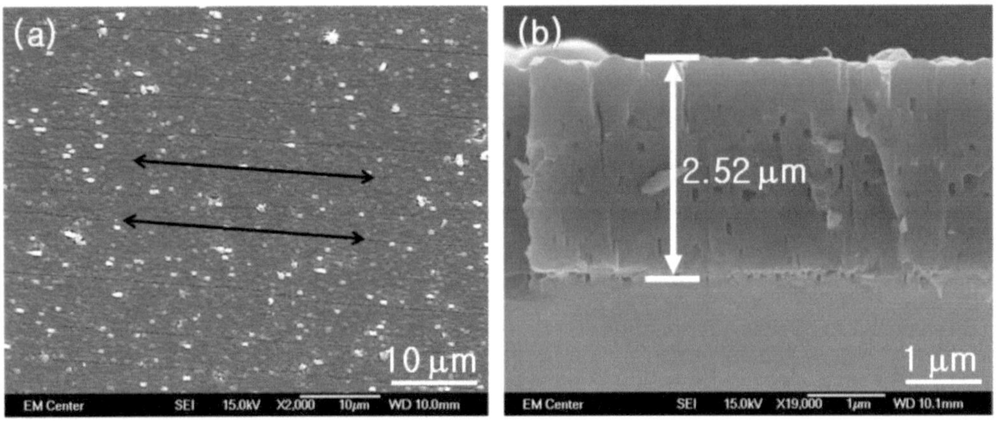

Fig. 6. Scanning electron microscopy images of a BaM film. (a) Film surface. (b) Film cross section. (After Ref. [10])

Figure 7 shows two magnetic induction ($4\pi M$) vs. magnetic field (H) hysteresis loops measured with the fields applied along the in-plane easy and hard axes, as indicated. The easy axis is along the *c* axis defined by the XRD measurements. The dashed lines indicate the extrapolations used to determine the effective anisotropy field H_a. Three results are evident in Fig. 7. (1) The film has an extremely well defined in-plane uniaxial anisotropy with the easy axis along the *c* axis. (2) The data indicate an anisotropy field H_a of about 16.9 kOe, a saturation induction $4\pi M_s$ of about 3.9 kG, and an easy-axis coercive force of about 200 Oe. These values are close to those for high-quality BaM films as reported previously.[7,35] (3) The film has an M_r/M_s ratio of 0.99, which is very close to unity and is the highest value ever obtained for BaM films. These results clearly confirm the in-plane orientation of the *c* axis and demonstrate the near ideal in-plane uniaxial anisotropy for this film.

Figure 8 gives representative FMR data. Figure 8(a) shows three FMR absorption derivative profiles at different frequencies, as indicated. The circles show the data, and the curves show fits to a Lorentzian derivative trial function. The lines and arrows indicate the linewidths based on these fits. The circles in Fig. 8(b) show the measured FMR field vs. frequency data. The line shows theoretical FMR fields based on Eq. (2) and the H_a and $4\pi M_s$ values cited above. These data show four results. (1) The film has a narrow FMR linewidth. The values indicated are four times lower than those for previous PLD films.[8,9] (2) There is a very good match between the experimental FMR profiles and the Lorentzian fits. (3) The theoretical FMR fields match nicely with the experimental values. This match confirms the H_a and $4\pi M_s$ values obtained from the hysteresis loop measurements. (4) The FMR frequency-field curve

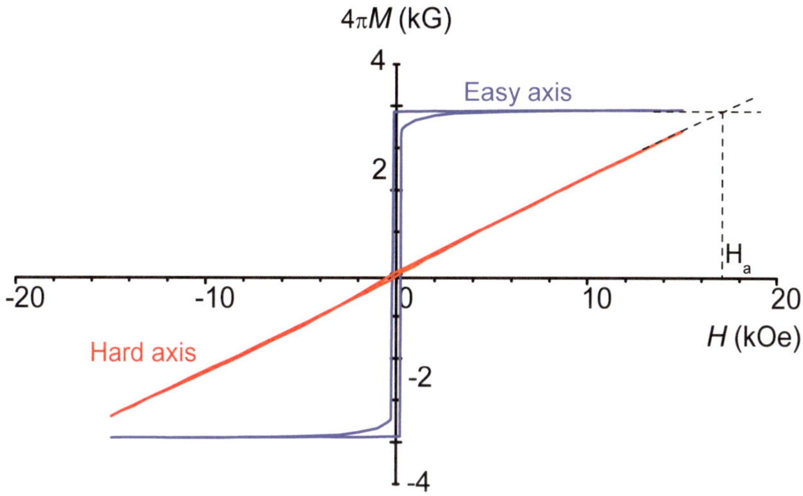

Fig. 7. Magnetic induction $4\pi M$ - field H loops of a BaM film measured with the fields along the in-plane easy and hard axes. The dashed lines indicate the determination of the anisotropy field H_a. (After Ref. [10])

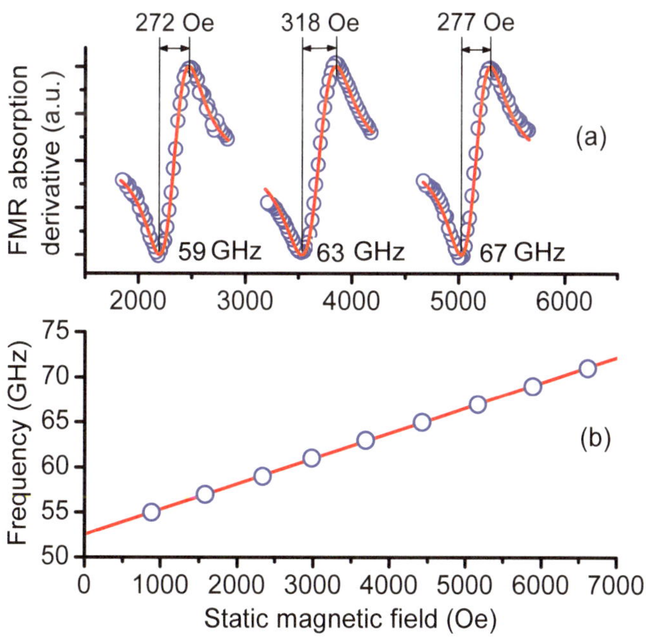

Fig. 8. FMR results for a BaM film. (a) FMR absorption derivative profiles measured at different frequencies. (b) FMR frequency as a function of static magnetic field. In both (a) and (b), the open circles show the data. The curves in (a) show Lorentzian fits. The line in (b) shows theoretical results. (After Ref. [10])

shows an almost linear response. This linear dependence results from the fact that the anisotropy field H_a is significantly larger than the induction $4\pi M_s$.

These results clearly demonstrate the feasibility of the basic PLD growth of BaM films with both high remanent magnetizations and low losses. Future work on the development of BaM films with similar quality but different anisotropy fields is of great interest. This can be realized, for example, through Sc or Al doping in BaM films.[51,52]

3.2 Deposition of BaM thin films on metallic electrodes

Single-crystal sapphire (α-Al_2O_3) has been the substrate of choice for the growth of BaM films by PLD and sputtering.[7,8,9,10,31,35,36] Sapphire is chosen because it has a rhombohedral crystal structure (a=5.128 Å and α=55°22') which is close to the hexagonal structure of BaM;[31] and the mismatches of the lattice parameters and thermal expansion coefficients between sapphire and BaM materials are relatively small.[7,31,36] Certain device applications, however, require the growth of BaM films on conductive substrates. In coupled-line and stripline-type devices, for example, a ground plane is needed underneath the active layer.

In 2010, Nie *et al.* succeeded in the fabrication of out-of-plane *c*-axis oriented BaM thin films on platinum (Pt) electrodes through metallo-organic decomposition (MOD) techniques.[15] The films were prepared on 300 nm-thick (111)-oriented Pt layers which were sputtered on Si wafers. The fabrication processes include three main steps, (1) spin coating of a precursor onto a substrate, (2) annealing at different temperatures (150-450 °C) to remove solvents and realize metallo-organic decomposition, and (3) rapid thermal annealing (RTA) at high temperatures (850-900 °C) to facilitate the formation of a proper structure. It was found that the RTA step was very critical for the realization of high-quality BaM films. It was demonstrated, for example, that the temperature significantly affects the crystalline structure of the film, and the type of the process gas strongly affects both the remanent magnetization and FMR linewidth.

Figure 9 shows the magnetic moment vs. field hysteresis loops measured for three film samples for both in-plane (| |) and out-of-plane (⊥) fields. The films were obtained with different RTA processes. For films #1 and #2, the RTA processes were done in O_2 and N_2, respectively. For film #3, the RTA process was made in N_2 first and then in O_2. The data for films #1, #2, and #3 are shown in Figs. 9(a), 9(b), and 9(c), respectively. One can clearly see that (1) these films show significantly different magnetic properties and (2) film #3 has the highest out-of-plane remanent magnetization, with $M_r/M_s \approx 0.93$, and the lowest out-of-plane coercivity, with H_c=4.5 kOe. Note that high remanent magnetizations are desirable from the point of view of device applications, as discussed in Section 3.1.

Nie *et al.* reported that film #3 not only showed the largest M_r/M_s ratio, but also had the lowest loss. Figure 10 presents the FMR linewidth vs. frequency data obtained for film #3. One can see that the film had a linewidth of about 338 Oe at 60 GHz and a slope of 4.2 Oe/GHz over the 45-60 GHz range. This slope corresponds to an effective damping constant α of about 0.006. These values are about one order of magnitude larger than those of the out-of-plane *c*-axis oriented BaM films grown by PLD on sapphire substrates.[7] Nevertheless, they are reasonably low in terms of practical applications.

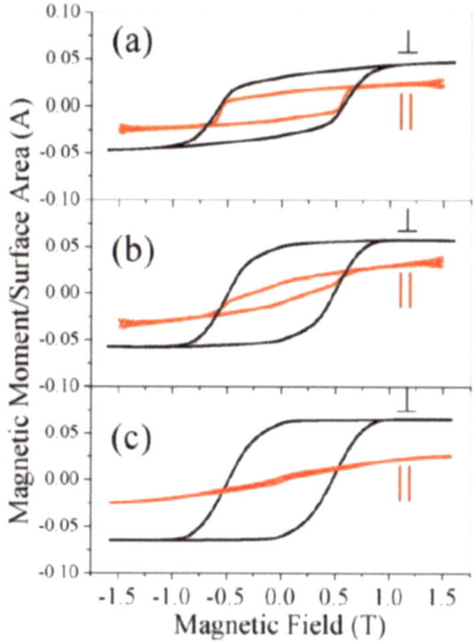

Fig. 9. Magnetic hysteresis loops obtained for three BaM films for both in-plane (| |) and out-of-plane (⊥) external magnetic fields. (After Ref. [15])

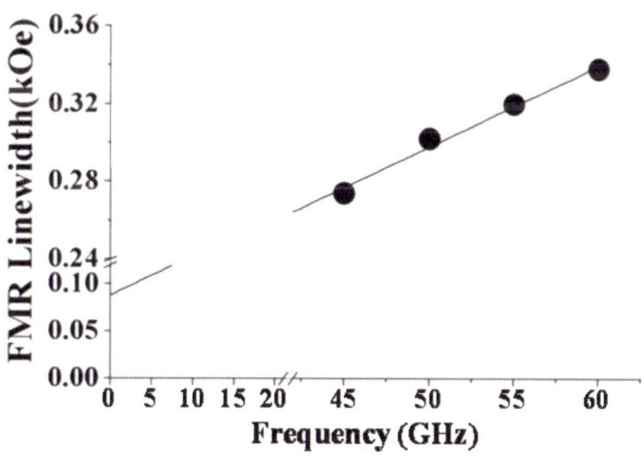

Fig. 10. FMR linewidth vs. frequency data for a BaM film. (After Ref. [15])

Two points should be highlighted. First, the work by Nie *et al.* clearly demonstrated the feasibility of the use of MOD techniques to fabricate on metallic electrodes BaM thin films with high M_r/M_s ratios and reasonable low losses.[15] Second, in comparison with PLD techniques, the MOD techniques are relatively inexpensive and can be used to make relatively large films. It should also be noted that for the data shown in Figs. 9 and 10, the films are about 200-500 nm thick. Future work on the use of MOD to make much thicker BaM films is of great interest. In principle, this can be done by the repetition of those processing steps mentioned above. Future work on the fabrication of BaM films on other "more conventional" electrodes, such as copper, is also of great interest. Note that the deposition of BaM films on copper can be challenging because one might need to face the issues associated with copper diffusion or oxidation at high temperatures.

3.3 Deposition of BaM Thin films on semiconductor substrates

The growth of high-quality BaM films on semiconductor substrates can allow for the development of BaM-based devices that are compatible with monolithic integrated circuits. In 2006, Chen *et al.* succeeded in the deposition of BaM thin films on single-crystal 6H-SiC substrates by PLD techniques.[11] The 6H-SiC substrates were chosen not only because they are promising wide band-gap semiconductor materials for next-generation electronics, but also because they have the same hexagonal crystal structure as the BaM materials, with a lattice mismatch of 4.38% in the *c* plane.[13] The films showed good crystal textures and effective anisotropy fields higher than 15 kOe. The loss of the films, however, was very high, with ΔH_{pp}>1 kOe. Such broad linewidths resulted from the random orientation of the grains in the BaM films and the diffusion of silicon from the substrates into the BaM films. Through the introduction of an interwoven MgO/BaM multilayered buffer into the film-substrate interface, Chen *et al.* were able to significantly improve the quality of the films and reduce the ΔH_{pp} value down to 500 Oe at 55 GHz.[12] The MgO/BaM multilayered buffer was also grown by PLD techniques. This buffer not only reduced the lattice mismatch between the film and the substrate, but also suppressed silicon diffusion at high temperatures.

In 2007, Chen *et al.* made use of a hybrid process and succeeded in the deposition of BaM films on 6H-SiC substrates with even lower losses.[13] The process involved three steps, (1) growth of an MgO (111) buffer layer on a 6H-SiC (0001) single crystal wafer by the MBE method, (2) growth of a BaM film onto the MgO layer by the PLD techniques, and (3) post-annealing in air. The MgO layer was grown at a substrate temperature of 150 °C. The BaM film was deposited at a substrate temperature of 915 °C. The post-annealing was done at 1050 °C. The thickness of the MgO layers varied from 2 nm to 12 nm, and the BaM films were 400 nm thick. The samples showed an abrupt MgO/BaM interface, with no Si or Fe diffusions. The films had out-of-plane *c*-axis orientation and showed an out-of-plane uniaxial anisotropy. The effective anisotropy field and saturation induction were 16.9 kOe and 4400 G, respectively, which were both close to the values for single-crystal BaM bulks (see Table III). The FMR measurements on these films indicated linewidths that were significantly narrower than those of previous BaM films grown on 6H-SiC substrates.[11,12] Before the post-annealing, the films showed a ΔH_{pp} value of about 220 Oe at 53 GHz. After the post-annealing, this value was reduced to about 100 Oe. Figure 11 shows an FMR profile measured at 53 GHz for a BaM film sample where the MgO buffer layer is 10 nm thick. This profile indicates a ΔH_{pp} value of 96 Oe.

The samples discussed above involved the MBE growth of MgO buffer layers and the PLD growth of BaM films. In 2010, Cai *et al.* demonstrated the MBE growth of both MgO buffer layers and BaM thin films on 6H-SiC substrates.[14] The MgO layers were grown at a substrate temperature of 150 °C. For the growth of BaM films, the substrate temperature was set to 800 °C. The oxygen pressure was optimized to allow for the growth of BaM films with near-perfect structures. As in Ref. [13], the 6H-SiC substrates were single-crystal (0001) wafers, the MgO layers had (111) orientation, and the BaM *c* axis was out-of-plane. Figure 12 shows the induction vs. field hysteresis loops obtained for a 200 nm-thick BaM film grown on a

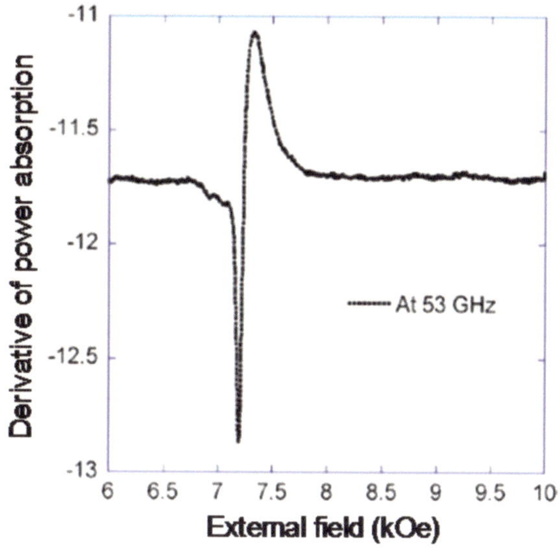

Fig. 11. FMR power absorption derivative profile for a BaM thin film grown on a single-crystal 6H-SiC substrate. (After Ref. [13])

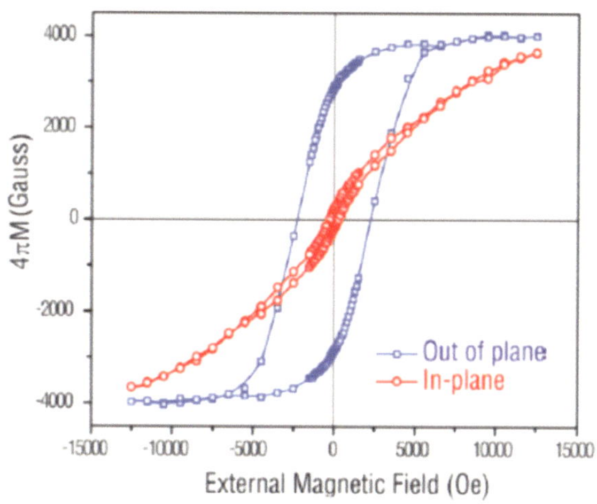

Fig. 12. Hysteresis loops for a BaM thin film grown on a single-crystal 6H-SiC substrate by the MBE method. The squares and circles show the data for external fields applied out of the film plane and in the film plane, respectively. (After Ref. [14])

6H-SiC substrate with a 10 nm-thick MgO buffer layer. The data confirm the out-of-plane anisotropy of the film and indicate an effective anisotropy field of 16.2 kOe, a saturation induction of 4100 G, and a M_r/M_s ratio of 0.7. These results, together with X-ray photoelectron spectroscopy and atomic force microscopy results, clearly indicate the feasibility of the use of MBE to prepare high-quality BaM thin films on 6H-SiC substrates.

There are two points to be mentioned. First, the low-loss films reported in Ref. [13] had very low remanent magnetizations. Future work on the use of the same process to grow in-plane *c*-axis oriented BaM films is of great interest. Such films have a near zero demagnetizing factor along the in-plane *c* axis and, therefore, are expected to exhibit near unity M_r/M_s ratios,[10] as discussed in Section 3.1. Second, the MBE growth of BaM thin films has the potential to achieve tight stoichiometric control and near-perfect crystal structures needed for the realization of extremely low losses.[14] Future work on the use of MBE thin films as seed layers for the fabrication of high-quality thick films is also of great interest.

3.4 Fabrication of BaM thick films on semiconductor substrates

For certain applications, high-power handling or strong coupling require the use of BaM thick films, with a desired thickness range of several hundreds of microns. In 2008, Chen *et al.* demonstrated the feasibility of the fabrication of polycrystalline BaM thick films on Si wafers by the screen printing method.[16,17] The fabrication process involved five main steps, (1) deposition of an amorphous Al_2O_3 buffer layer on a Si (001) substrate by magnetron sputtering, (2) spreading of a BaM/binder paste over a stencil onto the substrate, (3) drying of the "wet" film on a hot plate (300-400 °C), (4) sintering at 1100 °C in air for 2 hours, and (5) annealing at 1150 °C in air for 10 minutes. The Al_2O_3 layer was about 1 μm thick and served as a barrier to reduce the diffusion of Si into the BaM film during the sintering process. The printing paste consisted of 70-75 wt% BaM particles and 25-30 wt% binder, and the BaM particles had an average size of 0.6 μm. The sintering at 1100 °C facilitated the growth of BaM grains and the formation of a continuous film. The annealing at 1150 °C allowed for a moderate level of diffusion of aluminum ions into the BaM lattice near the interface. This moderate diffusion enhanced the adhesion of the film to the substrate.

The polycrystalline BaM film wafers with a thickness range of 50-200 μm and a diameter of 1 inch were fabricated.[16,17] Representative hysteresis loops are shown in Fig. 13.[17] The dashed and solid loops show the data obtained at 5 K and 300 K, respectively. The data indicate a $4\pi M_s$ value of 2090 G at 5 K and a value of 1477 G at 300 K. These values are smaller than

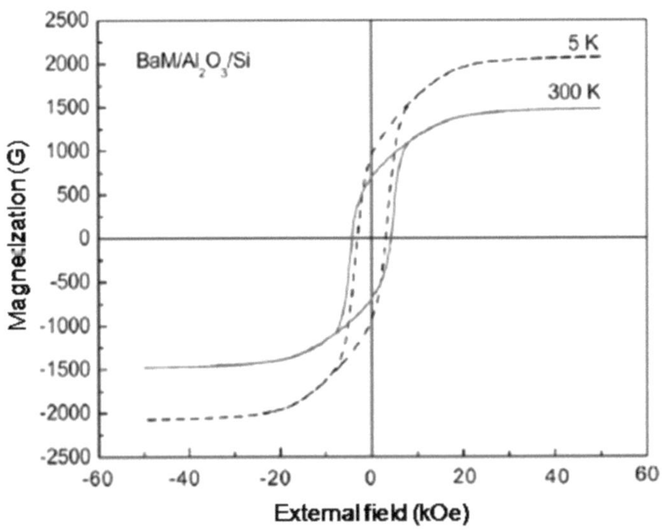

Fig. 13. Hysteresis loops for a BaM thick film grown on a Si substrate by screen printing. The dashed and solid loops show the data measured at 5 K and 300 K, respectively. (After Ref. [17])

that for single-crystal BaM bulks. The main reason for this lies on the relatively low density of the film, which is intrinsic to the screen printing technique. The data also indicate a M_r/M_s ratio of about 0.44 at 5 K, which is close the expected value of 0.5 for randomly oriented polycrystalline samples.

Figure 14 shows the FMR linewidth vs. frequency response of a BaM thick film.[16] One sees nearly linear behavior. All the linewidth values are larger than 1 kOe. Possible reasons for such large values include the following. (1) The magnetic fields used in FMR measurements were less than 20 kOe. As a result, the magnetization in the BaM film was not completely saturated. (2) The grains in the film were randomly oriented. (3) The porosity in the film was relatively high. Future work on the optimization of the screen printing process for BaM thick films with much narrower FMR linewidths is of significant interest.

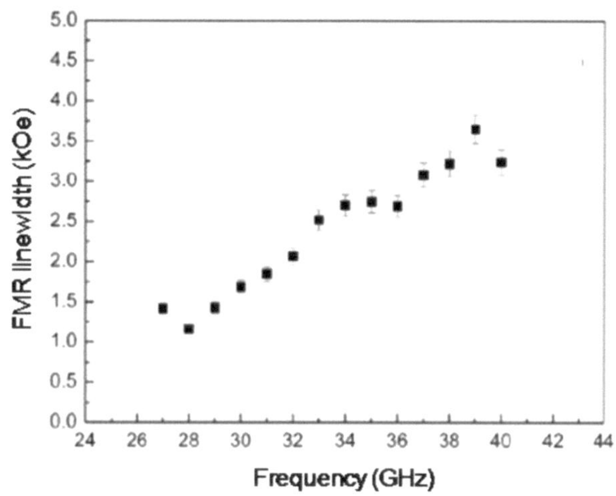

Fig. 14. FMR linewidth as a function of frequency for a BaM thick film fabricated on a Si substrate by screen printing. (After Ref. [16])

4. BaM thin film-based millimeter wave devices

Great progress has been made in recent years in the development of BaM-based planar mm-wave devices.[10,18-24] Such devices include (1) notch filters made of BaM thin films and microstrip geometry, (2) notch filters made of BaM slabs and strip-line geometry, (3) self-biased notch filters using high-M_r BaM thin films and coplanar waveguide (CPW) geometry, (4) notch filters based on the excitation of confined magnetostatic waves in narrow BaM strips and CPW configurations, and (5) phase shifters using BaM thin films and CPW geometry. This section describes the device configurations and responses of (3), (4), and (5). One can refer to Refs. [22] and [23] for details on (1) and (2).

4.1 Self-biased millimeter wave notch filters

Song *et al.* demonstrated a self-biased BaM notch filter in 2010.[10] The device consisted of a high-M_r BaM film element positioned on the top of a CPW structure. The alternating magnetic field produced by the CPW signal line is spatially non-uniform. This non-uniform field excites magnetostatic waves (MSW) in the BaM film.[1,3] Such waves propagate along the direction transversal to the CPW signal line and decay during the propagation. The net effect is a band-stop response of the device in a certain frequency range which corresponds to the bandwidth of the MSW excitation.

Figure 15 shows the structure and responses of such a notch filter. Figure 15(a) shows the device structure. Figure 15(b) shows a transmission profile of the device for zero external fields. Figure 15(c) shows transmission profiles measured at different external fields (Oe), as indicated. The BaM film element was 4.3 mm by 2.3 mm. Its properties are discussed in Section 3.1. The longer side of the film element, which was also the easy axis direction, was along the CPW signal line. The width of the CPW signal line was 50 μm. For the data in Fig. 15(c), the fields were applied along the CPW signal line.

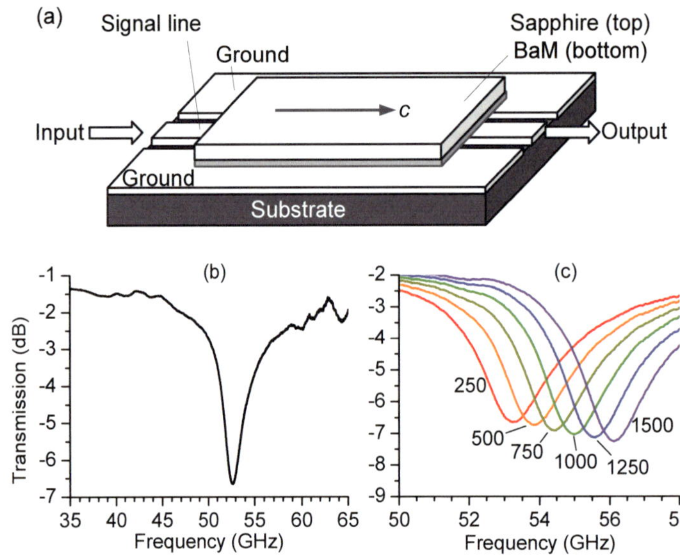

Fig. 15. (a) Diagram of a BaM thin film-based mm-wave notch filter. (b) Transmission profile of the filter for zero fields. (c) Transmission profiles of the filter for nonzero fields (Oe), as indicated. (After Ref. [10])

The data in Fig. 15(b) show a band-stop response, with a maximum absorption of 6.7 dB and an insertion loss less than 2 dB. The center frequency is 52.69 GHz, which is slightly higher than the zero field FMR frequency of 52.50 GHz. This difference results from the fact that the frequency of the magnetostatic waves transverse to the field is above the FMR frequency.[1,3] The data also show a 3 dB linewidth of about 2.52 GHz. This value is much larger than the frequency-swept FMR linewidth, Δf_{FMR}, of the film, about 1.46 GHz; this indicates that the filtering response is not due to the FMR effect. Note that the linewidth Δf_{FMR} was obtained by

$$\Delta f_{\mathbf{FMR}} = \left(\frac{\partial f_{\mathbf{FMR}}}{\partial H} \bigg|_{H=0} \right) \sqrt{3} \Delta H_{\mathbf{pp}} \tag{3}$$

with ΔH_{pp}=300 Oe. These responses resulted from the self-biased operation of the device. The data in Fig. 15(c) indicate that the filter can be tuned for higher frequency operations with relatively low fields. Note that, for a YIG-based notch filter, an operation at 53 GHz requires an external field of 18 kOe, which is significantly higher than the fields indicated in Fig. 15(c). It should also be noted that the insertion loss at the shoulders of the absorption dip results mainly from the ferromagnetic resonance. This is demonstrated by the decrease of the insertion loss at 52 GHz with the increase of the field shown in Fig. 15(c).

4.2 Notch filters based on excitation of confined magnetostatic waves

The notch filter discussed in Section 4.1 showed a maximum absorption of only 6.7 dB. In terms of practical applications, filters with much larger maximum absorption are desirable. In 2011, Lu *et al.* demonstrated that one could significantly increase the absorption level through the replacement of the wide BaM film element with a narrow BaM film strip.[24] This significant increase in absorption results from the excitation of confined MSW modes in the BaM strip, which is explained in detail below.

Figure 16 shows the configuration of such a notch filter. The BaM film strip has its substrate facing up and its length along the CPW signal line. The easy axis of the BaM strip is along the strip length. The CPW-produced non-uniform magnetic field excites MSW modes in the BaM strip. These waves propagate along the strip width direction and are confined by the edges of the strip. The net effect is that the modes that satisfy the phase constraint $2kw=n2\pi$ are relatively strong, while other modes are weak. In the phase condition, k is the wavenumber, w is the width of the BaM strip, and n is an odd integer. The waveforms for the first three modes are shown schematically in Fig. 16. The excitation of these confined modes results in reduced output power and a band-stop filtering response which is discussed below.

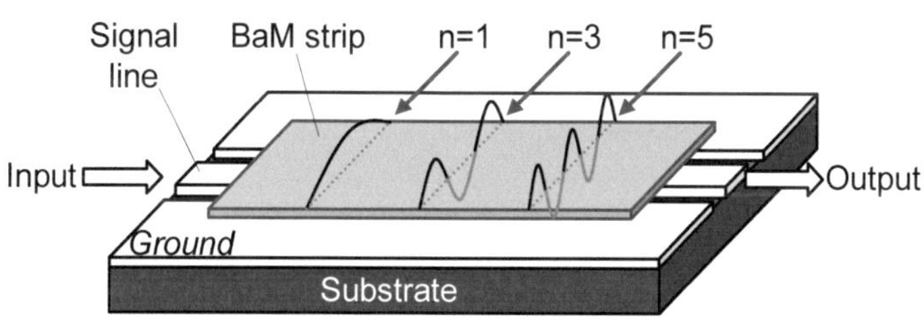

Fig. 16. Diagram of a confined magnetostatic wave-based BaM notch filter. (After Ref. [24])

There are two important points to be noted. (1) The even modes (n=2, 4, 6…) cannot be excited. This is because the in-plane components of the CPW signal line-produced alternating magnetic field is symmetric across the signal line. (2) Only the n=1 mode will be strongly excited and other modes will be very weak. This is due to the fact that the MSW excitation efficiency decreases significantly with the wavenumber k for the geometry considered here.

For the data described below, the coplanar waveguide has a 50 μm-wide signal line, a signal line-to-ground separation of 25 μm, and a nominal impedance of 50 Ω. A wide BaM film, which was the same as the films discussed in Sections 3.1 and 4.1, was cut into a narrow strip with a length of 4.30 mm. The width of the BaM strip was reduced to different values through polishing and ranged from 1.30 mm to 0.24 mm. The ΔH_{pp} values for the 1.30 mm strip and the 0.24 mm strip were 306 Oe and 307 Oe, respectively, at 56 GHz. The closeness of these two values indicates a good uniformity of the initial wide BaM film. With Eq. (3)

and the parameters cited in Section 3.1, one can convert these ΔH_{pp} values to a Δf_{FMR} value of about 1.49 GHz.

Figure 17 shows representative data on the self-biased operation of the device. Figure 17(a) gives the transmission profiles for filters with BaM strips of different widths, as indicated. Figures 17(b) and 17(c) give the maximum absorption and bandwidth data, respectively, for the filtering responses shown in Fig. 17(a). The bandwidth was taken at the transmission level 3 dB higher than the minimum transmission. Three important results are evident in Fig. 17. (1) The devices all show a self-biased band-stop response at about 52.7 GHz. (2) With a reduction in the BaM width, the maximum absorption increases substantially while the bandwidth decreases significantly. (3) In spite of the significant changes in absorption and bandwidth, there is only a slight increase in the device insertion loss. For all the measurements, the insertion loss is less than 2 dB on the low-frequency side and less than 4 dB on the high-frequency side.

The fact that a reduction in the BaM strip width leads to an increase in absorption and a decrease in bandwidth can be explained as follows. In a BaM strip which is relatively wide, the magnetostatic wave decays during its propagation along the width of the strip, and the MSW confinement across the strip width is weak. As a result, broadband magnetostatic waves are excited, and the filter shows a broad bandwidth and a small maximum absorption. In contrast, in a very narrow BaM strip, the effect of the MSW decay is insignificant and the MSW propagation is confined by the strip edges. This geometry confinement yields a phase constraint on the magnetostatic waves. As a result, narrowband magnetostatic waves with $k \approx \pi / w$ are excited, and the filter shows a much narrower bandwidth and a much larger absorption.

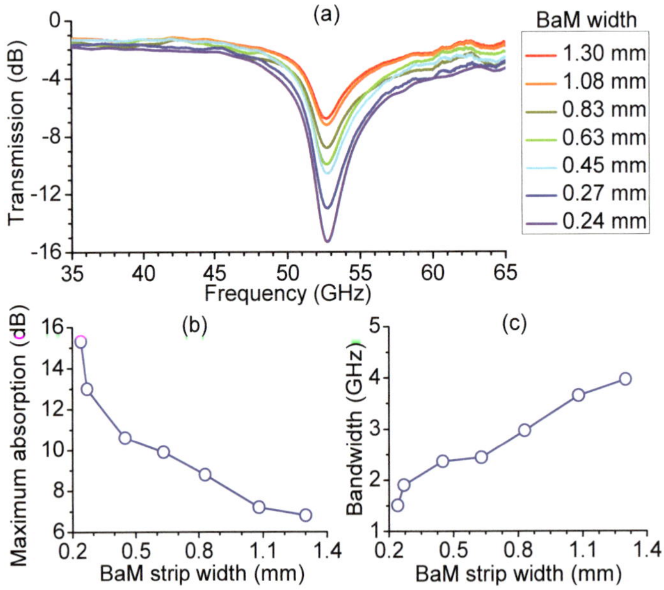

Fig. 17. (a) Transmission profiles for filters with BaM film strips of different widths. (b) Filter maximum absorption vs. BaM strip width. (c) Filter bandwidth vs. BaM strip width. (After Ref. [24])

The above explanation yields three expectations as follows. (1) As broadband magnetostatic waves are excited in wide BaM strips, the bandwidth of the filter using a wide BaM strip should be larger than the FMR linewidth Δf_{FMR} of the BaM strip. (2) The field/film configuration considered here supports the propagation of surface MSW modes, for which the frequency f is higher than f_{FMR} and increases with the wavenumber k. As a result, when one shifts from a regime where broadband magnetostatic waves are excited and the maximum absorption is at f_{FMR} to a regime where only the confined modes are excited and the center frequency is at $f=f_{k=\pi/w}$, one should see a shift of the maximum absorption frequency to a higher value. (3) One should also expect that, with a reduction in BaM strip width, the absorption profile is narrowed from the high frequency side, not the low frequency side, as the frequencies of the broadband MSW excitations are all above f_{FMR}.[3]

These expectations are all confirmed by the data in Fig. 17. (1) The bandwidths of the filters with wide strips are all larger than Δf_{FMR}, while that of the filter with the 0.24 mm strip is just slightly larger than Δf_{FMR}. (2) When the width was reduced from 1.30 mm to 0.24 mm, the maximum absorption frequency increased from 52.72 GHz to 52.74 GHz. The net increase agrees well with the theoretical value, which is evaluated as $f_{k=\pi/w} - f_{FMR}$ = 19 MHz. (3) With a reduction in strip width, the 3 dB bandwidth of the absorption profile is reduced on the high frequency side.

In addition to the self-biased operation described above, the filter is also tunable with low fields. Figure 18 demonstrates such low-field tuning. Figure 18(a) shows the transmission profiles for different external fields for a device with the 0.24 mm-wide strip. Figures 18(b) and 18(c) show the maximum absorption frequency and bandwidth, respectively, as a function of field for the filtering response. As for the data in Fig. 17(c), the bandwidth was taken at the transmission level 3 dB higher than the minimum transmission. For all the measurements, the fields were applied along the BaM strip. The curve in Fig. 18(b) shows MSW frequencies calculated as[3]

$$f(H)=|\gamma|\left[(H+H_a)(H+H_a+4\pi M_s)+\frac{1}{4}(4\pi M_s)^2\left(1-e^{-2\pi d/w}\right)\right]^{\frac{1}{2}} \tag{4}$$

where d is the film thickness. The calculations used the parameters cited above and those in Section 3.1. The only exception is that a value of 16.94 kOe was used for H_a.

The data in Fig. 18 show four things. (1) The filter is tunable with low fields. (2) The operation frequency increases almost linearly with field and matches almost perfectly with the theoretical value. (3) The bandwidth decreases slightly, with an overall change of only 7% in the entire field range. (4) The insertion loss remains almost constant. All of these results are critical for practical applications. The data in Fig. 18(a) also show a slight increase in maximum absorption with field. The fact that the absorption is lower and the bandwidth is larger at low fields may be a result of low-field loss effects.[53]

4.3 Millimeter wave phase shifters

The devices in both Sections 4.1 and 4.2 made use of CPW geometry. With similar geometry, it is also possible to make planar BaM phase shifters. This possibility was demonstrated by Wang *et al.* in 2010.[21] The phase shifter was made of the same structure as

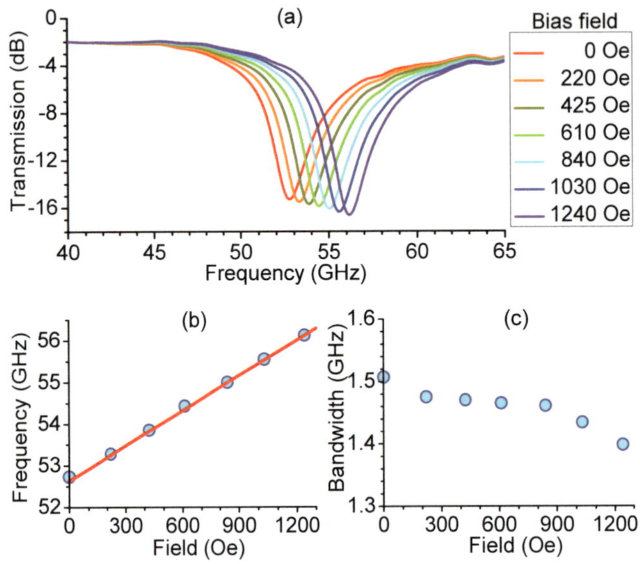

Fig. 18. (a) Transmission profiles measured at different fields, as indicated, for a filter with a 0.24 mm-wide BaM strip. (b) Filter center frequency as a function of field. (c) Filter bandwidth as a function of field. The curve in (b) shows theoretical frequencies. (After Ref. [24])

that shown in Fig. 15(a). The coplanar waveguide had a length of 5.0 mm and an impedance of 50 Ω. The signal line had a width of 100 μm, and the separation between the signal line and grounds was 50 μm. The BaM film element was 4 mm long, 2 mm wide, and 5 μm thick. It was positioned on the top of the coplanar waveguide with its substrate side facing up and its short edge parallel to the signal line. The film was grown on a *c*-plane sapphire substrate through PLD techniques.[7] It showed out-of-plane *c*-axis orientation, an effective out-of-plane anisotropy field of 16.5 kOe, a saturation induction of 4300 G, and a 60 GHz peak-to-peak FMR linewidth of 340 Oe. For the measurements of the phase shifter, an external field was applied perpendicular to the film plane to tune the phase of the signal.

Figure 19 shows the responses of the device.[21] Figures 19(a) and 19(b) present the amplitude and phase, respectively, of the transmission parameter S_{21} for four different fields. Figures 19(c) and 19(d) give the central frequency and 3 dB linewidth of the dips in the transmission profiles, four of which are shown in Fig. 19(a). The line in Fig. 19(c) shows the theoretical FMR frequencies of the BaM film calculated with Eq. (1), the parameters given above, and a small field correction of +0.14 kOe. It should be noted that previous theoretical calculations for microstrip geometry had shown that the position of the dip in transmission did not necessarily occur at the same frequency as the maximum absorption in a FMR experiment.[20] The field correction found here is consistent, in magnitude and sign, with the shift found in Ref. [20].

The data in Fig. 19 show four results. (1) As shown in Fig. 19(a), the transmission of the device shows a clear dip response, which is tunable with the field. (2) In the dip regime, the phase changes notably with the field. (3) There is a good agreement between the dip central frequencies and theoretical FMR frequencies. This match indicates that the transmission dip and phase change responses originate from the FMR effect in the BaM film. (4) The linewidth is one order of magnitude higher than and shows a frequency dependence opposite to that reported in Ref. [7].

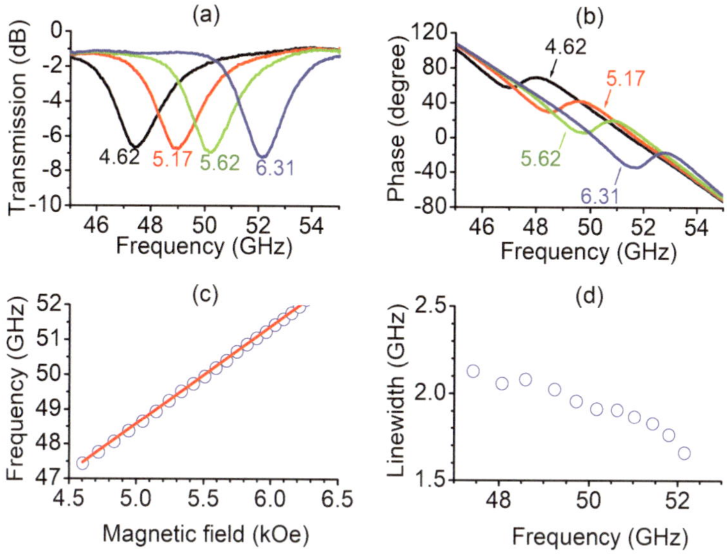

Fig. 19. Responses of a BaM-based mm-wave phase shifter. (a) and (b) show the transmission and phase profiles, respectively, which were measured for four different fields (kOe), as indicated. (c) and (d) show the frequency and 3 dB linewidth for the dips in the transmission profiles shown representatively in (a). (After Ref. [21])

Figure 20 shows representative data on the phase shift properties. Figures 20(a) and 20(b) show the phase shift and insertion loss, respectively, as a function of field measured at 50 GHz. The phase shift is the phase of the device output signal relative to that measured at H=5.35 kOe. Figures 20(c) and 20(d) show the data for 54 GHz. The phase shift in Fig. 20(c) is the phase relative to that measured at 4.60 kOe. For the field ranges in Fig. 20, the 50 GHz point is within the on-resonance regime, while the 54 GHz point is in the below-resonance regime.

The data in Figs. 20(a) and 20(b) show a negative phase shift which decreases almost linearly with H and an insertion loss which is almost constant over the entire field range. The phase shift decreased from 0 to -30° as the field was increased from 5.35 kOe to 5.70 kOe. This corresponds to a phase tuning rate of 43°/(mm·kOe). The insertion loss changed in a narrow range of 6.3±0.5 dB, and this corresponds to a loss rate of about 3.1 dB/mm. The responses at 54 GHz are significantly different. Specifically, both the phase shift and insertion loss are smaller than those for 50 GHz. The linear phase tuning range is 11°. The phase tuning rate is 3.2°/(mm·kOe). The insertion loss is only 0.7 dB/mm.

The above results clearly demonstrate the feasibility of the use of FMR effects in BaM thin films to develop planar phase shifters for mm-wave signal processing. Three points should be mentioned. (1) It is the built-in high anisotropy field that facilitates the operation of the phase shifter at 50 GHz for relatively low bias fields. (2) Both the linear tuning of the phase and the flatness of the loss curves are critical for practical applications. The maximal phase shift ranges are much wider than those shown in Fig. 20. In these wider ranges, however, the phase tuning is nonlinear and the loss is not constant. (3) Strictly speaking, the modes in the BaM film were MSW modes, as those discussed in Section 4.1, rather than the uniform

FMR modes. In fact, as the CPW signal line was relatively wide and the BaM film had a relatively broad FMR linewidth, one can view those MSW modes as quasi-FMR modes.

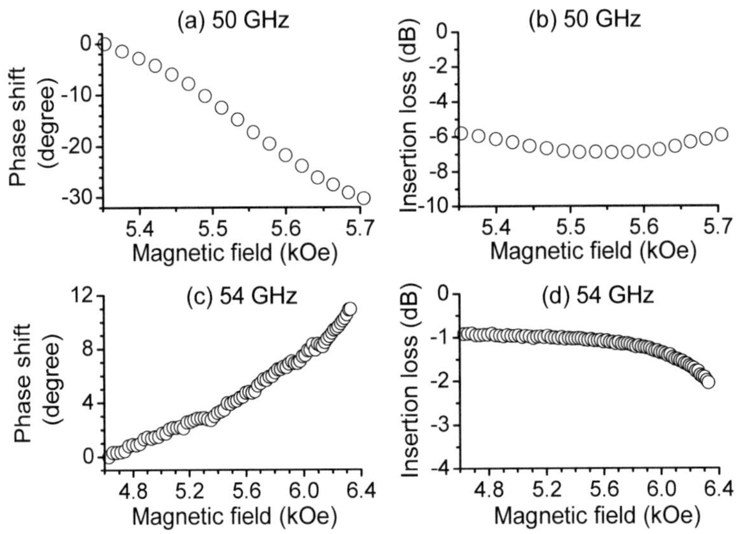

Fig. 20. Phase shift and insertion loss of a BaM-based phase shifter as a function of bias field. The data in (a) and (b) were measured at a fixed frequency of 50 GHz. The data in (c) and (d) were measured at 54 GHz. (After Ref. [21])

4.4 Theoretical analysis of millimeter wave phase shifters

Camley *et al.* carried out theoretical studies on the feasibility of BaM thin film-based mm-wave devices.[20,21,22] Figure 21 shows representative theoretical results obtained for a phase shifter that had the same structure and parameters as the one described in Section 4.3.[21] Figures 21(a) and 21(b) give the transmission and phase profiles, respectively, of the device for different fields. Figures 21(c) and 21(d) give the phase shift as a function of field for different frequencies. For a better comparison with experimental data, these theoretical data are shown in the same format as those shown in Figs. 19 and 20. The calculations were carried out with an effective medium theory[54] and the parameters cited in Section 4.3. The permeability for the BaM film was given in Ref. [20]. The FMR linewidth of the BaM film was taken to be $\Delta H_{FMR} = \Delta H_0 + 2\alpha f/|\gamma|$, where ΔH_0 was the frequency independent linewidth, chosen as 400 Oe, α was the damping constant, chosen as 0.0042, and f was the frequency. The dielectric constant for the BaM film and the substrates was chosen as 10. The effective medium was composed of a 5 μm-thick BaM film, a 150 μm-thick sapphire layer, and an air gap of 5 μm between the BaM film and coplanar waveguide. The air gap was considered because the BaM film was positioned, not deposited, on the coplanar waveguide.

The data in Figs. 21(a) and 21(b) indicate the following results. (1) The device shows a field-tunable resonance response. (2) The transmission dip is slightly asymmetrical at low fields. (3) The minimum transmission is about -7 dB and decreases slightly with frequency. (4) The width of the transmission dip decreases slightly with frequency. (5) The phase increases about 33° during the resonance. All of these results agree very well with the experimental

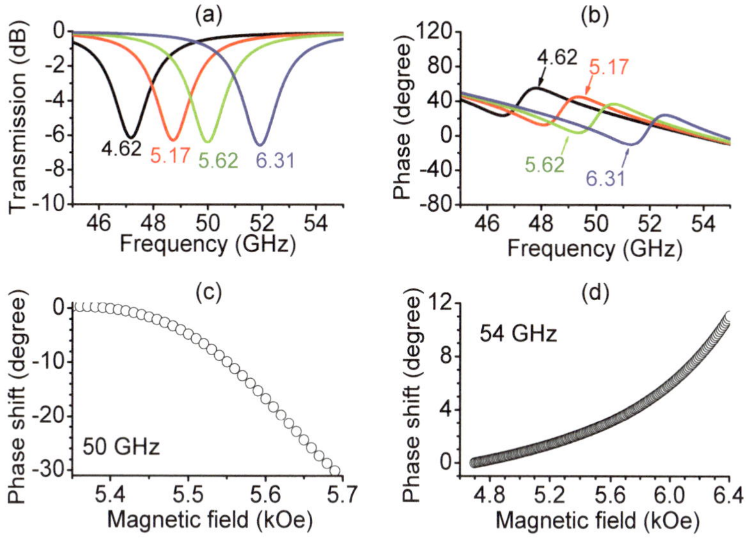

Fig. 21. Simulation of a BaM-based mm-wave phase shifter. (a) and (b) show the device transmission and phase profiles, respectively, for four different fields (kOe). (c) and (d) show the phase shift as a function of field for 50 GHz and 54 GHz, respectively. (After Ref. [21])

results shown in Fig. 19. However, both the off-resonance insertion loss and the entire phase change are smaller than experimental values. This is because the calculations considered only the short active portion of the device - the portion with the BaM film, not the 5 mm-long coplanar waveguide. The data in Figs. 21(c) and 21(d) show the phase shift ranges and tuning rates of the device that perfectly match the experimental values shown in Fig. 20.

One can see that the theoretical results agree with the measurements of the actual phase shifter. In addition to the analysis on the phase shifter, Camley *et al.* also conducted theoretical analysis on microstrip-type BaM notch filters.[20,22] The results, which are not discussed here, were very close to those measured experimentally.[22]

5. Outlook

Recent years have witnessed significant progress in the development of low-loss, self-biased BaM thin films, the deposition of high-quality BaM thin films on metallic and semiconductor substrates, the fabrication of BaM thick films on semiconductor substrates, and the development of BaM thin film-based planar mm-wave devices. Although not reviewed in this chapter, significant advances have also been made in the development of BaM/ferroelectric and BaM/piezoelectric heterostructures and the electrical tuning of magnetic responses therein, as well as in the study of multiferroic effects in BaM materials. These advances have far-reaching implications for the future development of BaM film-based devices for mm-wave signal processing. It is believed that BaM film-based devices will greatly impact the advancement of the mm-wave technology, just as YIG devices had impacted the microwave technology in the past decades.

In lieu of those significant advances, there exist many future research works that are of great interest and importance in terms of practical applications. These include, but not limited to, the following topics. Some of them have already been mentioned in previous sections.

1. Development of BaM thin films that have low loss and high remanent magnetization, as the films discussed in Section 3.1, but with either lower or higher anisotropy fields. This work is important for the development of self-biased devices for applications in different frequency regimes. It can be realized through the doping of scandium or aluminum into BaM films.[51,52]
2. Fabrication of high-quality BaM thin films in "conventional" metallic substrates. This effort is critical for the development of mm-wave devices where a ground plane is requisite.
3. Deposition of high-quality BaM films on semiconductor substrates through the use of MBE BaM seed layers.
4. Optimization of screen printing processes for the fabrication of high-quality, thick BaM films on semiconductor substrates.
5. Development of BaM thin film-based, planar, non-reciprocal devices, such as isolators and circulators. Such devices are critically needed by mm-wave signal processing and can be realized through the use of coupled-line geometry.[55,56]
6. Development of electrically tunable BaM mm-wave devices by the use of BaM/ferroelectric or BaM/piezoelectric heterostructures. Electrically tunable devices can have a number of advantages over devices that are tuned magnetically. These advantages include fast tuning, negligible power consumption, small size, and light weight.

6. Acknowledgments

The author wants to thank his colleagues Dr. Robert Camley, Mr. Lei Lu, Dr. Young-Yeal Song, Mr. Yiyan Sun, and Mr. Zihui Wang for meaningful scientific collaborations in the field of BaM thin films and devices. The author would like to particularly thank Dr. Young-Yeal Song for extensive useful discussions. The author also thanks Dr. Yan Nie, Mr. Michael Kabatek, and Mr. James Hitchman for proofreading this chapter. The author apologizes to those who contributed to this research field and whose results are not included in this chapter. This omission results from the author's inability to include all results into this chapter which has a length constraint. Support from the U.S. National Science Foundation, the U.S. Army Research Office, the U.S. National Institute of Standards and Technology, and Colorado State University is gratefully acknowledged.

7. References

[1] P. Kabos and V. S. Stalmachov, *Magnetostatic Waves and Their Applications* (Chapman & Hall, London, 1994).
[2] J. D. Adam, L. E. Davis, G. F. Dionne, E. F. Schloemann, and S. N. Stitzer, IEEE Trans. Microwave Theory Tech. 50, 721 (2002).
[3] D. D. Stancil and A. Prabhakar, *Spin Waves: Theory and Applications* (Springer, New York, 2009).
[4] F. Sammoura, *Micromachined Plastic Millimeter-Wave Radar Components* (VDM Verlag, Saarbrücken, 2008).
[5] D. Liu, U. Pfeiffer, J. Grzyb, and B. Gaucher, *Advanced Millimeter-Wave Technologies: Antennas, Packaging and Circuits* (Wiley, Chichester, 2009).
[6] U. Soerqel, *Radar Remote Sensing of Urban Areas* (Springer, New York, 2010).

[7] Y. Y. Song, S. Kalarickal, and C. E. Patton, J. Appl. Phys. 94, 5103 (2003).
[8] S. D. Yoon, C. Vittoria, and S. A. Oliver, J. Appl. Phys. 93, 4023 (2003).
[9] S. D. Yoon, C. Vittoria, and S. A. Oliver, J. Magn. Magn. Mater. 265, 130 (2003).
[10] Y. Y. Song, Y, Sun, L. Lu, J. Bevivino, and M. Wu, Appl. Phys. Lett. 97, 173502 (2010).
[11] Z. Chen, A. Yang, S. D. Yoon, K. Ziemer, C. Vittoria, and V. G. Harris, J. Magn. Magn. Mater. 301, 166 (2006).
[12] Z. Chen, A. Yang, Z. Cai, S. D. Yoon, K. Ziemer, C. Vittoria, and V. G. Harris, IEEE Trans. Magn. 42, 2855 (2006).
[13] Z. Chen, A. Yang, A. Gieler, V. G. Harris, C. Vittoria, P. R. Ohodnicki, K. Y. Goh, M. E. Mchenry, Z. Cai, T. L. Goodrich, and K. S. Ziemer, Appl. Phys. Lett. 91, 182505 (2007).
[14] Z. Cai, T. L. Goodrich, B. Sun, Z. Chen, V. G. Harris, and K. S. Ziemer, J. Phys. D: Appl. Phys. 43, 095002 (2010).
[15] Y. Nie, I. Harward, K. Kalin, A. Beaubien, and Z. Celinski, J. Appl. Phys. 107, 073903 (2010).
[16] Y. Chen, I. C. Smith, A. L. Geiler, C. Vittoria, V. Zagorodnii, Z. Celinski, and V. G. Harris, IEEE Trans. Magn. 44, 4571 (2008).
[17] Y. Chen, I. C. Smith, A. L. Geiler, C. Vittoria, V. Zagorodnii, Z. Celinski, and V. G. Harris, J. Phys. D: Appl. Phys. 41, 095006 (2008).
[18] N. Zeina, H. How, and C. Vittoria, IEEE Trans. Magn. 28, 3219 (1992).
[19] S. Capraro, J. P. Chatelon, M. Le Berre, T. Rouiller, H. Joisten, D. Barbier, and J. J. Rousseau, Phys. Status Solidi C 1, 3373 (2004).
[20] T. J. Fal and R. E. Camley, J. Appl. Phys. 104, 023910 (2008).
[21] Z. Wang, Y. Y. Song, Y. Sun, J. Bevivino, M. Wu, V. Veerakumar, T. Fal, and R. Camley, Appl. Phys. Lett. 97, 072509 (2010).
[22] R. E. Camley, Z. Celinski, T. Fal, A. V. Glushchenko, A. J. Hutchison, Y. Khivintsev, B. Kuanr, I. R. Harward, V. Veerakumar, and V. V. Zagorodnii, J. Magn. Magn. Mater. 321, 2048 (2009).
[23] Y. Y. Song, C. L. Ordóñez-Romero, and M. Wu, Appl. Phys. Lett. 95, 142506 (2009).
[24] L. Lu, Y. Y. Song, J. Bevivino, and M. Wu, Appl. Phys. Lett. 98, 212505 (2011).
[25] J. Das, B. Kalinikos, A. R. Barman, and C. E. Patton, Appl. Phys. Lett. 91, 172516 (2007).
[26] A. B. Ustinov and G. Srinivasan, Appl. Phys. Lett. 93, 142503 (2008).
[27] Y. Y. Song, J. Das, P. Krivosik, N. Mo, and C. E. Patton, Appl. Phys. Lett. 94, 182502 (2009).
[28] J. Das, Y. Y. Song, and M. Wu, J. Appl. Phys. 108, 043911 (2010).
[29] Y. Y. Song, J. Das, P. Krivosik, H. K. Seo, and M. Wu, IEEE Magn. Lett. 1, 2500204 (2010).
[30] Y. Tokunaga, Y. Kaneko, D. Okuyama, S. Ishiwata, T. Arima, S. Wakimoto, K. Kakurai, Y. Taguchi, and Y. Tokura, Phys. Rev. Lett. 105, 257201 (2010).
[31] S. R. Shinde, R.Ramesh, S. E. Lofland, S. M. Bhagat, S. B. Ogale, R. P. Sharma, and T. Venkatesan, Appl. Phys. Lett. 72, 3443 (1998).
[32] S. D.Yoon and C.Vittoria, J. Appl. Phys. 93, 8597 (2003).
[33] S. D.Yoon and C.Vittoria, IEEE Trans. Magn. 39, 3163 (2003).
[34] S. D.Yoon and C.Vittoria, J. Appl. Phys. 96, 2131 (2004).
[35] Y. Y. Song, J. Das, Z. Wang, W. Tong, and C. E. Patton, Appl. Phys. Lett. 93, 172503 (2008).

[36] M. S. Yuan, H. L. Glass, and L. R. Adkins, Appl. Phys. Lett. 53, 340 (1988).
[37] Z. Zhuang, M. Rao, R. M. White, D. E. Laughlin, and M. H. Kryder, J. Appl. Phys. 87, 6370 (2000).
[38] S. Capraro, J. P. Chatelon, M. Le Berre, T. Rouiller, H. Joisten, D. Barbier, and J. J. Rousseau, Phys. Status Solidi C 1, 3373 (2004).
[39] N. N. Shams, M. Matsumoto, and A. Morisako, IEEE Trans. Magn. 40, 2955 (2004).
[40] N. N. Shams, X. Liu, M. Matsumoto, and A. Morisako, J. Appl. Phys. 97, 10K305 (2005).
[41] S. Pignard, J. P. Senateur, H. Vincent, J. Kreisel, and A. Abrutis, J. Phys. IV France 7, C1-483 (1997).
[42] J. R. Hook and H. E. Hall, *Solid State Physics* (John Wiley & Sons, New York, 1991)
[43] C. Kittel, *Solid State Physics* (John Wiley & Sons, New York, 2005).
[44] J. Smit and H. P. J. Wijn, *Ferrites* (John Wiley & Sons, New York, 1959).
[45] W. H. von Aulock, *Handbooks of Microwave Ferrite Materials* (Academic Press, New York, 1965).
[46] K. Okumura, T. Ishikura, M. Soda, T. Asaka, H. Nakamura, Y. Wakabayashi, and T. Kimura, Appl. Phys. Lett. 98, 212504 (2011).
[47] R. Karim, K. D. McKinstry, J. R. Truedson, and C. E. Patton, IEEE Trans. Magn. 28, 3225 (1992).
[48] M. A. Wittenauer, J. A. Nyenhuis, A. I. Schindler, H. Sato, F. J. Friedlaender, J. Truedson, R. Karim, and C. E. Patton, J. Cryst. Growth 130, 533 (1993).
[49] M. Sparks, *Ferromagnetic-Relaxation Theory* (McGraw Hill, New York, 1964).
[50] A. G. Gurevich and G. A. Melkov, *Magnetization Oscillations and Waves* (CRC-Press, Boca Raton, 1996).
[51] X. Zuo, P. Shi, S. A. Oliver, and C. Vittoria, J. Appl. Phys. 91, 7622 (2002).
[52] A. B. Ustinov, A. S. Tatarenko, G. Srinivasan, and A. M. Balbashov, J. Appl. Phys. 105, 023908 (2009).
[53] S. S. Kalarickal, N. Mo, P. Krivosik, and C. E. Patton, Phys. Rev. B 79, 094427 (2009).
[54] T. J. Fal, V. Veerakumar, Bijoy Kuanr, Y. V. Khivintsev, Z. Celinski, and R. E. Camley, J. Appl. Phys. 102, 063907 (2007).
[55] L. E. David and D. B. Sillars, IEEE Trans. Microw. Theory Tech. 34, 804 (1986).
[56] J. Mazur, M. Solecka, M. Mazur, R. Poltorak, and E. Sedek, IEE Proc. Microw. Antennas Propag. 152, 43 (2005).

3

Tailoring the Interface Properties of Magnetite for Spintronics

Gareth S. Parkinson[1], Ulrike Diebold[1], Jinke Tang[2] and Leszek Malkinski[3]
[1]*Institute of Applied Physics, Vienna University of Technology, Vienna,*
[2]*Department of Physics and Astronomy, University of Wyoming, Laramie, WY*
[3]*Advanced Materials Research Institute and the Department of Physics, University of New Orleans, Lakeshore Dr., New Orleans, LA*
[1]*Austria*
[2,3]*USA*

1. Introduction

1.1 Spintronics and spintronic materials

The field of spintronics originates from the discovery of giant magnetoresistance (GMR) by Fert and Grünberg [1, 2] in 1988, for which they were awarded Nobel Prize in 2007. This effect, first observed in nanostructures comprised of two thin magnetic layers of Fe separated by a 1-2 nm thick Cr spacer, was both qualitatively and quantitatively different from the prior-known phenomenon of anisotropic magnetoresistance. GMR leads to magnetoresistance much larger than anisotropic magnetoresistance. Given the exciting nature of the effect, the underlying mechanism was promptly investigated and quickly understood [3]. Essentially, GMR can be described in terms of spin filtering; conduction electrons are polarized in one ferromagnetic layer, maintain spin memory while traveling through a thin spacer, and then enter the second magnetic layer. The scattering of electrons in this second magnetic layer depends on the direction of the magnetization relative to the first (polarizing) layer. The electrons pass through the two layers almost unperturbed if their respective magnetization is parallel. In contrast they experience enhanced scattering for an antiparallel magnetization configuration. This magnetization-dependent scattering potential can be explained through the availability (or non availability) of electron states in the second material around the Fermi level in the spin-up and spin down bands. Consequently, spintronics relies on materials in which a spin asymmetry exists in the density of states at the Fermi level. Such differences down in magnetically ordered materials arise from exchange interactions between magnetic atoms. The extreme case of energy band splitting occurs in half metals, where only one spin orientation can be occupied by electrons at the Fermi level. Therefore, half-metals should behave as a conductor for the electron current when electron spins match the direction of magnetization and as an insulator when the spin direction opposes the magnetization.

While the first observations of GMR were made in the Fe/Cr system of antiferromagnetically coupled Fe layers forming tri-layer spin valve structures or multilayers, it was quickly realized that the Ruderman-Kittel-Kasuya-Yosida (RKKY) type interaction between the Fe layers through the Cr spacer, responsible for this coupling, was not a necessary condition for the occurrence of the GMR effect. What really matters is that the antiparallel alignment of the magnetization in the magnetic layers in some range of the applied magnetic field. This effect could be achieved either using dissimilar magnetic layers (e.g. different thickness of the films or different materials constituting the layers) or by pinning the magnetization of the one of the layers using exchange interaction. This can be realized using a magnetically harder ferromagnetic or antiferromagnetic material, such that switching of the free layer occurs prior to the pinned (or harder) layer. One of the disadvantages of the antiferromagnetic coupling was the oscillation of its strength and sign (change from anti- to ferromagnetic) with thickness of the spacer layer, which limited the spacer thickness to 2 nm and imposed tough requirements on the technology. A new mechanism of switching the magnetization in the layers enabled the discovery that the spin diffusion length could be greater than just 2 nm, and allowed the building of structures with 4 to 6 nm spacer layers with performance nearly independent of fluctuations in the spacer thickness. A variety of materials have been tested and GMR, defined as $MR = (R_{max}-R_{min})/R_{min}$ (were R_{min} and R_{max} denote minimum and maximum resistance of the sample) exceeding 100% has been observed in the Co/Cu multilayers [4]. Shortly after the discovery of GMR it was demonstrated that this effect is not unique to layered structures, and significant magnetoresistance values were measured in granular systems of Co nanoparticles embedded in a Cu matrix [5].

The success of giant magnetoresistive structures encouraged researchers to study similar structures comprising two magnetic layers separated by an insulating barrier. Such architectures are similar to those in which Julliere discovered magnetoresistance at low temperatures back in 1975 [6]. It took Miyazaki [7] and Moodera [8] to realize that these structures can easily compete with the GMR structures. Transition metals with a 1 nm thick barrier of aluminum oxide or titanium oxide display a 40 to 80 % relative change of the resistance, exceeding the performance of spin-valves (of about 15%) by a factor of 5. Many of the critical features defining the magnetoersistance in tunneling junctions, or the tunneling magnetoresistance (TMR), remain similar to those important for GMR. The key difference of course, is that the electrons now tunnel from one material to another magnetic material through a barrier, replacing ballistic conduction.

The potential for using the GMR effect in magnetic field sensors and hard drive read heads was quickly recognized by companies in the information technology sector. Dynamic progress in research on GMR and TMR effects laid the foundations of a new discipline in electronics initiated by the publication of Prinz [9] in 1998, who suggested that spin dependent transport does not have to be limited to analog magnetic sensors, but can be developed into a branch of electronics that takes full advantage of electron spin. The name given to the emerging discipline evolved from magnetoelectronics, spin electronics and, finally, to spintronics, describing a new generation of nonvolatile electronic devices which use magnetization of spin valve structures to store or process digital information. More details can be found in several books and review articles on spintronics [10-12]

Current magnetoresistive devices such as magnetoresistive random access memories are based on transition metals and their alloys. The excellent performance of tunneling junctions with an MgO barrier, demonstrated by Yuasa [13] and Parkin [14] approached the theoretical limit of TMR for the transition metal based junctions. In order to compete with semiconducting switches and transistors, which change resistance by more than 6 orders of magnitude, the relative changes of resistance of spintronic devices must be improved by a few orders of magnitude. The only option to achieve this goal is to search for new magnetic materials that exhibit a higher degree of spin polarization than transition metals. While some researchers explore magnetic semiconductors as potential candidates for spintronic devices, half-metallic materials provide an attractive alternative path toward the challenging goal of further improving the performance of spintronic devices.

In addition to the more complex Heusler alloys, the binary compounds CrO_2 and Fe_3O_4 (magnetite) are primary half metal candidates for spintronic applications [15]. While spectroscopy measurements demonstrated 98.4% spin polarization of CrO_2 [16, 17], its performance in spintronic materials and devices has been rather disappointing. For the last decade many researchers have tried to understand this discrepancy and have attempted to improve the properties of the spintronic properties of CrO_2 based devices [18-31]. A similar problem has been found in Fe_3O_4-based structures, which have underperformed compared to expectations. In both cases, a reduction in the magnetoresistance has been attributed to surface modifications of these oxides occurring at the interfaces. Here, we will discuss how the surface properties of single crystal magnetite, particularly those of the (001) surface reconstruction, can produce an environment detrimental to the transport of a spin polarized current. Further, we discuss the possibility to preserve the half metallic character of this magnetic oxide on the surface of single crystals and nanoparticles by surface engineering. Our results show that molecular adsorption has a strong effect on the electronic structure of the interface, evidenced by spectroscopy results and enhanced magnetoresistance.

2. Structure and properties of Fe_3O_4

Spintronic devices, rely on materials in which a spin asymmetry can be established at the Fermi level, allowing the on/off state of a device to be controlled through the manipulation of a spin dependent charge transfer mechanism. An example of the principle of operation of a simple spintronic device, magnetic tunnel junction, where magnetic fields or spin current torques are used to switch the alignment of two ferromagnetic electrodes between a parallel and anti-parallel condition is illustrated in Fig. 1. With the spins aligned, a spin-polarized current can flow through the device as there are corresponding states available at the Fermi level. In contrast, with antiparallel alignment, there are no matching states at the Fermi level into which spin polarized electrons can flow and the device is in the off state. In the ideal device, the magneto resistance, i.e., the difference in resistance between the on and off states, is maximized. In recent years there has been much research into suitable materials with which to realize the spintronics dream, but it is clear that the optimum electrode materials should possess 100% spin polarization at the Fermi level, since this ensures that zero current flows in the "off" state.

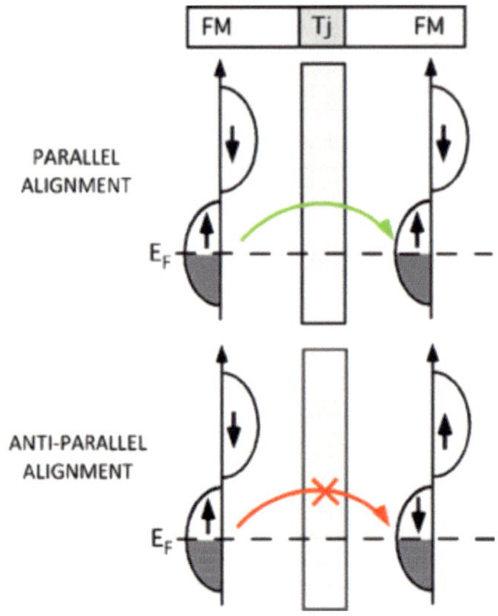

Fig. 1. Schematic of a magnetic tunnel junction. (top) Parallel alignment of two magnetic electrodes leads to tunneling of up-spin electrons through the insulating tunnel junction (Tj). (bottom) Anti-parallel alignment leads to high resistance since there are no matching states available at E_F

The considerations described above led to a surge of interest in half-metallic ferromagnets, including Fe_3O_4, which exhibits nearly 100% spin polarization at the Fermi level [32-35]. However in magnetite, which is particularly attractive to industry due to its ubiquity and inexpensive production, the expected benefits of 100% spin polarization have never been realized in prototype devices [36-41]. In fact, the highest magnetoresistance reported to date for a Fe_3O_4 based device is a mere 16 %. This disappointing turn of events is more often than not attributed to interface effects [42-45]. More specifically, it is thought that the problem lies in efficiently injecting the spin polarized current from a Fe_3O_4 electrode into a semiconductor. Unfortunately, the interface between Fe_3O_4 and a semiconducting buffer layer represents the most ill-defined region of such a device, and little is known about the factors that govern the electronic properties of the interface. This has led to a somewhat ad-hoc approach to performance improvement, i.e., constructing device prototypes using a variety of semiconductors and then measuring the resulting magnetoresistance and explaining the results afterwards.

In this section we discuss the alternative approach toward improving the performance of Fe_3O_4 based spintronics devices i.e. developing an understanding of the fundamental processes governing Fe_3O_4-semiconductor interface properties, and then using this knowledge to deterministically tailor suitable interfaces for certain tasks. More specifically, we describe recent progress toward understanding the properties of Fe_3O_4 through atomic scale studies of its' (001) surface, and how early experiments show that the surface properties can be affected by adsorbates. It is demonstrated herein that the electronic properties of the clean Fe_3O_4(001) surface diverge strongly from those of the bulk material as a result of a subtle surface reconstruction. Further, we demonstrate that surface engineering through adsorption represents a valid route toward tailoring the electronic structure, and consequently the spin transport properties at spintronic interfaces.

Fe_3O_4 crystallizes in the inverse spinel structure, which comprises a cubic close packed oxygen lattice with Fe cations occupying interstitial sites [46]. In the inverse spinel structure Fe^{3+} cations occupy 1/8 of the tetrahedral interstitial "A" sites and a 50:50 mixture of Fe^{3+} and Fe^{2+} cations occupy 1/2 of the octahedral "B" sites. The differing and opposed magnetic moment on the A and B sublattices results in ferrimagnetism (T_C = 858 K) with a magnetic moment of $4\mu_B$ per unit cell. In 1939, Verwey discovered a metal-insulator transition in Fe_3O_4 (T_V =123 K) [47,48] in which the conductivity drops by two orders of magnitude on cooling. Coincident with the electronic changes, the crystal structure goes from cubic to monoclinic. For the room temperature conductive phase, Verwey proposed that the conduction mechanism was electron hopping between Fe^{2+} and Fe^{3+} cations on the B sublattice [47,48]. However despite several decades of intense research, the conduction mechanism remains a controversial topic in the literature. Recent experiments have shown measured little difference in the charge between Fe(B) atoms, and today the charge is commonly written as $Fe^{2.5+}$ to reflect the extent of electron delocalization [48]. This makes Fe_3O_4 a particularly challenging system for modern theoretical calculations.

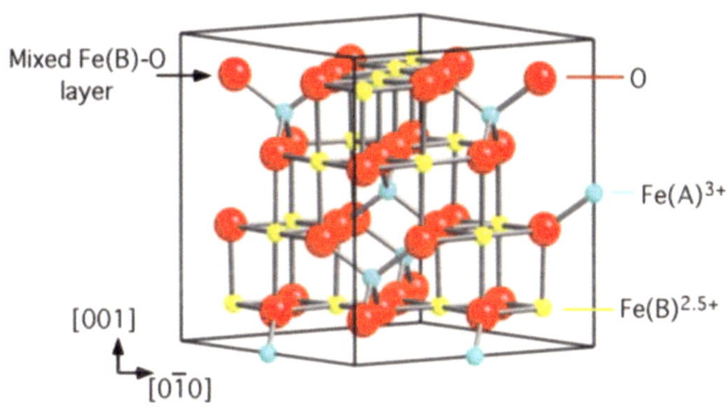

Fig. 2. Fe_3O_4 bulk unit cell (inverse spinel structure)

The interest in using Fe_3O_4 in spintronic applications arose in the 1980's, when band structure calculations predicted that the room temperature phase is half-metallic [49,50], with only minority spin electrons responsible for the conduction. This occurs because the 5-fold degenerate d levels of the Fe(B) cations are split by the crystal field into 3 t_{2g} orbitals and 2 e_g orbitals. For both the Fe^{3+} and Fe^{2+} cations the spin up levels are occupied. However, for the Fe^{2+} cations, the extra electron occupies the lowest lying t_{2g} orbital of the spin down state (marked by a gray arrow in Figure 3), which sits at the Fermi level. While this result has been confirmed by other theoretical calculations, experimental verification of this exciting prediction has not been forthcoming, despite significant effort.

The most common method for the absolute measurement of spin polarization, Andreev reflection [51], requires measurements at extremely low temperatures. Unfortunately this is not possible for the conducting phase of Fe_3O_4 as the Verwey transition occurs at 123 K [47,48]. Consequently, alternative methods to measure the Fermi level spin polarization were sought. Several groups performed spin resolved photoemission spectroscopy [52] (SP-PES)

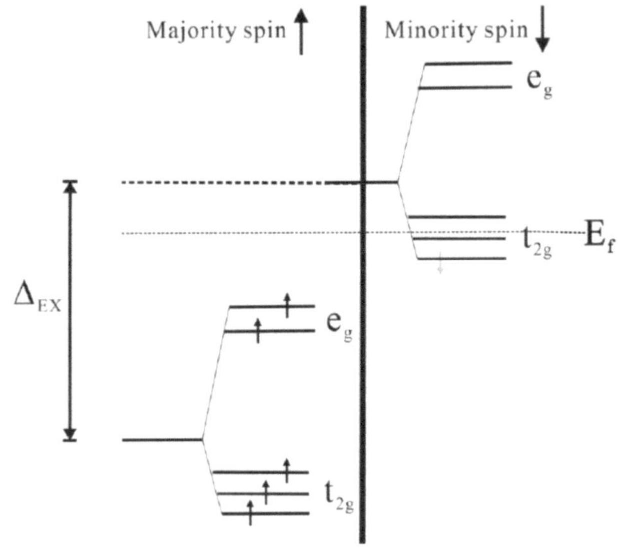

Fig. 3. Schematic representation of the energy levels on the Fe(B) sublattice in Fe_3O_4. Only down spin electrons are present at the Fermi level. Figure reproduced from ref. 33

experiments, where UV light excites photo-electrons directly from the valence band, which are analyzed by a spin sensitive electron analyzer. By including spin sensitive Mott detector [52], the fraction of spin-up and spin-down electrons at the Fermi level can be calculated. However, the measured values of the spin polarization in Fe_3O_4 vary wildly from group to group, and from sample to sample, with reported values ranging from 20-80% [53-56]. Thus, for several years there has been confusion and debate regarding the real spin polarization of Fe_3O_4.

In the light of the experimental results to be described in this chapter, it appears that the primary issue affecting the SP-PES measurements is that photoemission is an inherently surface sensitive technique. Since the mean free path of low energy electrons is extremely short in solids, the measured electrons emerge only from the first few atomic layers of the sample. Consequently, photoemission can only be reliably used to measure bulk properties if the surface layers are representative of bulk properties. If a material exhibits strong surface effects, the measurements reflect a superposition of the bulk and surface. In this regard, recent work has shown that Fe_3O_4 *does not* form simple bulk-truncated surfaces, and that Fe_3O_4 surface properties deviate dramatically from those of the bulk. While the inability to extract spin-polarized electrons from Fe_3O_4 surfaces provides a satisfactory explanation of the erroneous SP-PES results, it also provides a rationale for understanding the poor performance of Fe_3O_4 based spintronic devices.

3. Fe_3O_4 surfaces

The most energetically favorable Fe_3O_4 surfaces are the (111) and (001) planes. In recent years the surface science method, where single crystal samples are studied in a highly controlled UHV environment, has been applied to characterize these surfaces and the most common terminations have been determined. At the $Fe_3O_4(111)$ surface, three distinct terminations are possible, exposing either O, Fe(B) or Fe(A) atoms [57]. However, often the different terminations are found to coexist on the surface, with their relative coverage strongly dependent on the oxidation conditions during sample preparation [57]. This makes quantitative analysis of the surface properties problematic and difficult to reproduce exactly.

In contrast, it is now well established that the energetically favorable termination of the (001) surface is a mixed Fe(B)-O layer across a wide range of O chemical potentials [34,58-61]. However, low energy electron diffraction (LEED) patterns exhibit a $(\sqrt{2}\times\sqrt{2})R45°$ symmetry, indicative of surface reconstruction (see fig. 4.a, the black square indicates the $(\sqrt{2}\times\sqrt{2})R45°$ unit cell). The generally accepted structural model for this surface was proposed on the basis of a combined experiment-theory study in 2005 [58]. Essentially, the surface undergoes a Peierls-like distortion, in which a small amount of energy is saved by doubling the periodicity along the surface Fe(B) rows (see fig. 4.b,c). This model is consistent with STM images (fig. 5) [32, 62,63], which show alternate pairs of Fe(B) atoms relaxing in opposite directions perpendicular to the Fe(B) row direction. As neighboring rows relax in antiphase to one another, the $(\sqrt{2}\times\sqrt{2})R45°$ surface unit cell is produced, as indicated by the cyan square in figure 5.

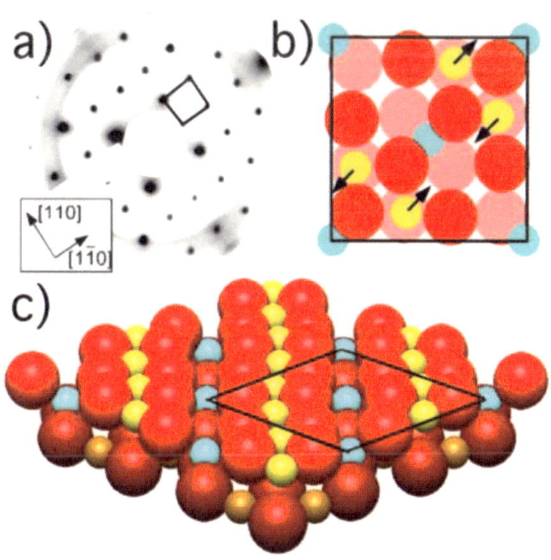

Fig. 4. a) LEED pattern recorded for 90 eV electron energy. (b) $Fe_3O_4(001)$ surface unit cell (c) perspective view of $Fe_3O_4(001)$ in which the undulations of the surface reconstruction are visible. Figure adapted from ref. 32

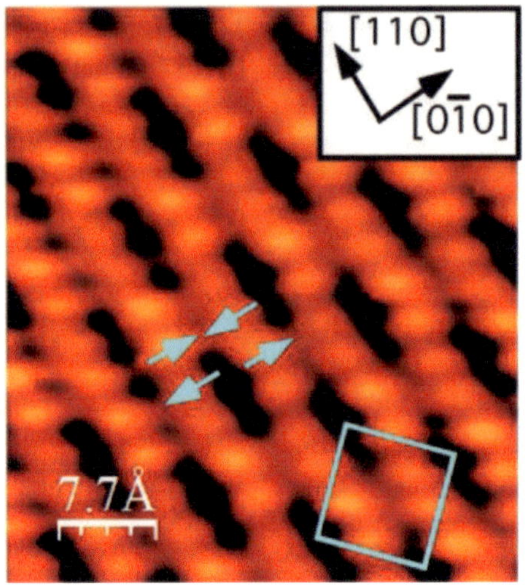

Fig. 5. 4x4 nm² STM image of the $Fe_3O_4(001)$ surface (V_{SAMPLE}=1.2 V, I_{TUNNEL}=0.3 nA). Adapted from Ref. 32

Further DFT calculations of the $Fe_3O_4(001)$ surface soon followed, extending on their predecessors by including a Hubbard "U" parameter to take account of electron correlation, which is significant in Fe_3O_4 [34,59]. These calculations reveal that charge order in the subsurface layers couples to the lattice distortion, resulting in the shift of the t_{2g} orbitals above the Fermi level, and the opening of a small bandgap in the surface layers. The surface band gap was experimentally measured at 0.2 eV in scanning tunneling spectroscopy experiments [33] (see figure 6b). These results taken together provide ample evidence that Fe_3O_4 is a material exhibiting strong surface effects. From this viewpoint, it is possible to

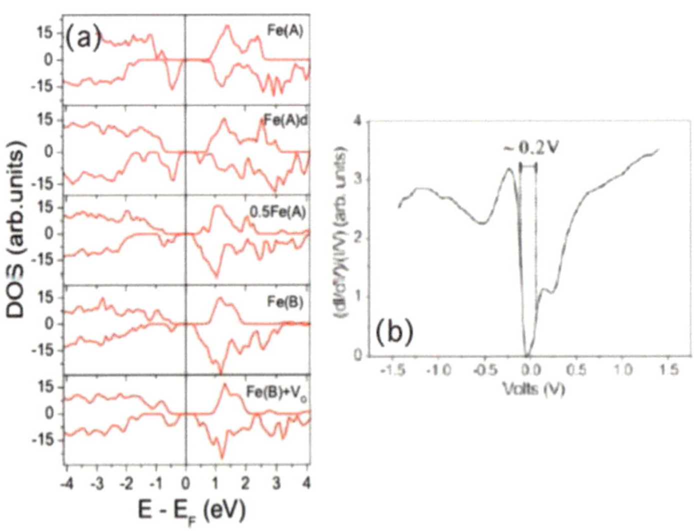

Fig. 6. DOS calculations for the $Fe_3O_4(001)$ surface which show that a surface band gap exists irrespective of the surface termination model. Reproduced from ref. 34 (b) STS results which demonstrate a band gap of 0.2 eV at the $Fe_3O_4(001)$ surface. Reproduced from ref. 33

explain the erroneous spin polarization measurements described above, and to rationalize the poor performance of Fe_3O_4 based spintronics devices.

The Fe_3O_4(001) surface represents one of the best characterized examples of the effect of surface reconstruction on electronic properties. Consequently, this system provides a perfect testing ground for experiments into the possibility of utilizing surface engineering to try and recover the bulk transport properties in the surface. Since the primary characteristic that distinguishes a surface from the bulk is the under coordination of surface atoms compared to their bulk counterparts, it is plausible that the addition of new bonds through the adsorption of molecules could modify the surface properties.

4. Fe_3O_4(001)–H – Recovery of surface half-metalicity

The simplest possible adsorbate that one can add to a metal-oxide surface is a single H atom. Atomic H has the advantage of being both computationally tractable and relatively simple to deposit cleanly in ultra-high vacuum experiments. Essentially, one simply backfills the vacuum system with a partial pressure of H_2 (~ 10^{-6} mbar) and then dissociates molecules close to the sample surface using a hot W filament. A fraction of the reactive atomic H are then directly incident to the sample surface, where they stick with unit probability. By changing the H_2 pressure and/or the exposure time it is possible to systematically vary the sample exposure.

The natural adsorption site for H atoms on a metal-oxide surface is the surface O atom, leading to the formation of surface hydroxyl groups. This expectation is confirmed by peaks associated with surface hydroxyl groups visible in both valence band (UPS) and core level (XPS) photoemission measurements upon saturation atomic H deposition (XPS data not shown).

However, the most interesting photoemission data occurs at the Fermi level (inset, figure 7a), where a huge increase in the density of states is observed, i.e. a metallization of the surface. Clearly, this result shows that adsorption is capable of inducing massive changes to the electronic structure at Fe_3O_4 surfaces. Independent but complementary measurements of the same system using a spin sensitive probe (metastable He-scattering) have shown that the

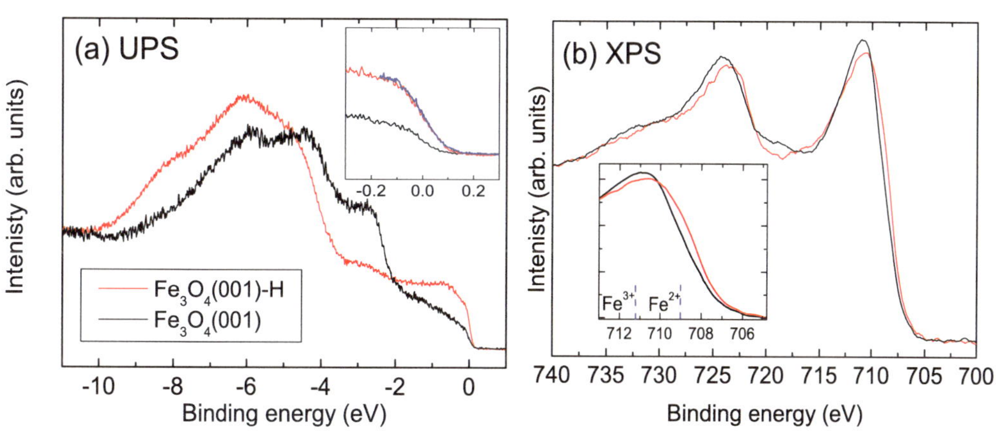

Fig. 7. (a) UPS spectra of the Fe_3O_4(001) valance band before and after saturation atomic H adsorption. (b) XPS spectra for the Fe $2p$ core levels before and after atomic H exposure. Figure reproduced from ref. 32

H-induced density of states at the Fermi level are highly spin polarized (> 50 %) [64], exactly the result required for spintronic applications.

5. $Fe_3O_4(001)$-H – Mechanism

While the possibility to remove the $Fe_3O_4(001)$ surface bandgap via atomic H adsorption is an important discovery, it is crucial that we understand the fundamental processes that underlie the macroscopic electronic changes. This is particularly important for the case of the H-$Fe_3O_4(001)$ system because it is highly unlikely that a monolayer (ML) of H atoms could be stabilized in a device architecture. Therefore, one needs to understand as much as possible about the mechanism responsible for the half-metallization in order to be able to reproduce the effect with other, more technologically relevant adsorbates.

To this end we investigated the adsorption of atomic H on the $Fe_3O_4(001)$ surface using several complementary surface science techniques. Firstly, the XPS measurements presented in figure 7b show a significant shift in the Fe *2p* peaks toward lower binding energy, consistent with a change in the valance of the surface Fe atoms from Fe^{3+} to Fe^{2+} character. This suggests that the atomic H has a significant impact on the properties of the surface Fe cations, despite the fact that it bonds directly only to the oxygen atoms. This is very important since conduction is thought to occur on the Fe(B) sublattice.

Typically, the first technique applied in a surface science experiment aimed at investigating structure is low energy electron diffraction. This technique provides instant qualitative information regarding changes in the surface symmetry and an assessment of the quality of the surface order can be made. Monitoring the $Fe_3O_4(001)$ LEED pattern as a function of exposure, it was observed that the diffraction spots related to the surface reconstruction fade and eventually disappear at the highest exposures, resulting in a (1×1) LEED pattern (see figure 8b). This suggests that the surface reconstruction is lifted by atomic H adsorption, and therefore that the surface may be representative of the $Fe_3O_4(001)$ bulk.

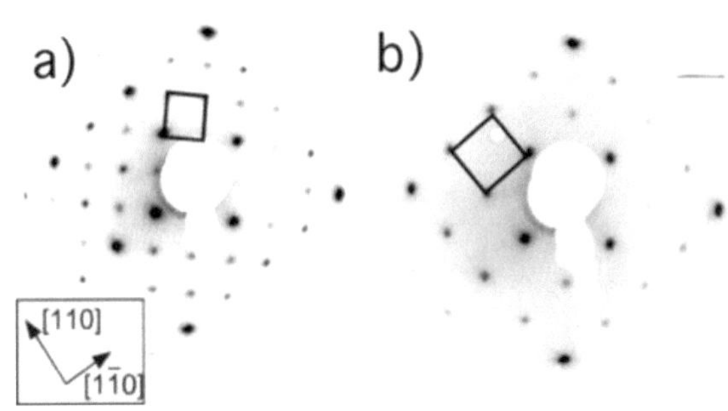

Fig. 8. $Fe_3O_4(001)$ LEED pattern before (a) and after (b) saturation atomic H exposure. The symmetry changes from $(\sqrt{2}\times\sqrt{2})R45°$ to (1×1). Figure adapted from Ref. 32

To extend upon the LEED analysis we utilized scanning tunneling microscopy (STM) [65] to investigate atomic H adsorption at the atomic scale. In an STM experiment, an atomically sharp STM tip (typically W) is brought very close to a sample surface, close enough that electrons can tunnel between them. With the application of a sample bias (of the order 1 V), a measurable current (of the order 1 nA) is observed. Scanning the tip across the surface and utilizing a feedback loop to keep the tunneling current constant, a topographical image of the surface can be constructed. It is important to note that the image is not a direct measure of the surface topography alone, rather it is effectively a contour plot of the surface electronic density of states. This results, for example, in the low lying Ti atoms being imaged brighter than the protruding O atoms in images of the $TiO_2(110)$ surface [66,67].

In the STM images of the hydroxylated $Fe_3O_4(001)$ surface (fig. 9), a similar case of electronic contrast is observed. The H atoms bound to the surface O are not directly observed, but their presence leads to a change in the density of states at the neighboring Fe(B) atom pair, resulting in an enhancement of their contrast in STM. When the H atom jumps the small distance to the symmetrically equivalent O atom within the unit cell, the electronic effect is transferred to the opposite Fe(B) row. This back and forth diffusion between the two sites is observed frequently at room temperature in STM movies (i.e. series of images collected on the same sample area). Two frames from such a movie are shown as figure 9a,b.

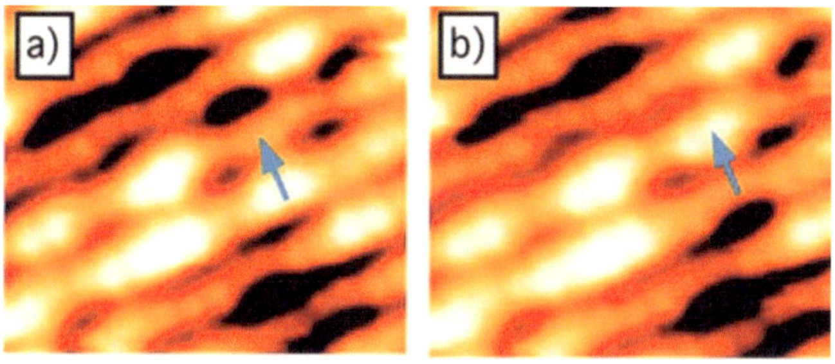

Fig. 9. Sequential STM images of the Fe_3O_4 following atomic H exposure (4x3.5 nm^2, V_{SAMPLE}=1.2 V, I_{TUNNEL}=0.3 nA). Between the two images one bright pair jumps to the opposite Fe(B) row (indicated by the blue arrow). Adapted from Ref. 32

From these STM images it is also possible to discern that atomic H has a strong preference for bonding to one particular O atom site within the surface unit cell; where the Fe(B) rows relax towards each other, hereafter called the O_{NARROW} site (see fig. 10 for a schematic model). The observed preference for the O_{NARROW} site is likely related to the bonding environment in the unit cell or the different local environment produced by the charge

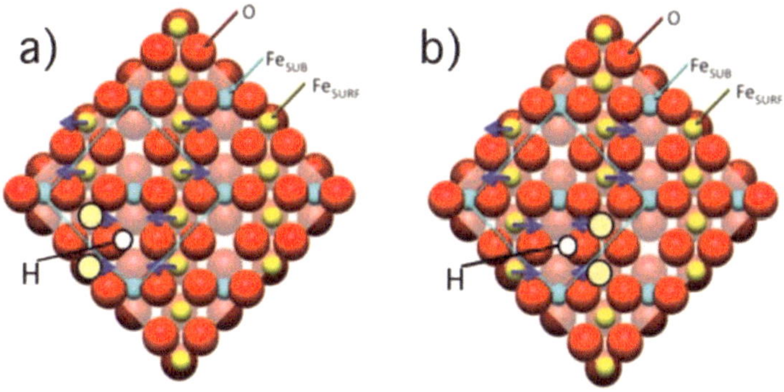

Fig. 10. Schematic model of the adsorption geometry of atomic H on the $Fe_3O_4(001)$ surface. The H atoms preferentially bind to O atoms where the Fe(B) relax together (O_{NARROW} site) and the neighboring Fe(B) become brighter. In (b) the H has jumped to the opposite O_{NARROW}, and the Fe(B) on the opposite Fe(B) pair becomes bright. Figure adapted from Ref. 32

ordering in the subsurface layers. At 1/8 ML coverage, one O_{NARROW} site per unit cell is hydroxylated, and a ($\sqrt{2}\times\sqrt{2}$)R45° ordering is observed amongst the adsorbates in STM [32].

When the H coverage is increased past 1/8 ML the O_{WIDE} sites begin to be occupied. Figure 11 shows an STM image in which the total coverage is approximately 1/6 ML, but locally there are some rows with slightly higher and lower occupation. In some places one can still clearly see the undulating rows of Fe(B) atoms (green arrow), but there are also rows in which long sections are affected by the H atoms. In such rows (one is indicated by the blue dashed line in figure 11) the Fe(B) atoms appear to be straight. The straightening of the Fe(B) rows with H adsorption is consistent with the observation from LEED that the saturated surface reverts to (1×1) symmetry.

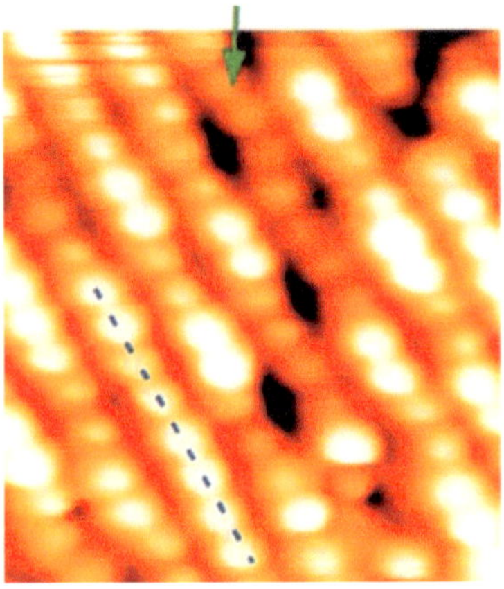

Fig. 11. STM image (4x4 nm^2, V_{SAMPLE}=1.2 V, I_{TUNNEL}=0.3 nA) showing the straight Fe(B) rows (blue dash line) formed following atomic H adsorption on the Fe_3O_4(001) surface. The green arrow indicates an area of reconstructed surface. Adapted from Ref. 32

The atomically resolved structural studies clearly show that atomic H causes the surface Fe rows to revert to a bulk-like arrangement. However, it is important to understand whether the surface metallization occurs because the surface is bulk-like, or if the adsorption of atomic H leads creates a surface distinct from the bulk with its own unique properties.

The method most commonly used to investigate crystal properties from a theoretical standpoint is density functional theory (DFT) [68], for which Walter Kohn was awarded the Nobel Prize for chemistry in 1998. In DFT, the ground state energy of a system is calculated through the minimization of an energy functional, which depends only on the electron density in the system. The energy functional contains terms related to the kinetic energy of the electrons, their interaction with the atomic nuclei, and their interaction with each other. The use of periodic boundary conditions allows an infinite crystal to be calculated using a small slab of atoms, typically numbering less than 100. The calculations yield the electron density and with it the optimum geometry of the system. This can be used to shed light on physics underlying phase transitions, electrical, magnetic and optical behavior. It should be noted that the accuracy of DFT relies heavily on the energy functional used, and that the electron-electron interactions represent the most difficult aspect, particularly in highly correlated systems such as Fe_3O_4. In the calculations that follow, a Hubbard "U" parameter of 5 eV was included to account for this, more details can be found in ref. 59.

Calculations relating to the H-Fe_3O_4(001) system confirm the experimentally determined adsorption site (fig. 12a), but find that the H sits slightly off the atop site at the O_{NARROW} atom, forming a hydrogen bond to the opposite O_{NARROW} atom. This explains the facile diffusion observed in the STM movies. Moreover, the calculations show that the Fe atom pair closest to the H atom relax back to bulk terminated positions and become Fe^{2+}, which explains the change in contrast observed in STM (see fig. 9). The yellow lobes in fig. 12(b) are indicative of occupied t_2g minority orbitals, and hence Fe^{2+} character. At saturation H coverage (all surface O atoms hydroxylated) all Fe atoms in the outermost 2 layers are

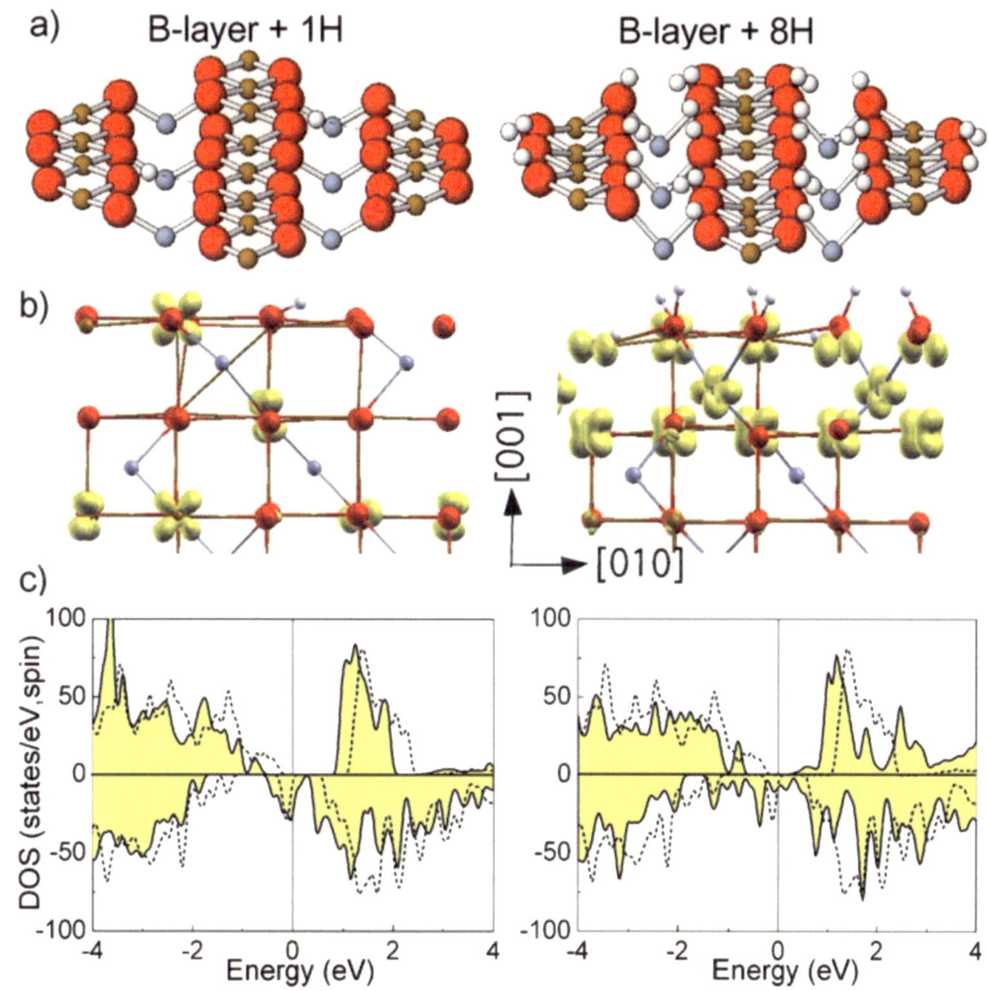

Fig. 12. Modification of an Fe_3O_4(001) surface with (left)1 and (right) 8 Hydrogen atoms per surface unit cell. O atoms are red/large, Fe(B) are gold/gray and Fe(A) are blue/light gray. (a) Adsorption geometry. (b) Side view showing the occupation of the minority t_{2g} orbitals at the Fe(B) sites (i.e., ions with Fe^{2+} character); electron density integrated between −1.3 eV and *EF*. (c) Total density of states (solid black line with yellow filled area) showing the characteristics of half-metallic system. For comparison the DOS of the clean surface (modified B layer; black dashed line), shows a band gap of 0.3 eV. Figure reproduced from Ref. 32

converted to Fe^{2+}, a finding consistent with the XPS results presented in figure 7b. Finally, the calculations reveal that the fully hydroxylated surface is half-metallic, that is, has only minority spin electrons present at the Fermi level, in agreement with the metastable He scattering results. Interestingly, the calculations show that a surface with 0.5 ML atomic H (not shown), i.e., with only the O_{NARROW} and O_{WIDE} sites hydoxylated, is the closest to the bulk material, and this also exhibits half-metallicity.

6. Prospective solutions

Overall, the results described here show that the adsorption of atomic H atoms leads to localized modifications of the structural and electronic properties of the $Fe_3O_4(001)$ surface. For H coverages in excess of 0.5 ML, the surface band gap is removed and half-metallicity is restored to the surface region. These results demonstrate that the adsorption is a valid route toward tailoring the properties of Fe_3O_4 surfaces for spintronics devices. Recent work by Pratt et al. [69] has shown that a similar half-metallization of the surface occurs with the adsorption of the organic molecule benzene. In recent years the potential of organic spintronics has been championed [70-75] as organics exhibit extremely favorable spin transport properties. If the right combination of molecule-Fe_3O_4 can be found that combines high performance with ease of fabrication and low cost, it is possible that Fe_3O_4 can play an important role in spintronics devices. In the next section we demonstrate the effect of polymer and organic coatings of magnetite nanoparticles on their magnetoresistance.

7. Tunneling magnetroresistance and spin polarization in Fe_3O_4 nanoparticles with organic coatings

In principle, a high spin polarization should result in large tunneling magnetoresistance (TMR), since the latter is proportional to the spin polarization of the tunneling electrons [76-78]. Various studies have focused on the MR ratio in Fe_3O_4 of different forms including epitaxial and polycrystalline films, powders, and tunnel junctions [39,79-82]. In early reports, some groups have claimed a large MR response on breaking contact of two microscaled single crystals of magnetite [83] and thin film structure composed of a few stacked monolayers of organically encapsulated magnetite nanocrystals [84]. However, in most cases the MR ratio is much smaller than expected, especially at room temperature. In fact, it is well known that in polycrystalline specimens and powder compacts of Fe_3O_4, the surfaces or interfaces at the grain boundaries have rather different magnetic properties and reduced spin polarization than the bulk [85-88]. This is a consequence of off-stoichiometry, surface reconstruction, oxidation, defects, and bonding effects located at or close to the surfaces and interfaces. Recently, some investigations have focused on improving MR performance of Fe_3O_4 [89-93]. Hao Zeng *et al.* reported 35% MR at 60 K for ordered three-dimensional arrays of Fe_3O_4 nanoparticles with annealing in high vacuum [89]. Rybchenko *et al.* have shown enhancement in MR of bulk granular magnetite by annealing in paraffin wax [91]. Lu *et al.* found relatively large low-field MR in ultrathin Fe_3O_4 nanocrystalline films by rapid thermal annealing at 800 °C for 120 s in pure nitrogen [92]. These works have all used passive annealing process to prevent the surfaces or interfaces from oxidation. Under ordinary conditions, the surface of Fe_3O_4 is oxidized and contains Fe^{3+}-rich oxide [94]. Combined with surface reconstruction mentioned earlier, these two factors are believed to be the main reasons for the unsuccessful attempts to observe high spin polarization in Fe_3O_4. To reveal the true spin polarization of Fe_3O_4, the reconstruction and the Fe^{3+}-rich oxide on the surface should be lifted or removed. We have selected polymers: polystyrene (PS) [90], polycarbonate (PC) [90], poly(methyl methacrylate) (PMMA), [90] fullerene (C_{60}), hexabromobenzene (C_6Br_6) [95], and polytetrafluoroethylene (C_nF_{2n+2}, also called Teflon) as a coating layer to modify the surfaces of Fe_3O_4 nanoparticles. These coatings also serve as good insulating barriers between the Fe_3O_4 nanoparticles.

To coat magnetite nanoparticles with polymer and organic compound, α-Fe_2O_3 nanoparticles and PS, PC, PMMA, Teflon, C_{60} and C_6Br_6 with various weight ratios were mixed together by first dissolving polymer/organic compound in solvents, then adding Fe_2O_3 particles and stirring, and finally evaporating the solvent. The samples were annealed at 200 °C in pure hydrogen flow, and then pressed into pellets. The pellets were again annealed at temperatures ranging from 200 to 320 °C in pure hydrogen flow, depending on the polymers/organic compounds used.

The structural analysis was done by x-ray diffraction (XRD) and transmission electron microscopy (TEM). XRD patterns indicate that there is a complete phase transformation from α-Fe_2O_3 to inverse spinel Fe_3O_4 after annealing in hydrogen. TEM images show that the Fe_3O_4 particles are generally spherical/ellipsoidal and their size is between 10-30 nm (Fig. 13(a and c)). They are dispersed in the polymer/organic matrix. Some particles are close to each other but separated by a 1-3 nm thick layer of polymer/organic compound (Fig. 13(b)), which allows the electrons to tunnel from one Fe_3O_4 nanoparticle to another, achieving intergranular tunneling of electrons in these nanocomposites.

The transport properties of the samples were measured using the four point method in a Physical Property Measurement System (PPMS) from Quantum Design. The temperature dependence of the resistance (R) follows $\ln R \propto T^{-1/2}$, and the I-V curves are nonlinear (Fig. 14(a,b), insets). These results suggest that the electron transport in the samples is via intergranular tunneling, which has been broadly accepted as exhibiting $\exp(T^{-1/2})$ behavior in the resistance [78, 96]. We have tested the $\exp(T^{-1/4})$ form expected for variable range hopping resistance and found that R could not be described by such a law. Similar behaviors are seen in other samples.

The magnetoresistance of the samples is shown in Fig. 14. The MR of the Fe_3O_4 samples coated with PC (Fig. 14(a)) is similar to the reported data on pressed Fe_3O_4 powders and polycrystalline films [81, 97-99], which have typical MR ratio of 4-7% or lower at room temperature. The Fe_3O_4 samples coated with PMMA have almost identical results to that coated with PC. On the other hand, the sample coated with PS exhibits a MR ratio of 22.8% in an applied field of 14 T at room temperature. The maximum MR of 40.9% is obtained at 110 K (Fig. 14(b)). The MR curves show rapid change in low fields but are not completely saturated in a field of 14 T. In low field region, butterfly shaped hysteresis in MR curves have been observed, and the coercivities in the MR coincide with that in the magnetization curves for both low and high temperatures measured with either a superconducting quantum interference device (SQUID) system or PPMS (Fig. 14 (c) and (d).

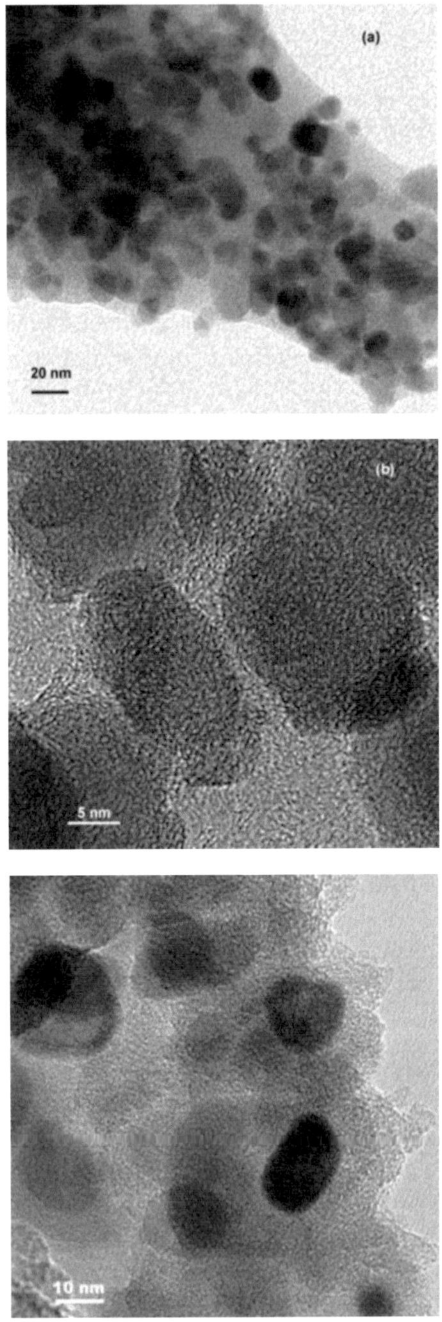

Fig. 13. (a) TEM image for the PS coated Fe_3O_4 sample indicating that spherical Fe_3O_4 nanoparticles are embedded in polymer matrix. (b) High resolution TEM image showing that the Fe_3O_4 particles are separated by a thin PS polymer layer of a few nanometers, forming tunnel barrier. (c) TEM image of a sample where Fe_3O_4 nanoparticles are dispersed in C_{60} matrix.

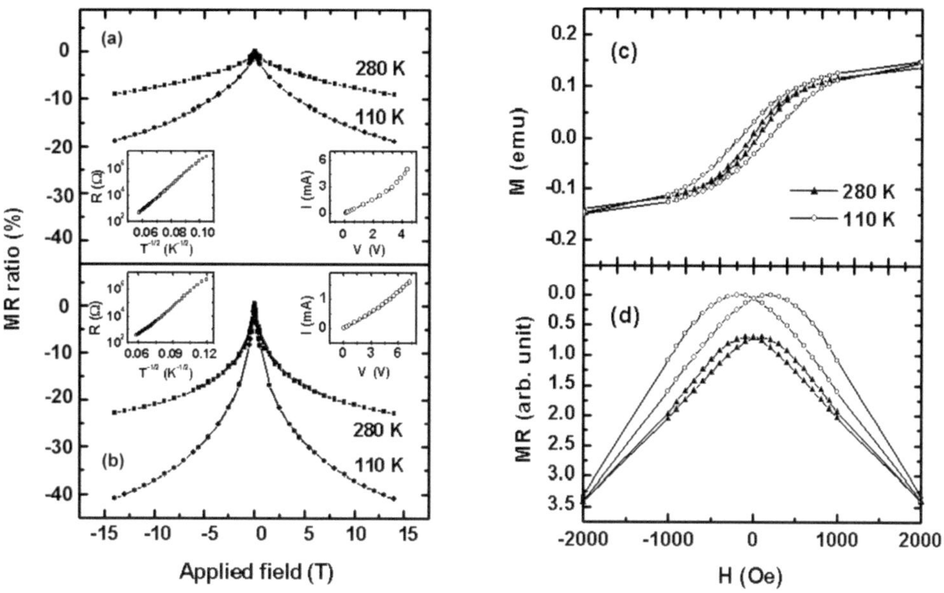

Fig. 14. MR ratio in an applied field of 14 T at 280 and 110 K for (a) PC coated Fe_3O_4 and (b) PS coated Fe_3O_4. The weight ratio of Fe_3O_4 to polymers is 1 : 1 for both. The MR ratio of PMMA coated Fe_3O_4 is similar to that of PC coated samples. (Left insets) resistance as a function of temperature: log R versus $T^{-1/2}$ curves exhibit linear relation; (Right insets) I-V curves at room temperature which have nonlinear behavior. These are consistent with the transport mechanism of intergranular tunneling. (c) Magnetic hystereses and (d) butterfly-shaped MR hyestereses in low fields for PS coated Fe_3O_4.

Several models can be used to calculate the spin polarization P from the MR data. Inoue and Maekawa have derived a simple relationship between MR and P for intergranular tunneling,[78]

$$MR = \frac{P^2 m^2}{1 + P^2 m^2} \tag{1}$$

where m is the relative magnetization of the system and m^2 = ‹cosθ›. In saturated state, $m = 1$, then

$$P = \sqrt{\frac{MR_S}{1 - MR_S}} \tag{2}$$

where MR_S is saturated MR ratio. A modification of the model is to consider serial connection of the grains in addition to parallel connections used in the model [100-102]. The difference between these models becomes very small when the three-dimensional nature of the network of grains is considered. This is true even for relatively high spin polarization P > 0.5. [101,102] Our composite pellet samples fall well within the 3-d regime, and Eqn. (2) was used to calculate the spin polarization. It should be mentioned that the models proposed by Slonczewski [103] and MacLaren [104] suggest that the nature of the barrier is a factor influencing the effective spin polarization. Such an effect diminishes with increasing barrier height. The polymer/organic barriers used here are in general of very large band gap, therefore we believe the effect of the barrier on the spin polarization is small in this case. More importantly, as long as one understands that the P obtained from Eqn. (2) is the effective spin polarization, the result is valid.

The spin polarization P of Fe_3O_4 derived from the MR values according to Eqn. 2 is 83% at low temperature and 54% at room temperature. These values are much higher than the reported experimental results [105-107] and higher than a recent theoretical calculation after taking into the consideration of modified surface state [58], Fe_3O_4 indeed belongs to the category of highly spin polarized half-metals. The more than 50% value for spin polarization at room temperature is significant because this is the first time that such a high spin polarization has been observed in Fe_3O_4 at room temperature in a spin dependent transport measurement, which has both practical and scientific implications. Coulomb blockade effect is believed to be another factor that sometimes contributes to the enhancement of MR, however it occurs only at very low temperature [76,108] and should not play a significant role here. We believe these P values only set lower limits for Fe_3O_4 and the actual values of its spin polarization can be higher, especially at room temperature for the following reasons. There may exist spin independent conductance channels due to the imperfections, defects, and impurities in the barrier in our samples, which reduce the tunneling MR ratio [30]. Bulk magnons and surface magnons will reduce the MR via magnon assisted tunneling [109]. Although they may also reduce the spin polarization itself [110], theoretical studies indicate that the MR ratio will decrease more rapidly with temperature than P[111]. Even an applied field of 14 T may not be high enough to completely align the magnetic moments of the Fe_3O_4 particles of 10-30 nm in size in our samples, especially those on the surface. It should be noted that using MR data taken at 14 T to calculate P is justified because the intrinsic magnetoresistance of Fe_3O_4 is very small in such a field [81]. We have assumed that the relative magnetization $m = 1$ in the calculation of P using Eqn (2), but it takes a reduced value at high temperatures.

It ought to be noted that there are different ways to define spin polarization [112]. Some definitions measure the spin polarization of the density of states, while others measure the transport current density. The spin polarization obtained from our intergranular tunneling experiments is the spin polarization of the tunneling current. We argue that the spin polarization of density of states is also high for Fe_3O_4. The difference between the two definitions becomes significant when there are "heavy" (e.g., d-electrons) and "light" electrons (e.g., s-electrons) co-existing at the Fermi level, which is not the case for Fe_3O_4. For Fe_3O_4, t_{2g}(Fe) electrons form small polarons and hop among the B-sites of the inverse spinel structure in a fully spin-polarized spin-down band. Therefore, the two numbers, 54% and 83%, may also represent the approximate values for the lower limits of the spin polarization of the density of states at room temperature and 110 K, respectively.

It is necessary to study the temperature dependence of MR since Fe_3O_4 undergoes a Verwey transition, which is characterized by an increase in the resistivity by about two orders of magnitude at the transition temperature $Tv \sim 120$ K. This transition is associated with an order–disorder transition from a charge-ordered state of the Fe ion on the B sites at low temperature to a statistical distribution at high temperature. A sharp narrow negative MR peak is normally observed at the Verwey point in single crystal F_3O_4 [79,113]. In our samples, the MR ratio continuously increases with decreasing temperature before the Verwey transition (Fig. 15(a)). After the transition, the MR ratio exhibits a plateau between 80 and 120 K. We cannot acquire MR data below 80 K because the resistance of the samples becomes too high to measure with our setup. According to the zero-field-cool (ZFC) and field-cool (FC) magnetization curves shown in Fig. 15(b), the Verwey transition is quite sharp and occurs in a relative narrow temperature range of 110 - 120 K in our samples. This suggests that MR observed in our samples can be used to calculate the spin polarization because it is not a part of the sharp peak ordinarily associated with the Verwey transition. The latter does not arise from the spin polarization but is a critical phenomenon at the phase transition and thus cannot be used for deriving spin polarization.

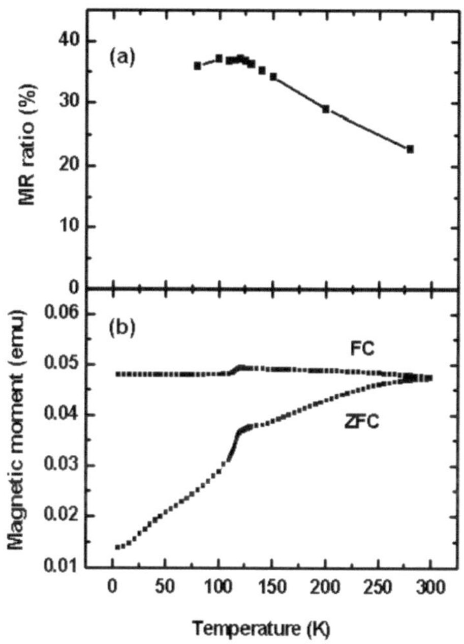

Fig. 15. (a) Temperature dependence of MR ratio in an applied field of 14 T for a PS coated sample. (b) ZFC-FC curves with an applied field H = 200 Oe, which shows sharp Vervey transiton in the range of 110 - 120 K in our samples.

The Verwey transition does not significantly change the tunneling MR and the spin polarization, consistent with reported results [105,106]. To understand the temperature dependence of MR ratio in our samples, we propose the following model. There exist two channels of the conductance. One is the intergranular spin-independent channel and the other is a spin-independent channel because of thermal excitation or inelastic hopping through localized states due to imperfections in the barrier, *etc.* Above the Verwey transition, the current of spin independent channel rapidly decreases with decreasing temperature, whereas the current of the spin dependent channel decreases relatively slowly. This results in the enhancement of tunneling MR ratio with the decrease of temperature. Below the transition, the resistance of Fe_3O_4 increases rapidly with decreasing temperature. The number of carriers available for tunneling decreases accordingly, which greatly diminishes the spin-dependent tunneling current. At the same time, the spin independent current decreases with temperature as well. The plateau in the MR ratio below the Verwey point is the combined effect of these two channels.

XPS is one of the most powerful tools to obtain information about the electronic structure of a solid's surface. Figure 16 shows the XPS Fe 2p core-level spectra for pure Fe_3O_4 powders and PS, PMMA and PC coated Fe_3O_4 samples, which contain contributions from the top 15 layers of the surface. Different from the polymer coated Fe_3O_4 samples, the lineshape of the pure Fe_3O_4 sample exhibits shake-up satellite at a binding energy of ~719 eV and a little narrow peak of the Fe $2p_{3/2}$, which is characteristic of Fe^{3+} oxide [114,115]. This clearly demonstrates that the surface of Fe_3O_4 is Fe^{3+} oxide once it is exposed to air. The lineshapes of PS, PMMA and PC coated samples reveal the characteristics of Fe_3O_4 and there is no obvious difference among them, suggesting the coating prevented the oxygen infusion into the Fe_3O_4 particles. However, the very top layers of the surface may be different from the top 15 layers, the latter of which are probed by XPS, depending on the coating materials. Studies have indicated that the top two layers of surface are rich in Fe^{3+} compared to the top 15 layers in Fe_3O_4 films MBE-grown in ultra-high vacuum [94]. Polymer PMMA and PC contain oxygen whereas PS is oxygen free. For the PMMA and PC coated samples, it is possible to form bonding between the oxygen in polymer and Fe^{2+} ion on the surface, which will result in the presence of Fe^{3+}-rich oxide on the top one or two layers of Fe_3O_4 surface, which will greatly diminish the spin polarization as it does in pure Fe_3O_4 powders or polycrystalline film. Since PS contains no oxygen element, the Fe_3O_4 state can be preserved and the high spin polarization can survive on the surface. This greatly enhances the spin-dependent tunneling MR.

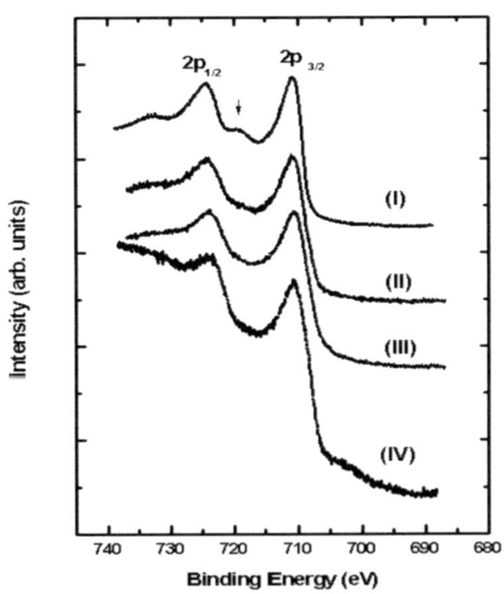

Fig. 16. Al Kα-exited Fe 2p core level photo emission spectra for (I) pure Fe_3O_4 powder sample, (II) PC coated Fe_3O_4, (III) PMMA coated Fe_3O_4 and (IV) PS coated Fe_3O_4. The arrow indicates the characteristic shake-up satellite associated with Fe(3+) ion photoemission at a binding energy of ~719 eV.

While the nanoparticle nature of the Fe_3O_4 investigated here does not allow for examination of the surface reconstruction and its potential lifting by the polymer cover layer, we have been able to study a number of other polymer/organic coatings and verify that the spin polarization of Fe_3O_4 can be preserved to a large extent when they contain no oxygen atoms.

Figure 17(a) shows the MR = $(R_H-R_0)/R_0$ of a C_{60} coated Fe_3O_4 sample annealed at 280 °C. Giant negative MR was observed at room temperature (280 K) and the MR ratio is over 11.4 % in an applied field of 5 T. The MR ratio is higher than 20 % at 150 K, however it slightly decreases to 17.6% at 75 K. These MR values are higher than reported data in pressed Fe_3O_4 powders and polycrystalline films [81,97-99], which have MR ratios typically near 4-5 % at room temperature. In our sample, C_{60} is coated on the surface of Fe_3O_4, and C_{60} is a good insulator and contains no oxygen. It may help prevent the oxidation of the surface of Fe_3O_4, which alters the half-metallic state at its surfaces.

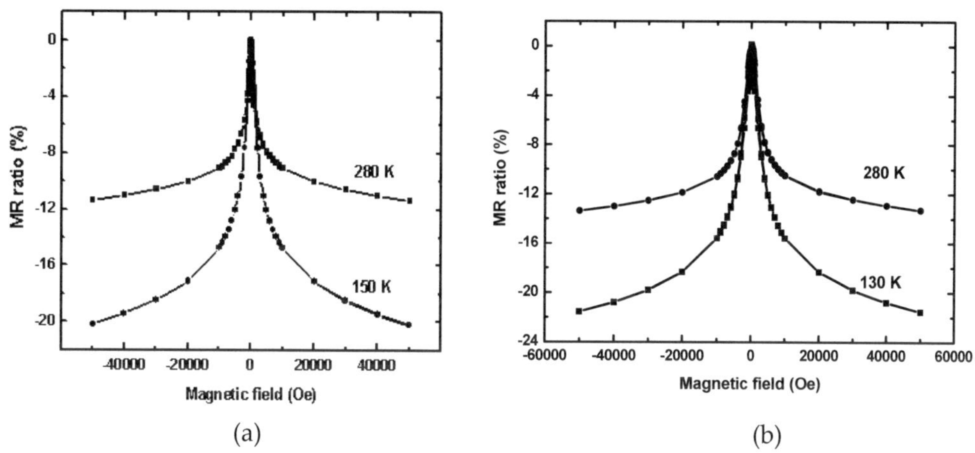

Fig. 17. (a) Magneroresistance as a function of magnetic field at 150 and 280 K, respectively for a C_{60} coated Fe_3O_4 sample annealed at 280 °C. (b) That of a C_6Br_6 coated Fe_3O_4 annealed at 250 °C measured at 130 and 280 K.

Nevertheless, the MR ratio in the magnetite/fullerene nanocomposite system is relatively low compared with the results for the magnetite/polystyrene system. Polystyrene is an excellent insulator with very high resistivity (about 10^{16} Ωm). On the contrary, there exists a wide range of data for the resistivity of C_{60}, and the highest is about 10^{14} Ωm [116-118]. More importantly, many elements can be doped into C_{60}, which results in a drastic decrease of the resistivity. In our magnetite/fullerene nanocomposites, it is possible that the defects and Fe doping in C_{60} will results in an increased hopping conductance, which gives rise to an increased spin-independent current and thus reducing the MR ratio.

Figure 18 shows the room temperature MR ratio in an applied field of 5 T versus annealing temperature. There is no obvious change for the MR ratio when the annealing temperature is between 220 and 300 °C. However, for the samples annealed at 320 °C in hydrogen, the MR ratio sharply drops to about 2%. Correspondingly the resistivity also decreases rapidly and exhibits metallic behavior. X-ray diffraction pattern indicates that there is precipitation of pure Fe in the samples. The precipitation and percolation of the iron precipitates should be responsible for the observed behaviors.

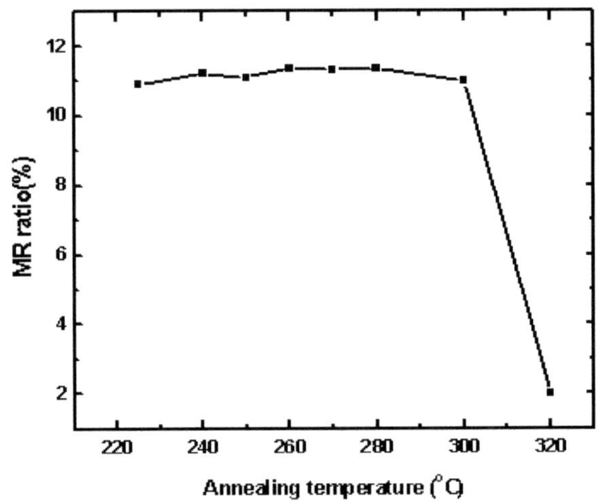

Fig. 18. Room temperature MR ratio versus annealing temperature for C_{60} coated Fe_3O_4. When the annealing temperature exceeds 300 °C, the MR ratio sharply drops to about 2% owing to the precipitation of pure Fe in the samples.

Organic compound C_6Br_6 coated Fe_3O_4 nanoparticles exhibit similar behaviors (see Fig. 17(b)). Giant negative MR was observed near room temperature (280 K) and the MR ratio is 13.4 % in an applied field of 5 T. The MR ratio is 21.5 % at 130 K, however it slightly decreases to 19.4% at 85 K.

Another oxygen-free insulating polymer, polytetrafluoroethylene (C_nF_{2n+2}), also called Teflon, was chosen as the tunnel barrier in a Fe_3O_4 intergranular tunneling experiment. A MR ratio about 16.6 % at room temperature in an applied field of 5 T was observed for a sample annealed at 320 °C in hydrogen. The temperature dependence of the resistivity exhibits characteristics of intergranular tunneling in the samples. Again, the enhancement of the MR ratio is attributed to that the Teflon can act as barrier material and, more importantly, prevent the oxidation of the surface of Fe_3O_4, which is believed to alter the half-metallic state at the surface. Our results suggest that there is high degree of spin polarization at room temperature for half metallic Fe_3O_4.

8. Magnetite nanowires

In addition to 2-dimensional epitaxial films and zero-dimensional nanaoparticles it has recently been demonstrated that fabrication of 1-dimensional nanowires, nanorods and nanotubes is possible using porous polymer or anodized alumina templates [119-121]. These structures are far less explored than the film or nanoparticulate systems, and only few reports exist on magnetoresistance of magnetite nanowires which is of the order of 7% at room temperature [122,123]. Therefore, it is believed that there is a large field for improvement of their magneto-electronic properties by functionalizing their surface, in analogy to nanoparticle systems. The advantage of the template method is that one-dimensional magnetite structures can be fabricated in the form of regular arrays of nanopillars, which suits CPI (Current Perpendicular to the Plane) geometry of giant magnetoresistive devices. In contrast, self-assembly of nanoparticles or patterning of films is required to build spintronic nanodevices.

Below, we present a new method of fabricating magnetite nanowires by oxidation of pure Fe metallic wires [124]. Conventional mild anodized AAM with about 60 nm pore diameter and 100 nm interpore distance was prepared by two-step anodizing process. The Al substrate was removed by electrochemical process in a 1:1 mixture solution of $HClO_4$ and CH_3CH_2OH at 45 V (10 °C). Fe nanowires were grown in the alumina pores by electrodeposition using a solution containing 240 g/L $FeSO_4 \cdot 7H_2O$, 45 g/L H_3BO_4, and 1 g/L of Ascorbic acid with the current of -0.9 mA for several minutes using a Princeton Applied Research VMP2 instrument. In order to convert Fe metal to magnetite nanowires (while still embedded in the porous alumina) two-step oxidation process was used. First, Fe nanowires were annealed at 500°C for 2 h in pure oxygen flow which resulted in the formation of nanowires with mixed Fe_3O_4 and Fe_2O_3 phases. Subsequently the wires were annealed at 350 °C in pure hydrogen flow for 2 h. This annealing reduced oxygen content and transformed the nanowires into Fe_3O_4 phase. An example of polycrystalline magnetite wires in alumina pores is presented in Fig.19. The nanowires exhibit high coercivity of 730 Oe and large saturation field of 9300 Oe. The length of the wires can be easily controlled in the range from a fraction of micrometer to several microns by adjusting electrodeposition time.

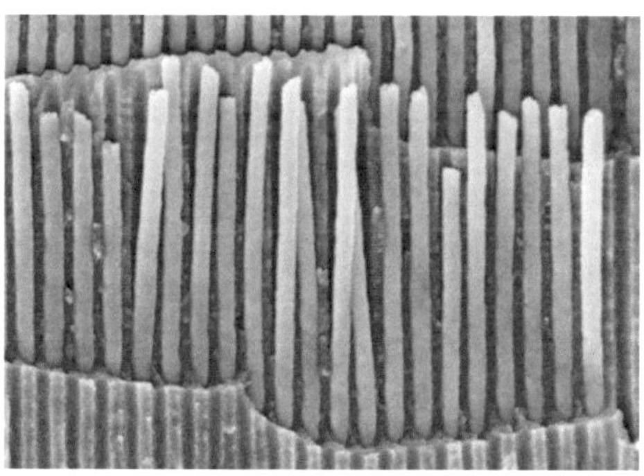

Fig. 19. Scanning electron microscopy image of Fe_3O_4 nanowires inside porous alumina template made by two step oxidation of Fe nanowires. Reproduced from ref. 124.

9. Concluding remarks

Ti-substituted magnetite ore, called loadstone, is the most common magnetic mineral on the Earth. It was also the first magnetic material discovered by humans over 2500 year ago, sparking the perpetual interest of mankind in magnetic phenomona, and resulting in first applications of magnetism in technology. Will history write another chapter for magnetite? In the era of digital compasses and Fe-Nd-B or Co-Sm permanent magnets (which eclipse the performance of magnetite by orders of magnitude), new opportunities are opening up for magnetite (in its purest form) in at least two emerging branches of nanotechnology. One of them is biomedicine, where Fe_3O_4 nanoparticles with their excellent biocompatibility are preferred for targeted drug delivery schemes. The second avenue, spintronics, emerges from the predicted half-metallicity properties and resulting high degree of spin polarization.

Our experiments on single crystal surfaces clearly demonstrate that the presence of the most simple adsorbate, H atoms, on the Fe_3O_4 surface lifts the insulating surface reconstruction and restores half-metallic character of the surface. There remain however, many unsolved issues to be investigated, especially in reference to nanostructures which exhibit less well-defined surfaces. In particular, the role of high curvature of the surfaces in nanoparticles and nanowires is unclear. Should we expect that similar surface reconstructions occur on curved heavily stepped surfaces. To date little work has been done to assess such questions.

The recent progress in magnetoresistance of nanoparticle systems has been very encouraging. Drastic enhancements of the MR ratio clearly suggest that there is indeed a high degree of inherent spin polarization at both low and room temperatures in (theoretically) half metallic Fe_3O_4. The improvement was achieved by controlling the surface of Fe_3O_4 through surface engineering using oxygen-free insulating barriers. Based on these result, it is possible that a simple tunnel junction made of Fe_3O_4 exhibits large MR in a relatively small field, pointing toward potential application as an effective spin injector. Knowing that the surface effects can be alleviated by adsorbing appropriate adlayers, Fe_3O_4 has the potential to play an important role in spintronic devices.

10. Acknowledgments

This work was supported by grants LEQSF(2007-12)-ENH-PKSFI-PRS-04 from Louisiana Board of Regents Support Fund, EPS-1003897 through contract NSF(2010-15)-RII-UNO and National Science Foundation (DMR-0852862). GSP and UD acknowledge partial support by the Center for Atomic-Level Catalyst Design, an Energy Frontier Research Center funded by the U.S. Department of Energy, Office of Science, Office of Basic Energy Sciences under Award Number #DE-SC0001058.

11. References

[1] M.N. Baibich, J.M. Brote, A. Fert, F. NguyenVan Dau, F. Petroff, P. Etienne, G. Crauzet, A. Friederich and J. Chazelas, *Phys. Rev. Lett.* 61, 2472 (1988)
[2] G. Binasch, P. Grünberg, F. Saurenbach and W. Zinn. *Phys. Rev. B* 39, 4828 (1989)
[3] J. Barnaś, A. Fuss, R.E. Camley, P. Grunberg and W. Zinn, *Phys. Rev.*, B. 42, 811 (1990)
[4] S.S.P. Parkin, *Phys. Rev. Lett.* 71, 1641 (1993)
[5] A.E. Berkowitz, J.R. Mitchell, M.J. Carey, A.P. Young, S. Zhang, F.E. Spada, F.T. Parker, A. Hutten and G. Thomas, *Phys. Rev. Lett.* 68 (25) 37745 (1992)
[6] M. Julliere, *Phys. Lett.*, 54 A, 225-226 (1975)
[7] J. S. Moodera *et al. Phys. Rev. Lett.* 74 (16): 3273–3276 (1995)
[8] T. Miyazaki and N. Tezuka *J. Magn. Magn. Mater.* 139: L231–L234 (1995)
[9] G. A. Prinz, *Science 282*, 1660 (1998)
[10] S. Maekawa, Concepts in Spin Electronics, Oxford University Press, New York, 2006
[11] H. Kronmüller and S. Parkin, Handbook of Magnetism and Advanced Magnetic Materials, vol. 5: Spintronics and Magnetoelectronics, John Wiley and Sons Ltd., USA, 2007
[12] M. Ziese and M.J. Thornton, Spin Electronics, Springer Verlag, Germany, 2001

[13] S. Yuasa, T. Nagahama, A. Fukushima, Y. Suzuki, and K. Ando *Nat. Mat.* 3 (12): 868–871 (2004)
[14] S. S. P. Parkin et al. *Nat. Mat.* 3 (12): 862–867 (2004)
[15] M. Bibes and A. Barthélémy, *IEEE Trans. On Electron Dev.*,54, (5), 1003-1016 (2007)
[16] Y. Ji, G. J. Strijkers, F. Y. Yang, C. L. Chien, J. M. Byers, A. Anguelouch, G. Xiao, and A. Gupta, *Phys. Rev. Lett.* 86, 5585 (2001).
[17] A. Anguelouch, A. Gupta, G. Xiao, D. W. Abraham,Y. Ji, S. Ingvarsson and C.L.Chien, *Phys. Rev. B* 64, 180408 (R) (2001).
[18] L. Yuan, Y. Ovchenkov, A. Sokolov, C.-S. Yang, B. Doudin, and S. H. Liou *J. Appl. Phys.* 93, 10, 6850-6852 (2003)
[19] H. Y. Hwang and S.-W. Cheong, Science 278, 1607 ~1997!.
[20] A. Gupta, X. W. Li, and G. Xiao, *J. Appl. Phys.* 87, 6073 (2000)
[21] S. M. Watts, S. Wirth, S. von Molna´r, A. Barry, and J. M. D. Coey, *Phys. Rev.* B 61, 9621 (2000)
[22] K. Suzuki and P. M. Tedrow, *Phys. Rev. B* 58, 11597 (1998)
[23] K. Suzuki and P. M. Tedrow, *Appl. Phys. Lett.* 74, 428 (1999)
[24] S. S. Manoharan, D. Elefant, G. Reiss, and J. B. Goodenough, *Appl. Phys. Lett.* 72, 984 (1998)
[25] S. J. Liu, J. Y. Juang, K. H. Wu, T. M. Uen, Y. S. Gou, and J.-Y. Lin, *Appl. Phys. Lett.* 80, 4202 (2002)
[26] P. A. Stampe, R. J. Kennedy, S. M. Watts, and S. von Molna´r, *J. Appl. Phys.* 89, 7696 (2001)
[27] J. M. D. Coey, A. E. Berkowitz, L. Balcells, F. F. Putris, and A. Barry, *Phys. Rev. Lett.* 80, 3815 (1998)
[28] A. Sokolov, C.-S. Yang, L. Yuan, S. H. Liou, Ruihua Cheng, B. Xu, C. N. Borca, P. A. Dowben, and B. Doudin, *J. Appl. Phys.* 91, 8801 (2002)
[29] J. Dai, J. Tang, H. Xu, L. Spinu, W. Wang, K. Wang, A. Kumbhar, M. Li, and U. Diebold, *Appl. Phys. Lett.* 77, 2840 pp. 1-3(2000)
[30] J. Dai and J. Tang, *Phys. Rev., B* 63, 064410 pp.1-5 (2001)
[31] L. Dai and J. Tang, *Phys. Rev., B* 63, 054434 pp.1-4 (2001)
[32] G. S. Parkinson, *et al. Phys. Rev. B*, 125413 (125415 pp.) (2010).
[33] K. Jordan, *et al. Phys. Rev. B* 74, 085416, (2006).
[34] Z. Lodziana, *Phys. Rev. Lett* 99, 206402, (2007).
[35] J. M. D. Coey and C. L. Chien, *Mrs Bulletin* 28, 720-724 (2003).
[36] C.Park, J. G. Zhu, Y. G. Peng, D. E. Laughlin, and R. M. White, *IEEE Trans. on Magn.* 41, 2691-2693, (2005).
[37] R. Mantovan, A. Lamperti, M. Georgieva, G.Tallarida and M. Fanciulli, *J. Phys. DAppl. Phys.* 43, 11, (2010).
[38] K. S.Yoon, *et al., J. Magn. and Magn. Mater.* 285, 125-129, (2005).
[39] W. Eerenstein,T. T. M. Palstra, S. S. Saxena, and T. Hibma, *Phys. Rev. Lett* 88, (4), 247204 (2002)
[40] S. S. A. Hassan, *et al. IEEE Trans. on Magn.* 45, 4360-4363, (2009)
[41] Y. B. Xu, *et al. J. Magn. Magn. Mater.* 304, 69-74, (2006).
[42] G. Schmidt, D. Ferrand, L. W. Molenkamp, A. T. Filip, and B. J. van Wees, *Phys. Rev. B* 62, R4790-R4793, (2000).
[43] S. A. Wolf, *et al.* Spintronics: *Science* 294, 1488-1495, (2001).

[44] I. Zutic, J. Fabian, and S. Das Sarma, Spintronics: *Reviews of Modern Physics* 76, 323 410, (2004).
[45] V. Dediu, M. Murgia, F. C. Matacotta, C. Taliani, and S. Barbanera, *Sol. State Communications* 122, 181-184, (2002).
[46] R. M. Cornell, & U. Schwertmann, *The Iron Oxides: Structure, Properties, Reactions, Occurrences and Uses*. (Wiley-VCH, 2003).
[47] E. J. W. Verwey, *Nature* 144, 327, (1939).
[48] J. Garcia, & G. *J.* Subias, *Phys. Condens. Matter* 16, R145-R178, (2004).
[49] A. Yanase, & N. Hamada, *J. Phys. Soc. Jpn* 68, 1607, (1999).
[50] A. Yanase, & K. Siratori, *J. Phys. Soc. Jpn.* 53, 312-317, (1984).
[51] R. J.Soulen, *et al. Science* 282, 85-88, (1998).
[52] J. Osterwalder, Spin-polarized photoemission. *Magnetism: A Synchrotron Radiation Approach* 697, 95-120, (2006).
[53] M. Fonin, Y. S. Dedkov, R. Pentcheva, U. Rüdiger and G. Güntherodt, *J. Phys. Condens. Matter* 19, 315217, (2007).
[54] M. Fonin, Y. S. Dedkov, R. Pentcheva, U. Rüdiger, and G. Güntherodt, *J. Phys. Condens. Matter* 20, 142201, (2008).
[55] M. Fonin, *et al., Phys. Rev. B* 72, 104436, (2005).
[56] J. G. Tobin, & et al. *J. Phys. Condens. Matter* 19, 315218, (2007).
[57] W. Weiss, and W. Ranke, *Progress in Surface Science* 70, 1-151, (2002).
[58] R. Pentcheva, *et al. Phys. Rev. Lett* 94, 126101, (2005).
[59] N. Mulakaluri, , R. Pentcheva, , M. Wieland, , W. Moritz, & M. Scheffler, *Phys. Rev. Lett* 103, 176102, (2009).
[60] G. Tarrach, D. Burgler, T.Schaub, R. Wiesendanger and H. J. Guntherodt, *Surf. Sci.* 285, 1-14, (1993).
[61] R. Wiesendanger, *et al. Science* 255, 583-586, (1992).
[62] B. Stanka, W. Hebenstreit, U. Diebold, and S. A. Chambers, *Surf. Sci.* 448, 49-63, (2000).
[63] S. F.Ceballos, *et al., Surf. Sci.* 548, 106-116, (2004).
[64] M. Kurahashi, X. Sun, & Y. Yamauchi, *Phys. Rev. B* 81, (2010).
[65] C. J. Chen, *Introduction to Scanning Tunneling Microscopy*. (Oxford University Press, 2007).
[66] U. Diebold, *Surf. Sci. Rep.* 48, 53-229, (2003).
[67] U. Diebold, J. F. Anderson, , K. O. Ng. & D. Vanderbilt, *Phys. Rev. Lett.* 77, 1322-1325, (1996).
[68] R. G. Parr, & W. Yang, *Density-Functional Theory of Atoms and Molecules*. (Oxford University Press, 1989).
[69] A. Pratt, M. Kurahashi, X.Sun, & Y. Yamauchi, *J. Phys. D-Appl. Phys.* 44, (2011).
[70] C. Barraud, *et al. Nature Physics* 6, 615-620, (2010).
[71] V. A. Dediu, , L. E. Hueso, , I. Bergenti, & C. Taliani, *Nat. Mater.* 8, 707-716, (2009).
[72] A. R. Rocha, *et al. Nat. Mater.* 4, 335-339, (2005).
[73] P. Ruden, *Nat. Mater.* 10, 8-9, (2011).
[74] S. Sanvito, & A. R. Rocha, *J. Comput. and Theoret. Nanoscience* 3, 624-642, (2006).
[75] Z. H. Xiong, D. Wu, Z. V. Vardeny, & J. Shi, *Nature* 427, 821-824, (2004)
[76] S. Mitani, S. Takahashi, K. Takanashi, K. Yakushiji, S. Maekawa, and H. Fujimori, Phys. Rev. Lett. 81, 2799 (1998).
[77] X. W. Li, A. Gupta, G. Xiao, W. Qian, and V. P. Dravid, Appl. Phys. Lett. 73, 3282 (1998).

[78] J. Inoue and S. Maekawa, Phys. Rev. B 53, R11927 (1996).
[79] V. V. Gridin, G. R. Hearne, and J. M. Honig, Phys. Rev. B 53, 15518 (1996).
[80] W. Eerenstein, T. T. M. Palstra, T. Hibma, and S. Celotto, Phys. Rev. B 66, 20110(R) (2002).
[81] J. M. D. Coey, A. E. Berkowitz, L. Balcells, F. F. Putris, and F. T. Parker, Appl. Phys. Lett. 72, 734 (1998).
[82] D. L. Peng, T. Asai, N. Nozawa, T. Hihara, and K. Sumiyama, Appl. Phys. Lett. 81, 4598 (2002).
[83] J. J. Versluijs, M. A. Bari, and J. M. D. Coey, Phys. Rev. Lett. 87, 026601 (2001).
[84] P. Poddar, T. Fried, and G. Markovich, Phys. Rev. B 65, 172405 (2002).
[85] J. M. De Teresa, A. Barhélémy, A. Fert, J. P. Contour, F. Montaigne, and P. Seneor, Science 286, 507 (1999).
[86] J. H. Park, E. Vescovo, H.-J. Kim, C. Kwon, R. Ramesh, and T. Venkatesan, Phys. Rev. Lett. 81, 1953 (1998).
[87] H. Dulli, E. W. Plummer, P. A. Dowben, J. Choi, and S. H. Liou, Appl. Phys. Lett. 77, 570 (2000).
[88] S. I. Rybchenko, Y. Fujishiro, H. Takagi, and M. Awano, Phys. Rev. B 72, 054424 (2005).
[89] Hao Zeng, C. T. Black, R. L. Sandstrom, P. M. Rice, C. B. Murray, and Shouheng Sun, Phys. Rev. B 73, 020402(R) (2006).
[90] W. Wang, M. Yu, M. Batzill, J. He, U. Diebold, and J. Tang, Phys. Rev. B 73, 134412 (2006).
[91] S. I. Rybchenko, Y. Fujishiro, H. Takagi, and M. Awano, Appl. Phys. Lett. 89, 132509 (2006).
[92] Z. L. Lu, M. X. Xu ,W. Q. Zou, S. Wang, X. C. Liu, Y. B. Lin, J. P. Xu, Z. H. Lu, J. F. Wang, L. Y. Lv, F. M. Zhang, and Y. W. Du, Appl. Phys. Lett. 91, 102508 (2007).
[93] K. Mohan Kant, K. Sethupathi, and M. S. Ramachandra Rao J. Appl. Phys. 103, 07F318 (2008).
[94] S. A. Chambers, S.Thevuthasan and S. S. Joyce, Surface Science 450, L273 (2000).
[95] W. Wang, J. He and J. Tang, *J. Appl. Phys.*, 105, (2009) 07B105.
[96] P. Sheng, B. Abeles and Y. Arie, *Phys. Rev. Lett.*, 31, 44 (1973).
[97] J. Tang , K.-Y. Wang and W. Zhou, *J. Appl. Phys.* 89, 7690(2001).
[98] H. Liu, E. Y. Jiang, H. L. Bai, R. K. Zheng, H. L. Wei, and X. X. Zhang, *Appl. Phys. Lett.* 83, 3531 (2003).
[99] D. Serrate, J. M. De Teresa, P. A. Algarabel, R. Fernández-Pacheco, J. Galibert, and M. R. Ibarra, *J. Appl. Phys.* 97, 084317 (2005).
[100] M. Ziese, Appl. Phys. Lett., 80, 2144 (2002).
[101] L. P. Zhou, S. Ju and Z. Y. Li, J. Appl. Phys., 95, 8041 (2004)
[102] S. Ju, K. W. Yu and Z. Y. Li, Phys. Rev. B, 71, 014416 (2005)].
[103] J. C. Slonczewski, Phys. Rev. B, 39, 6995 (1989).
[104] J. M. MacLaren, X. G. Zhang, and W. H. Bulter, Phys. Rev. B 56, 11827 (1997).
[105] G. Hu and Y. Suzuki, *Phys. Rev. Lett.* 89, 276601 (2002).
[106] M. Ziese, U. Köhler, A. Bollero, R. Höhne and P. Esquinazi, *Phys. Rev. B* 71, 180406(R) (2005).
[107] S. A. Morton, G. D. Waddill, S. Kim, Ivan K. Schuller, S. A. Chambers and J. G. Tobin, *Surface Science* 513, L451 (2002).
[108] T. Zhu and Y. J. Wang, *Phys. Rev. B* 60, 11918-11921 (1999).

[109] F. Guinea, *Phys. Rev. B* 58, 9212 (1998).
[110] P. A. Dowben and R. Skomski, *J. Appl. Phys.*, 95, 7453(2004).
[111] H. Itoh, T. Ohsawa, and J. Inoue, *Phys. Rev. Lett.* 84, 2501(2000).
[112] J.M.D. Coey and S. Sanvito, *J. Phy. D: Appl. Phys.* 37, 988 (2004).
[113] M. Ziese and H.J. Blythe, *J. Phys.: Condens. Matter* 12 13(2000).
[114] N.S. Mcintyre and D. G. Zetaruk, *Analytical Chemistry* 49, 1521(1977).
[115] T. Fujii, F. M. F. de Groot, G. A. Sawatzky, F. C. Voogt, T. Hibma, and K. Okada, *Phys. Rev. B* 59, 3195(1999).
[116] C.W. Chang, B.C. Regan, W. Mickelson, R.O. Ritchie and A. Zettl, Solid State Communications 128, 359-363 (2003).
[117] C. Wen, J. Li, K. Kitazawa, T. Aida, I. Honma, H. Komiyama, and K. Yamada, Appl. Phys. Lett. 61, 2162 (1992).
[118] G. B. Alers, Brage Golding, A. R. Kortan, R. C. Haddon, F. A. Theil, Science 257, 511 (1992).
[119] T. Wang, Y. Wang, F.S. Li, C.T. Xu, D. Zhou, *J. Phys.: Condens. Matter* 18 (47) 10545-10551 (2006)
[120] L. Zhang,and Y. Zhang, *J. Magn. Magn. Mater.,* 321 L15–L20 (2009)
[121] D.S. Xue, L.Y. Zhang, X.F. Xu, A.B. Gui, Appl. Phys. A–Mater. Sci.& Process 80, 439, (2004)
[122] Z.-M. Liao, Y.-D. Li, J. Xu, J.M. Zhang, K. Xia and D.-P. Yu, Nanoletters 6 (6) 1087-1091 (2007)
[123] M. Abid, J.-P. Abid, S. Jannin, S. Serrano-Guisan, I. Palaci and J.-Ph. Ansermet, J. Phys.: Condens. Matter 18 6085–6093(2006)
[124] S. Min, J.-H. Lim, L. Malkinski and J. B. Wiley, *J. Spintronics and Magnetic Nanomaterials,* 1, 52, (2012).

4

Biomedical Applications of Multiferroic Nanoparticles

Armin Kargol[1], Leszek Malkinski[2] and Gabriel Caruntu[2]
[1]*Department of Physics, Loyola University, New Orleans, LA*
[2]*Advanced Materials Research Institue, University of New Orleans, New Orleans LA*
USA

1. Introduction

Magnetic nanoparticles with functionalized surfaces have recently found numerous applications in biology, medicine and biotechnology [1,2]. Some application concern in vitro applications such as magnetic tweezers and magnetic separation of proteins and DNA molecules. Therapeutic applications include hyperthermia [3], targeted delivery of drugs and radioactive isotopes for chemotherapy and radiotherapy and contrast enhancement in magnetic resonance imaging. In this chapter we describe an entirely new concept of using magnetic/piezoelectric composite nanoparticles for stimulation of vital functions of living cells.

The problem of controlling the function of biological macromolecules has become one of the focus areas in biology and biophysics, from basic science, biomedical, and biotechnology points of view. New approaches to stimulate cell functions propose using heat generated by hysteresis losses in magnetic nanoparticles placed in high frequency magnetic field (or magnetic nanoparticle hyperthermia) [5] and mechanical agitation of the magnetic nanoparticles which are attached to the cells using external low frequency magnetic field. In this chapter we propose a new mechanism to be explored which relies on localized (nanoscale) electric fields produced in the vicinity of the cells. Research in this field provides new insight into complexity, efficiency and regulation of biomolecular processes and their effect on physiological cellular functions. It is hoped that as our understanding of complex molecular interactions improves it will lead to new and important scientific and technological developments.

Magnetoelectric properties of nanoparticles

A magnetoelectric is a generic name for the material which exhibits significant mutual coupling between its magnetic and electronic properties. Especially important group in this category are multiferroics which simultaneously demonstrate ferroelectricity and ferromagnetism. Although there exist a few examples of single phase multiferroics [6-11], the multiferric composites [7] with superior magnetoelectric parameters are much more attractive for applications. These composites consist of mechanically coupled ferroelectric and ferromagnetic phases. The conversion of magnetic to electric energy takes advantage of piezoelectric (or more general electrostrictive) properties of the ferroelectric phase and piezomagnetic (or magnetostrictive) properties of the ferromagnetic phase. Magnetostrictive stresses generated in the magnetic phase by the variation of the applied magnetic field ΔH are transferred through the interface between the ferroic phases to the ferroelectric which in turn changes polarization ΔP and associated electric field ΔE due to the piezoelectric effect. Magnetoelectric coefficient α is considered to be the figure of merit which characterizes the performance of a multiferroic composites. In the simplest case, the coefficient α is defined as the ratio of the polarization change in response to the change of the field: $\alpha = \Delta P/\Delta H$. Alternatively, magnetoelectric voltage coefficient $\alpha_E = \Delta E/\Delta H$ can be used instead.

In general, the response of the electric flux density **D** in the multiferroic composites to the applied electric field **E**, external stress **S** and the magnetic field **H** vectors is determined by the permittivity tensor ε_{ij}, piezoelectric coefficients e_{ij} and the magnetoelectric coefficient tensor α_{ij}:

$$D_i = \sum_{j=1}^{j=3} \varepsilon_{ij} E_j + \sum_{j=1}^{j=6} e_{ij} S_j + \sum_{j=1}^{j=3} \alpha_{ij} H_j$$

The components of the magnetoelectric coefficient tensor depend on the piezoelectric coefficients of the ferroelectric, the piezomagnetic coefficients of the magnetic phase, geometry of the constituting components and the quality of the mechanical contact between the piezoelectric and magnetic phases. Examples of formulas for the magnetoelectric coefficients for laminated and particulate composites can be found in review papers by Nan et al. [7] and Bichurin et al. [12,13]. Vast majority of publications on multiferroic composites concerns particulate and laminar composites because of potential applications of these structures in magnetic-field sensors, magnetic-field-tunable microwave and millimeter wave devices, and miniature antennas, data storage and processing, devices for modulation of amplitudes, polarizations and phases of optical waves, optical diodes, spin-wave generation, amplification and frequency conversion.

So far, relatively little attention has been drawn to fabrication and characterization of composite nanoparticles. The multiferroic nanoparticles, which are useful for the particular application we propose, can be in the form of a spherical core-shell nanoparticles with a ferromagnetic core and a ferroelectric shell as depicted in Fig. 1 a), magnetic rods with a piezoelectric coating, as seen in Fig. 1b) (also concentric magnetic/piezoelectric tubes) as well as composite spheres of piezoelectric ceramics or piezopolymers with embedded magnetic nanoparticles (Fig.1c).

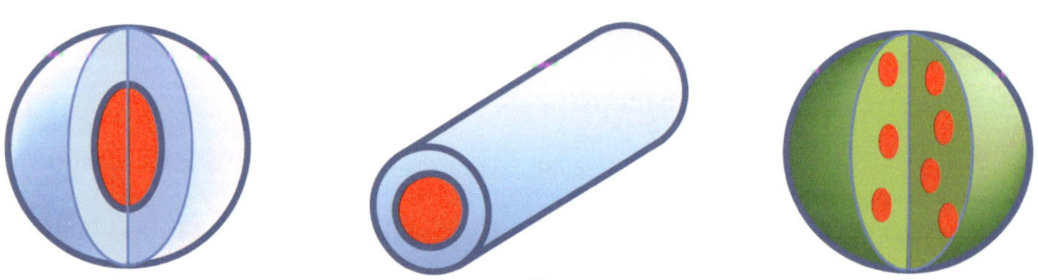

(a) (b) (c)

Fig. 1. Three types of the magnetoelectric nanoparticles: spherical core-shell nanoparticles with magnetostrictive core encapsulated in piezoelectric shell (a), cylindrical core-shell nanoparticle with magnetostrictive rod and piezoelectric coating (b) and a composite superparticle with magnetic nanoparticles embedded into piezo-polymer or piezo-ceramics (c).

The reasons for which these magnetoelectric particles are expected to have excellent magnetoelectric performance are as follows:

a. Very good mechanical contact at the interface between two synthesized ferroic phases (which is an issue for many particulate or laminar composites bonded by gluing, pressing or sintering).
b. Efficient transfer of magnetostrictive stresses to the piezoelectric phase. Because of very small thickness of the piezoelectric shell the magnetostrictive stress will be absorbed by the whole volume of the piezoelectric phase (not just by interfacial layer as it may happen in thicker film coatings or large grains). Here, it is important to mention that in our preliminary studies we did not observe any dramatic deterioration of piezoelectric or magnetostrictive properties of the constituting components of the nanoparticles as their dimensions decreased down to 25 nm.
c. No substrate clamping effect which restricts magnetostrictive or electrostrictive strains in some laminated or thin film structures.

Since a few existing articles about magnetoelectric core-shell nanoparticles [12-20] do not report on their magnetoelectric coefficients our estimate of the expected properties of multiferroic nanoparticles will be based on the coefficients of the laminated composites. At this point it is important to note, that the record values of magnetoelectric voltage coefficient exceeding 90 V/(cm·Oe) [7,21] have been measured for the laminated composites in magnetoacoustic resonance conditions.

Theoretical model by Petrov et al. [22] predicts even higher values (up to 600 V/cm Oe) of the α_E at electromechanical resonance for the free standing multiferroic nanopillar structures, which can be identified with our cylindrical core-shell nanoparticles. However, because of specific frequency range (0-5000 Hz) which corresponds to ion channel dynamics we expect in our application the values of magnetoelectric voltage coefficient α_E of the core-shell nanoparticles to be closer to the non-resonant values ranging from 1 to 5 V/ (cm·Oe).

The difference of potentials generated by the particle will depend on the particle size. It is easy to estimate that nanoparticles as small as 50 nm with magnetoelectric voltage coefficient of 5V/(cm·Oe) should be capable of generating voltage of 25 mV when exposed to the field of 100 Oe (or 0.01 T). This voltage is sufficient to control ion transport through ion channels and also to trigger action potential in nerves. By increasing the size of the particles to 1 micrometer (which is still a small fraction of the cell size) this voltage can be increased by more than 2 orders of magnitude.

Whereas the magnetoelectric performance of the multiferroic core-shell nanoparticles can be elucidated to a high degree, there is not enough literature data available for the piezopolymer based composite nanoparticle. However, these nanoparticles are intriguing for at least three reasons:

- These composites may use superparamagnetic nanoparticles embedded in the piezopolymer. In the absence of the external field they will not display effective magnetic moment which prevents potential problems with particle aggregation.
- The key difference between the multiferroics and these piezopolymer clusters is that they may use different mechanisms for generation of electric fields. Because of mechanically soft matrix, attractive dipole-dipole interactions between magnetic nanoinclusions in the presence of a magnetic field may be more effective than the

magnetostrictrive strains of individual magnetic particle. The strained piezopolymer will respond by changing its polarization. The primary factors which will affect the performance of these magnetoelectric nanocomposites will be: saturation magnetization of the nanoparticles, separation between the particles and the piezoelectric coefficients of the piezopolymer, which can be as good as those of piezoelectric solids [23]. Due to insufficient literature data, good performance of these novel structures can be predicted based on excellent performance of piezopolymers in laminated composites [24].

- Some polymer composites are biocompatible and have already been successfully used in cancer research. Therefore, this design of magnetoelectric nanoparticles may be the ultimate solution in therapeutic applications.

Voltage-gated ion channels

One of the primary examples of research on interaction of electromagnetic fields with biological macromolecules is the study of ion channels. Ion channels are membrane proteins that form pores for controlled exchange of ions across cellular membranes [24]. Their two main characteristics are their selectivity, i.e. the type of ions they flux (e.g. K^+, Na^+, or Ca^{++}) and their gating, i.e. the process of opening and closing in response to the so-called gating variable. Probably the most common are the voltage-gated ion channels which gate in response to the transmembrane electric field. They are central to shaping the electrical properties of various types of cells and regulate a host of cellular processes, such as action potentials in neurons or muscle contraction, including the heart muscle. Ion channel defects have also been identified as causes of a number of human and animal diseases (such as cystic fibrosis, diabetes, cardiac arrhythmias, neurological disorders, hypertension etc.) [25]. As such, ion channels are primary targets for pharmacological agents for therapeutic purposes. Cellular responses to various chemical stimuli, such as drugs and toxins, can be analyzed in terms of their effect on ion channels. From a biophysical standpoint ion channels are the primary examples of voltage-sensitive biomolecules.

The main aims of ion channel studies are to understand the channel gating kinetics and how it is affected by various external factors as well as to develop methods of controlling the channel gating. It is believed that this will lead to knowledge of corresponding molecular mechanisms of gating and eventually to a full understanding of the physiological role of the channels. Gating of voltage-gated ion channels results from rearrangements of the tertiary structure of the channel proteins, i.e. transitions between certain meta-stable conformational states in response to changes in transmembrane potential. Various conformations can be described by a finite set of states (closed, inactivated or open), connected by thermally activated transitions, and known as the Markov model of channel kinetics [26,27]. It can be mathematically described as a discrete Markov chain with transitions between the states given as:

$$\alpha(V) = \alpha(0)\exp(qV / kT) \tag{1}$$

where α is a generic transition rate, V - membrane voltage, k - Boltzmann constant, T - temperature, q - gating charge. The topology of Markov models (i.e. the number and connectivity of its discrete states) as well as the parameters for the transition rates are determined by fitting model responses to various sets of experimental data, mostly obtained from electrophysiological experiments. It needs to be understood, though, that Markov models are only a coarse approximation and more precisely the channel gating should be viewed as a motion of a "gating particle" in a certain energy landscape, subject to thermal fluctuations and governed by the Langevin or Fokker-Planck equation [28]. Models based on the Smoluchowski equation describing channel gating as a diffusion process in a one-dimensional energy landscape have also been developed [29-31]. However, a picture of channel gating as a discrete Markov chain with the discrete states corresponding to the minima in this energy landscape has been widely and very successfully used. A large proportion of scientific publications on ion channels is devoted to channel gating properties and the corresponding Markov models.

A functional model of a voltage-gated ion channel consists of three parts: the voltage sensor, the pore, and the gate. Commonly, the channels are made of four subunits (K^+ channels) or one protein with four homologous domains (Na^+ and Ca^{2+} channels). Each of these has six transmembrane segments S1-S6. It is believed that the voltage sensitivity of ion channel is based on the movement of the charged S4 domain (described as the movement of the "gating charge") that causes conformational changes of the molecule resulting in channel opening or closing [24].

Ion channel electrophysiology

The bulk of experimental data on ion channel function comes from patch-clamping technique [24,32] (Fig. 2), in which the voltage across a cell membrane is controlled by a feedback circuit that balances, and therefore measures, the net current. Variations of this technique include whole-cell recordings (measured from a large number of ion channels), single channel recordings (ionic currents through membrane patches containing only single channels), and gating currents. Electrophysiological experiments have been recently complemented by X-ray crystallographic methods, fluorescence methods (FRET), and other imaging techniques [33-35].

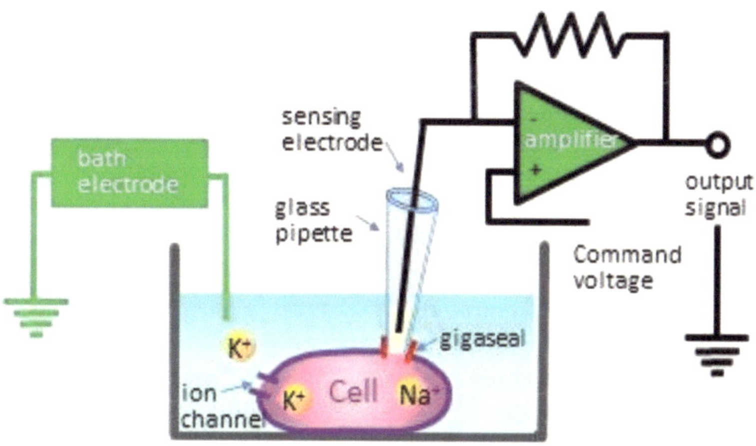

Fig. 2. A schematic illustration of the patch-clamp method

For electrophysiological experiments the channels can be either in their native cells or expressed in a suitable expression system (typically mammalian cells transfected with the channel DNA or Xenopus laevis oocytes injected with the RNA). A cell is placed in a grounded solution in a recording chamber under an inverted microscope. A glass pipette with another electrode is moved using a micromanipulator until it forms a tight seal (a gigaseal) with the cell membrane. A patch-clamp amplifier overrides the membrane potential that naturally exists across a membrane in every living cell, and measures the current flowing through the ion channels. The current reflects the kinetics of channel gating. In the whole-cell mode it is the total current from all channels present in the cell, while in the single-channel mode it is recorded from one or few channels in the excised membrane patch.

The paradigm of ion channel electrophysiology is a stepwise change of the potential from the holding hyperpolarized value. Membrane depolarization changes the transition rates, according to (1), and affects the stationary probability distribution among all Markov states. The transitions are thermally activated and are stochastic in nature. The time constant of channel relaxation to a new probability distribution depends on voltage and the temperature, but it typically is of the order of few ms. Experiments with time-variable, including rapidly fluctuating voltages have been also proposed for various purposes [36-40] and they only require sufficient bandwidth of the recording apparatus, faster than the time scale of channel gating kinetics. Bandwidth in excess of 5 kHz can be easily obtained with standard equipment. Based on these and other types of experimental data Markov models have been developed for many known channels and can be found in literature [36,37].

Ion channels in cell membranes can be found in any of their conformational states with certain probabilities and can "jump" from one state to another in a random fashion. This probability distribution among these different states is determined by the channel kinetics and in equilibrium states cannot be directly controlled. In other words, the molecules cannot be "forced" into any particular state. However, the transition rates between different states in a Markov model are voltage-dependent and as the voltage is varied the probability distribution for channel conformational states changes. One should distinguish here between equilibrium and non-equilibrium distributions. The former are achieved in response to static voltages while the latter are achieved for fluctuating voltage stimuli. The idea of remote channel control is to enhance a probability of finding channel molecules in a selected state (open, closed, or inactivated) by driving them into a non-equilibrium distribution with fluctuating electric fields [38,39] (Fig. 2). This is an idea that might be new to biologists but has been well-researched in statistical physics, both theoretically and experimentally [41-45].

Control of channel gating using multiferroic nanoparticles

While a significant progress has been made on the channel gating kinetics modeling, there hasn't been as much success in controlling the channel function. The primary approach has been the use of pharmacological agents that modify the gating kinetics or block the channel transiently or permanently [46]. This has been used both for research and therapeutic purposes. Ion channel function can be modified by many natural agents extracted from plants and animals as well as drugs developed for that specific purpose. This has been the basis for treatment for many ion channel-related disorders. On the other hand ion channels are very difficult targets for pharmacological agents since their function is state and voltage dependent and the interactions are highly nonlinear. In this chapter we discuss a different, innovative approach, based on using multiferroic nanoparticles introduced extra- or intracellularly to locally modify electric fields and invoke appropriate conformational changes of channel proteins. This method would allow remote control of ion channel gating by externally applied magnetic fields.

Electric and magnetic fields have been shown to influence biological systems. The effect depends on the intensity and frequency of the fields and may be beneficial, adverse, or neutral to the organism. Only relatively gross effects, such as cell damage by electroporation or tissue heating by microwaves, are reasonably well understood [47-52]. Otherwise little is known about the mechanism of the interaction, and even the significance of the effect can be questioned. Some issues, like the adverse effects of electromagnetic fields generated in the vicinity of power lines or of microwave radiation from cell phone use are quite controversial [47,49,52]. On the other hand there are therapeutic uses of electric and/or magnetic fields. For instance, electro-stimulation techniques such as transcutaneous electrical nerve stimulation (TENS) or deep brain stimulation (DBS) have evolved from highly experimental to well-established therapy used for treatment of pain, epilepsy, movement disorders, muscle stimulation, etc. [53,54]. Although they provide evidence for the interaction of electric fields with biological systems these methods have many restrictions and drawbacks. First of all, many of present electrostimulation methods use high voltage (sometimes above 200V) and expose large portions of tissues or whole organs to electric fields and currents. Moreover, some internal organs (e.g. brain) are not accessible to electrostimulation without surgery. Secondly, the best results in therapy and research have been achieved using percutaneous electrodes which are in direct contact with bodily fluids.

The new method we propose is based on remote generation of local electric fields in the proximity of the cells by nanoparticles of magnetoelectric composites. Therefore, individual cells or selected groups of cells can be targeted, rather than whole tissues, and the voltage required for their stimulation will be of the order of several mV, not hundreds of volts. Also, because of the remote way in which the stimulation will be performed, the functions of the cells in the internal organs, including brain, can be controlled. Moreover, because of the magnetic component the particles can be concentrated in a specific location (of the brain, for instance) using magnetic field gradient. Finally, the nano-electrostimulation can target specific types of cells or even certain parts of cell membranes. This selectivity of stimulation can be achieved by functionalizing the surface of the nanoparticles and binding them to cells through antigens.

Figure 3 illustrates the proposed method for controlling ion channel gating. Multiferroic nanoparticles will be either placed in extracellular medium or introduced internally. When an external magnetic field is applied these multiferroic particles will respond to it and convert it to localized electric fields. For particles placed near cell membranes containing ion channels this will lead to local membrane depolarization or hyperpolarization. The ion channels would respond by opening or closing accordingly. The stimulating magnetic field can be generated e.g. by Helmholtz coils and hence its properties can be very easily controlled. Moreover, the field can be applied globally (i.e. to the selected parts or to the entire organism) but the electric fields will be generated only locally, in the areas to which nanoparticles have been delivered. By modifying the properties of the stimulating magnetic field, such as its strength, duration, and spectral properties, as well as by controlling the delivery locations for the nanoparticles, we achieve a tight control over the localized electric fields. This will lead to very localized changes in ion channel gating. The resulting changes in ionic currents can be measured using standard electrophysiological techniques.

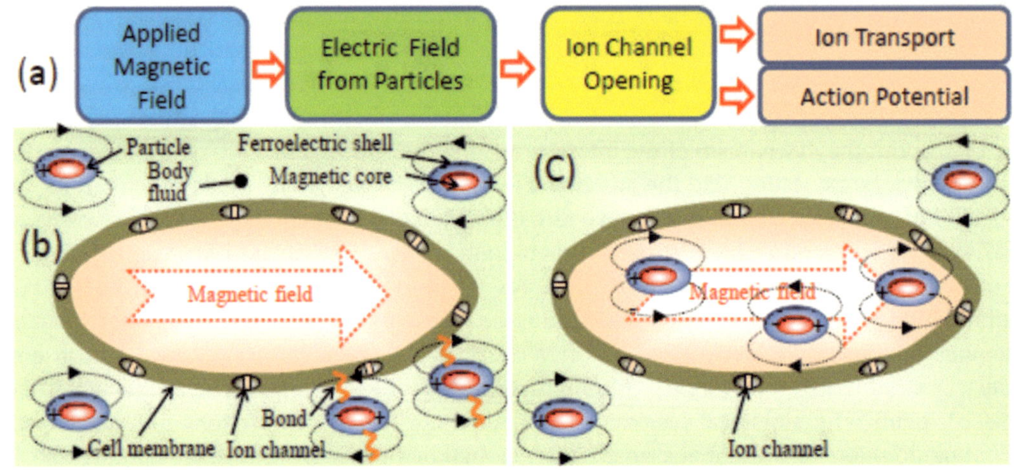

Fig. 3. Illustration of possible mechanisms of stimulation of ion channels by dipolar electric fields from magnetoelectric nanoparticles generated remotely by external magnetic field: (a) A chain of actions triggered by the applied magnetic field pulses. (b) External stimulation of the ion channels using nanoparticles in body fluid or particles bound to the cell membranes. (c) Internal stimulation of cells by uptaken nanoparticles.

The minimum strength of the signals used to control ion channels is of the order of mV and maximum frequency to which they respond should not exceed the typical bandwidth of the patch-clamping apparatus (few kHz). The purpose of the first experiments would be to determine the existence and the scale of this effect in mammalian cells in vitro, as well as its possible future biomedical applications.

Typical electrophysiology experiments include measurement of activation and "tail" ionic currents (Fig. 4). The former are recorded in response to stepwise voltage increases from the holding potential to a series of depolarizing values. The currents reflect the channels' opening as a result of the depolarizing stimulus. The "tail" currents are recorded when the potential is changed from a depolarized value to a series of repolarizing values and show the process of channel closing in response to a hyperpolarizing stimulus. Similar types of recordings can be performed in the presence of nanoparticles and we can observe how the process of channel gating (opening or closing) is affected by the nanoscale electric fields generated by magnetically stimulated nanoparticles. It is expected that if a sufficient number of nanoparticles is placed close to the ion channels, there will be observable changes in the value and the time course of the ionic currents. The next step will be to apply modulated magnetic fields to generate variable electric fields that would allow us to directly impact the channel gating using remote stimulation. Both extracellular and intracellular application of nanonparticles can be tested.

It is important to mention that both the extracellural and intracellural action of the particles can be investigated (Fig. 5). Extracellular delivery is very straightforward. The nanoparticles will be dispersed in the extracellular medium. Since the ferroelectric component can be charged the multiferroic particles in the vicinity of the cells will be attracted to the cell membranes. For intracellular delivery, two methods can be contemplated. One is the particle uptake by cells that has been reported in several studies for particles of size up to 1.5 μm [10].

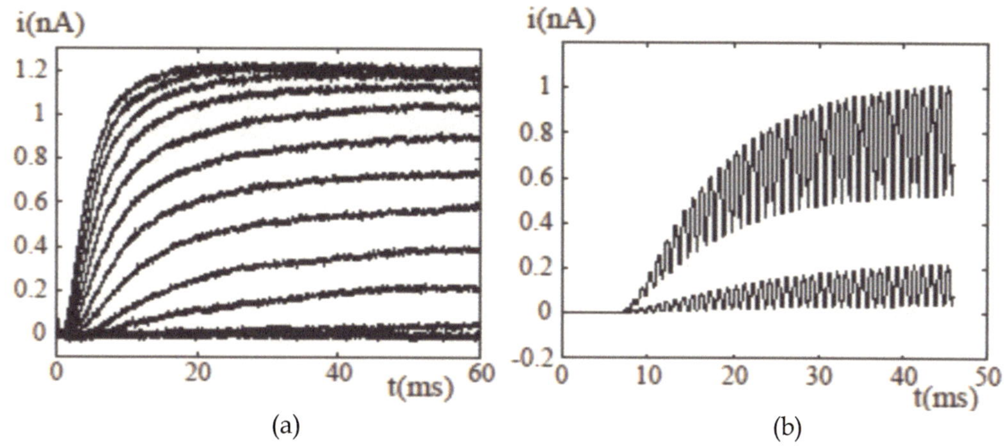

Fig. 4. Sample whole-cell currents recorded from Shaker K+ channels using patch-clamp technique. a) Currents obtained for stationary voltages. At t = 0 voltage was changed from a holding value of -70 mV to a series of values varying from -70 to 42 mV in 8 mV steps. b) Currents obtained for sinusoidally modulated voltages. The voltage amplitude was 90 mV peak-to-peak, the frequency 1 kHz, and the mean values: 0 mV and -30 mV. It is expected that a similar modulation of current can be observed when ion channels are exposed to multiferroic nanoparticles and a modulated external magnetic field is applied.

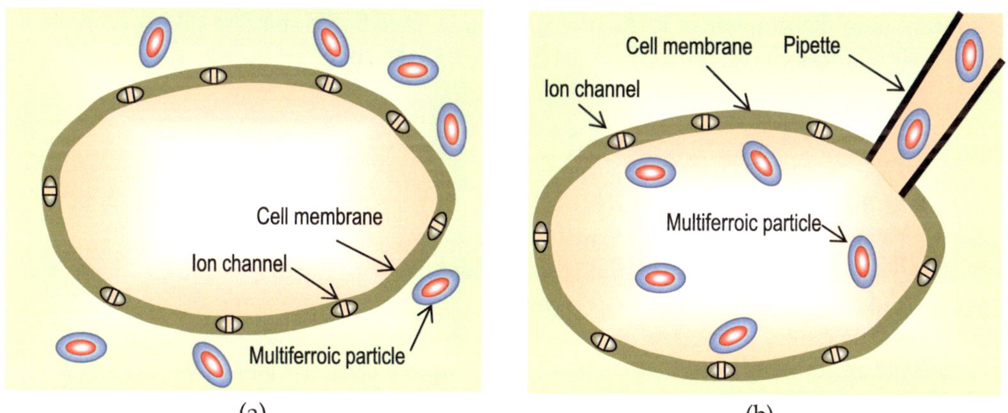

Fig. 5. Two possible methods for multiferroic nanoparticles: a) nonoparticles are dispersed in the extracellular medium.; b) patch-clamping pipette is filled with solution containing multiferroic nanonparticles. After the gigaseal is formed and the membrane patch at the tip of the pipette ruptured, the cell is perfused with the pipette solution, including the nanoparticles. The in-vivo uptake of nanoparticles into cells has been also reported after they are initially introduced extracellularly

Another method would make use of the recording apparatus used in patch-clamping. In the whole-cell mode the membrane patch at the tip of the recording pipette is ruptured, allowing a direct contact between the cell cytosol and the pipette solution. Since the volume of the pipette is significantly larger than the cell itself, the cell is quickly perfused with the pipette solution. If the nanoparticles are added to the pipette solution they quickly diffuse to the cell interior. It is a method commonly used for intracellular drug delivery or recording solution replacement in ion channel research. For a precisely timed and controlled cell perfusion with different solutions an electronic valve controller can be used.

An ultimate goal would be using nanoparticles tethered to, or covered by ligands that bind to the selected membrane protein, in this case ion channel, ensuring that the nanoparticle stays in the vicinity of the target molecule. Fluorescent labels can be used in order to detect and track the multiferroic particles inside the cells.

It seems that there is no alternative to the magnetoelectric composites for generation of highly localized strong electric field in biological systems. At this point it is worth to mention that the pulses of magnetic flux ranging from 10^3 to 10^6 T/s used in the brain studies can induce only weak electromotive forces of the order of nanovolts to microvolts on cellular level and are more effective only on the scale of whole organs [2].

One can question the choice of multiferroic particles for generation of electric fields on sub-cellular level over the inductive systems which can also remotely generate local electric fields. A simple estimate shows that it would be extremely difficult to fabricate inductive systems with the micrometer size which would be capable of generating voltages sufficient to affect ion channels. Such circuits could be made by means of nanolithography (electron beam writing or focused ion beam etching) or by coating the nanostructures which have chiral shape (for example some ZnO nanowires or carbon nanotubes, which can grow in the chiral shape). According to the Faraday's law the voltage ε generated by a miniature coil is proportional to the number of turns n of the coil, its area S and the rate dB/dt of magnetic flux density B through the coil in the axial direction of the coil:

$$\varepsilon = -\frac{\partial \Phi}{\partial t} = -nS\frac{\partial B}{\partial t} \qquad (2)$$

where Φ is the magnetic flux. Microscopic coils with n=10 turns and the size of $1\text{x}10^{-6}$ m (and the corresponding area $S=10^{-12}$ m^2) can be fabricated using current nanolithographic methods. When placed in the alternating magnetic field with the amplitude B=0.1 Tesla and frequency of 1 kHz, such coils are expected to generate voltage of the order of 10^{-9} V (or nanovolts rather than a fraction of volt) which is orders of magnitude smaller than that required for ion channel stimulation. This signal can be further enhanced by a factor of 1000 by inserting a magnetic core of magnetically soft material, but the signals generated by such inductors will still be below the useful range.

In contrast, multiferroic particles seem to be much more promising for generation of mV signals at micron- or nano-scale. As already mentioned, the strains of magnetostrictive material due to applied magnetic field produce deformation of the piezoelectric component which results in change of polarization of the particle. For the 1 micrometer magnetoelectric particles with medium magnetoelectric voltage coefficient of 1V/cmOe (measured in non-resonant conditions of 1 kHz) the same fields of 0.1 T can potentially generate voltage of 100 mV and even more than 10 times smaller particles can generate useful electric potentials for the stimulation of mammalian cells. The volume of these particles can be million times smaller than the size of a typical mammalian cell, thus they can easily be accommodated inside the cells. Also because of small size the nanoparticles can easily propagate with the bloodstream through the vascular system and with the static magnetic field they can be directed to specific organs. Cells of these organs and ion channels closest to the multiferroic particles will experience presence of electric field when alternating magnetic field with the frequency in the acoustic range is applied. These local fields are expected to affect opening and closing of the ion channels and thus to alter ion transport across cell membranes.

Synthesis of core-shell magnetoelectric architectures

The role of the interfaces in the strain-mediated magneto-electric coupling

A key requirement for achieving a sizeable ME response in composite materials is to combine phases with a robust piezomagnetic and piezoelectric properties. Perovskites ABO_3 (A=alkaline/alkali earth metal, Pb; B=Ti, Zr, Nb, etc.) and ferrites (either spinel MFe_2O_4 (M= transition metal) or M-type hexagonal ferrites $AFe_{12}O_{19}$ (A=alkali earth metal) are good candidates for the design of ME nanocomposites since they present robust ferroelectric and magnetic responses at room temperature, are structurally compatible, have high chemical, thermal and mechanical stability, can be fabricated by simple methods at relatively low costs and present immiscibility gaps which limit the number of secondary phases. Recent advances in the chemical synthesis of nanoscale spinel-perovskites magnetoelectric structures resulted in materials that are approaching the quality needed for practical devices.

Since the control of the ferroic order parameters in magnetoelectric composites is dependent on how efficiently the generated strain is transferred across the shared interfaces, the intrinsic surface stress, the curvature of the surface [55], the contact area and the characteristics of the interfaces at the atomic scale are critical for the enhancement of the ME coupling. Any imperfect interface will more or less reduce the transfer of the elastic strain between the magnetostrictive and ferroelectric phases, thereby leading to a decrease of the ME response of the nanocomposites. The common methods of bonding two constituent phases in bulk magnetoelectrics include co-sintering [56,57], tape-casting [58], hot pressing [59] or adhesive bonding [60], which inevitably results in losses at the interface [61]. Furthermore, the elastic coupling between the constituent phases can be hindered by the formation of undesired phases as a result of the interfusion and/or chemical reactions during the annealing, thermal expansion mismatch between the two phases, as well as the presence of voids, residual grains, phase boundaries, porosity, dislocations and clamping effects [61,62].

Nanostructuring, as mentioned before, is an efficient way to engineer the interphase boundaries in hybrid ME composites which will lead to a maximum transfer of the mechanical strain between the two phases and, in turn will allow for an optimum conversion between the magnetic field and the electric field. Interfacing a ferroelectric with a ferromagnetic material in core-shell geometry offers a much higher anisotropy and high surface contact between the components which can also result in a strong ME effect. The low processing temperatures associated characteristic to the solution-mediated routes are preferable to substantially reduce the diffusion pathways between the molecular species, thereby preventing the formation of unwanted secondary phases. Magnetoelectric nanocomposites formed by assembling ferrites as the magnetic phase with perovskites as the electrostrictive phase can present a large ME response due to the large magnetostriction of the ferrites and the large piezoelectric coefficients of the perovskites. Additionally, in core-shell geometry the magnetic grains are completely isolated from each other by the perovskite shells, which prevents the high dielectric loses related to electrical conduction, space-charge effects and Maxwell-Wagner interfacial polarization which are detrimental for the ME properties of the composites [64]. Last, but not least, the control of the size of the core and thickness of the shell enables the fine adjustment of the microstructure and phase fraction of the composites, resulting in materials with tunable properties and reproducible characteristics.

Chemical synthesis of core-shell ME ceramic nanocomposites

Soft chemistry-based chemical approaches represent a viable alternative to the classical "shake and bake" methods for the fabrication of metal oxide-based magnetoelectric nanocomposites. It is well known that the application of high temperatures in conventional sintering processes yields the formation of unwanted phases as a result of chemical reaction. For example undesired phases such as $BaFe_{12}O_{19}$, $BaCo_6Ti_6O_{19}$ or hexagonal $BaTiO_3$ could appear in the $BaTiO_3$-$CoFe_2O_4$ ferroelectric magnetostrictive composites being very detrimental to their physical properties. In addition, the interphase diffusion of the constitutional atoms, though difficult to prove experimentally, lowers the local eutectic point around the boundary region, thereby facilitating the formation of structural defects in high concentration as well as liquid phases. These phenomena are detrimental to the piezoelectric signal and/or magnetostriction of the constituent phases and the strain created at the interface between them. Moreover, the large thermal expansion mismatch between the perovskite and ferrite phases deteriorates the densification and leads to the formation of micro-cracks. Solution-based methods alleviate these problems since they have the potential to achieve improved chemical homogeneity at molecular scale which significantly increases the surface contact between the two phases and, in turn, induces a notable enhancement of the magnetoelectric voltage coefficient.

Last, but not least, when ME nanocomposites are fabricated by chemical routes, diffusion distances are reduced on calcination compared to conventional preparation methods owing to the mixing of the components at colloidal or molecular level that favors lower crystallization temperatures for multicomponent oxide ceramics. This will eliminate the intermediate impurity phases leading to materials with small crystallites and a high surface area. Another challenges that need to be overcome are those related to the coupling between the two phases and include: a) the mechanical coupling between two phases must be in equilibrium; b) mechanical defects, such as the pores or cracks should be as low as possible to ensure a good mechanical coupling between the two phases; c) no chemical reaction between the constituent phases should occur during the synthesis; d) the resistivity of magnetostrictive phase should be as high as possible and the magnetic grains should be isolated from each other to minimize the leakage currents and prevent the composites from electric breakdown during electric poling; e) the magnetostriction coefficient of piezomagnetic phase and piezoelectric coefficient of piezoelectric phase must be high; and f) the proper poling strategy should be adopted to get high ME output in composites. Due to their assembled structure, core-shell piezoelectric-ferro/ferrimagnetic nanostructures have been fabricated exclusively by chemical routes. In the following we will briefly review the most relevant approaches described in the literature, as well as explore some novel methods which can be extended to the fabrication of ME core-shell ceramic nanoparticles with controlled morphology and predictable multiferroic properties.

Synthesis of core-shell magnetoelectric ceramic nanocomposites and assemblies

The synthesis of core-shell spinel-perovskite nanoparticles (0-2 connectivity scheme) is conventionally carried out in two successive steps: (1) the precipitation of the ferrite nanoparticles and (2) the creation of a perovskite shell around each nanoparticle (Figure 6).

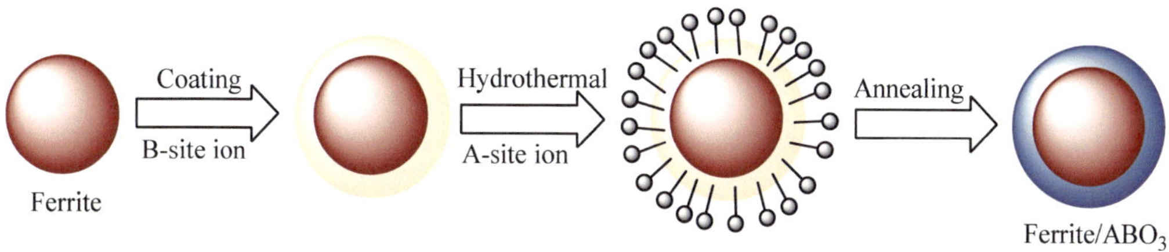

Fig. 6. Schematic of the two-step approaches used for the fabrication of core-shell ferrite/perovskite nanoparticles by soft-solution routes

Due to their low percolation threshold, the agglomeration of the ferrite nanoparticles should be avoided during the creation of the perovskite shells in order to prevent short-circuits during the measurement of their magnetoelectric properties. In general, the ferrite nanoparticles are obtained by classical alkaline precipitation of stoichiometric mixtures of Fe^{3+} and M^{2+} (M=Mn, Fe, Co, Ni, Cu, Zn) in aqueous solutions. Liu and coworkers prepared spinel ferrite-perovskite core-shell nanoparticles by combining a solvothermal route with conventional annealing [65]. In the first step of the synthesis spheroidal iron oxide microparticles (Fe_3O_4, γ-Fe_2O_3 and $CoFe_2O_4$) were obtained by treating hydrothermally solutions containing the metal ions and NaOH dissolved in ethylene glycol at 200 ^oC. The scanning electron microscopy (SEM) images shows that the as-prepared particles are spheroidal and have rough surfaces, are aggregate-free and have an average diameter of 280 nm. In the next step the ferrite nanoparticles were individually coated with a Ti hydroxide layer upon the slow hydrolysis and condensation during the aging of a solution of $Ti(SO_4)_2$ in which the particles were suspended. The Ti hydroxide layers have a thickness of several tens on nanometers (Figure 7f) and they render the surface of each ferrite microparticles very smooth (third image in Figure 6 and Figures 7b and c). Furthermore, the hydroxide layer can be converted into a uniform and dense perovskite coating by *an in-situ* hydrothermal reaction at 140 ^oC between the Pb^{2+} ions from the solution and the Ti^{4+} ions anchored onto the surface of the magnetic particles. Because the perovskite layer has a low crystallinity, the resulting magneto-electric core-shell nanoparticles were subjected to an additional annealing in air at 600 ^oC.

X-Ray diffraction analysis confirmed that the nanocomposites are polycrystalline and free of secondary phase and consist of a spinel and perovskite phase suggesting that the proposed methodology can be used for the fabrication of core-shell nanoparticles with different chemical compositions (Figure 8a). Changing the chemical identity of the M^{2+} ions in the ferrite structure allows the tuning of the magnetic properties of the nanocomposites in a wide range. In general the coercivity of the nanocomposites increases upon annealing with H_c values of the nanocomposites of 15, 38, 40 and 1200 Oe for the $Fe_3O_4/PbTiO_3$, γ-$Fe_2O_3/PbTiO_3$, γ-$Fe_2O_3/Pb(Ti, Zr)O_3$ and $CoFe_2O_4/PbTiO_3$ particles, respectively (Figures 8b and c).

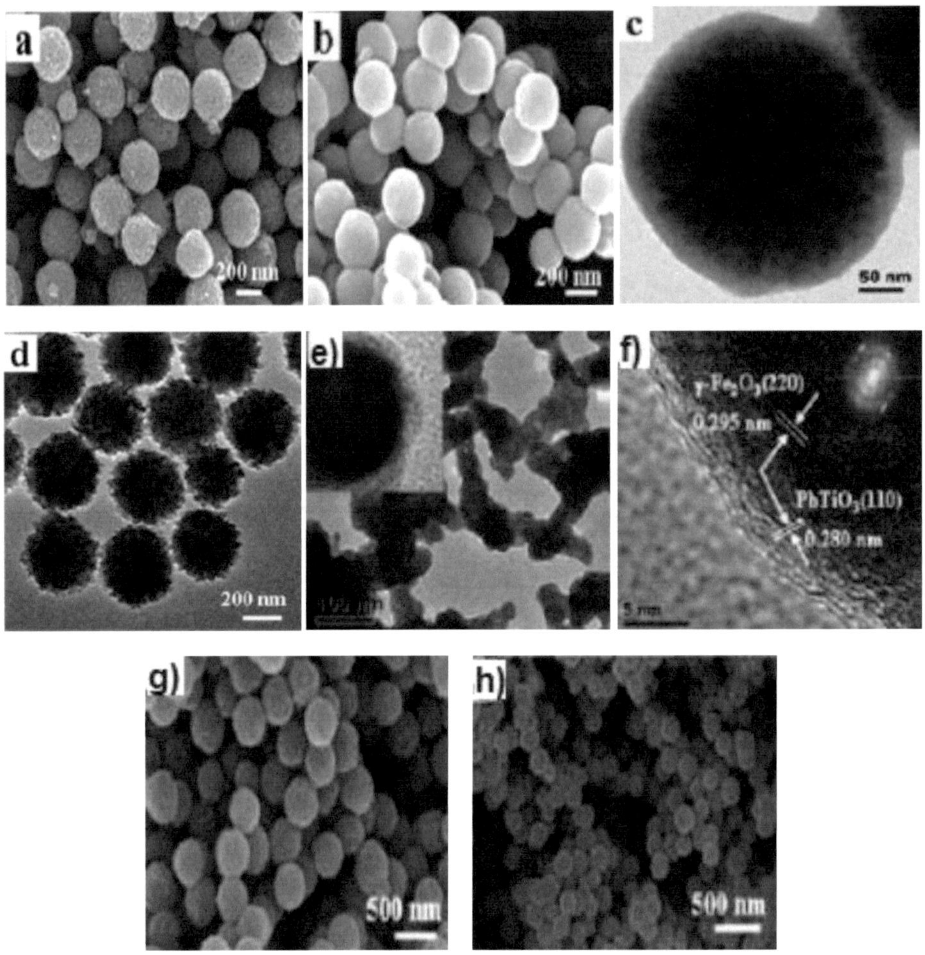

Fig. 7. SEM images of Fe_3O_4 nanoparticles obtained under solvothermal conditions from polyol solution (a); Fe_3O_4 microparticles coated individually with a layer of Ti hydroxide (b) and (c); $Fe_3O_4/PbTiO_3$ core-shell particles (d); (e) and (f) TEM and HRTEM images of chain-like aggregates of γ-$Fe_2O_3/PbTiO_3$ core-shell nanoparticles; SEM images of $CoFe_2O_4/BaTiO_3$ (g) and $CoFe_2O_4/PbTiO_3$ core-shell nanoparticles (h);

The values of the saturation magnetization were found to be in agreement with the molar fraction of the ferrite phase, as well as the average diameter of the magnetic core. These nanostructures are not dispersible in solvents, which somehow limits their use in biomedical applications. From the viewpoint of their magnetic and electric properties they were characterized in detail; however, no studies on their *magnetoelectric properties* were reported so far. A similar approach was proposed by Buscaglia and coworkers, who reported on the preparation of α-Fe_2O_3@$BaTiO_3$ [65] and $Ni_{0.5}Zn_{0.5}Fe_2O_4$@$BaTiO_3$ [66] core-shell submicron particles. Similar to the method reported by Liu *et al.* ferrite nanoparticles were prepared by co-precipitation and then a uniform layer of amorphous titania (TiO_2) was formed around each nanoparticle by treating a solution containing these magnetic grains with a peroxotitanium solution followed by a treatment of the nanoparticles with a solution of

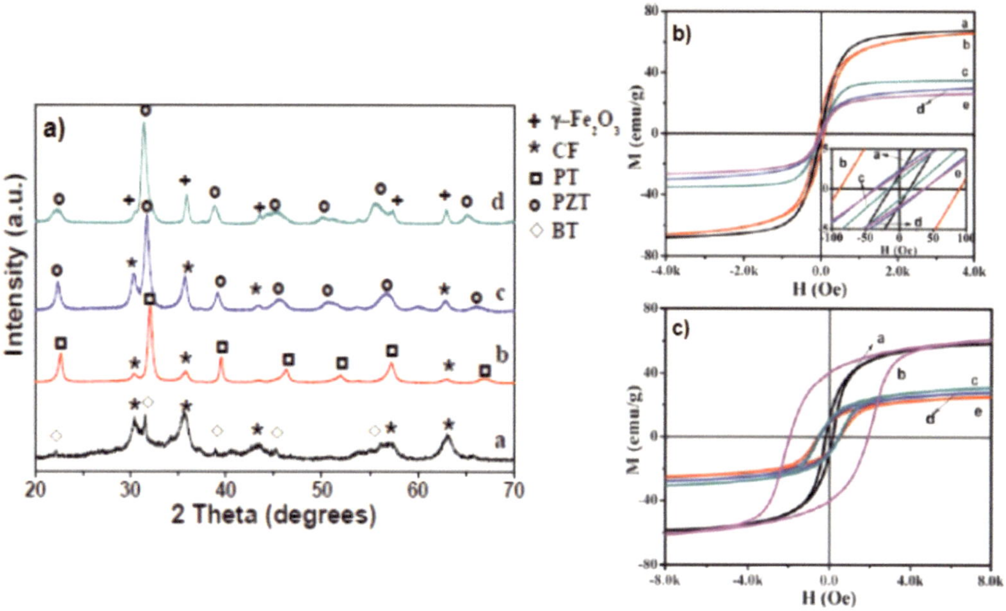

Fig. 8. XRD patterns of core-shell nanostructured powders of SEM images of $CoFe_2O_4$/$BaTiO_3$ (a); $CoFe_2O_4$/$PbTiO_3$; Fe_3O_4 (b); $CoFe_2O_4$/Pb(Ti, Zr)TiO_3 (c) and γ-Fe_2O_3/$PbTiO_3$ (d) nanoparticles. Magnetic hysteresis loops of Fe_3O_4 (a); γ-Fe_2O_3 (b); Fe_3O_4/$PbTiO_3$ (c); γ-Fe_2O_3/$PbTiO_3$(e); γ-Fe_2O_3/Pb(Ti, Zr)O_3 nanoparticles (upper panel) and the as-prepared $CoFe_2O_4$ nanoparticles (a) and after annealing at 600 °C (b) and $CoFe_2O_4$/$BaTiO_3$(c); $CoFe_2O_4$/$PbTiO_3$(d) and $CoFe_2O_4$/Pb(Ti, Zr)O_3 (e) nanopowders, respectively

$BaCO_3$ nanoparticles. A heat treatment at 700 °C in air ensured the formation of the perovskite layer on the surface of the magnetic nanoparticles. From a mechanistic viewpoint, it is believed that $BaCO_3$ selectively binds titania on the surface of the magnetic grains, thereby promoting the diffusion of the Ba^{2+} and O^{2-} ions from the $BaTiO_3$/$BaCO_3$ interface to the $BaTiO_3$/TiO_2 interface throughout the perovskite lattice with the preservation of the initial titanium-containing layer and preserving a good dispersion of the particles [67-69].

As seen in Figure 9a, the as-prepared α-Fe_2O_3 particles are spheroidal with narrow edges and an average diameter of 400-500 nm. These particles increase their sizes upon the coating with the perovskite layer (Figure 9d) yielding porous powders formed by 300-600 nm grains with a certain degree of inherent aggregation after the heat treatment (Figures 9b and c). Although this method allows the variation of the molar fraction of the nanostructured ferrites, as well as their chemical composition, the phase analysis revealed that the resulting two-phase composites are slightly contaminated with secondary phases, such as $BaFe_{12}O_{19}$, and $Ba_{12}Fe_{28}Ti_{15}O_{84}$ which can be potentially detrimental to their magnetoelectric properties.,

HRTEM microscopy experiments revealed that the hexagonal ferrites form as secondary phases when the amount of $BaTiO_3$ in the composite was decreased from 73% (wt.) to 53% (wt.), whereby the nanocomposites seem to be coated with a 20-30 nm thick $BaTiO_3$ glassy layer separated by a thin (3-5 nm) interdifussion region (Figure 9c). Such a behavior was ascribed to the formation of an eutectic during sintering, despite the fact that liquid phases were not found on the isothermal section (t=1200 °C) of the phase diagram. The microstructural

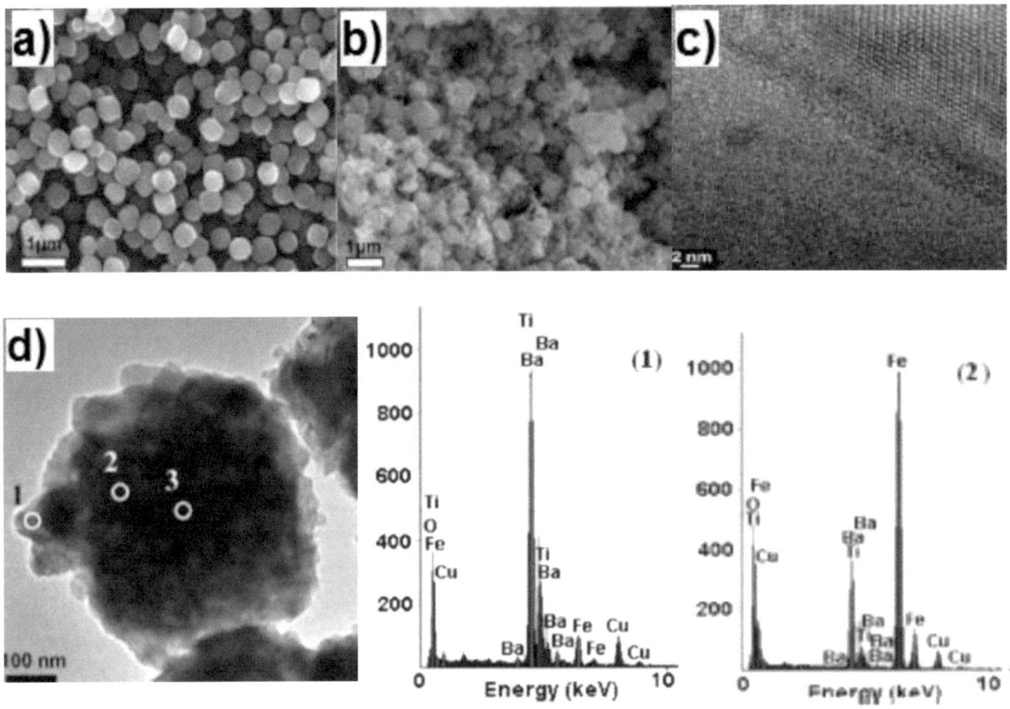

Fig. 9. SEM images of the α-Fe_2O_3 nanoparticles and α-Fe_2O_3/$BaTiO_3$ composite particles with 73% (b) and 53 % (wt.) (c) $BaTiO_3$ core-shell nanostructured powders after annealing at 1100 °C; (d); TEM image of a single α-Fe_2O_3/$BaTiO_3$ composite particle with 73 % BaTiO3 and the corresponding EDX spectra for points 1 and 2 in Figure 4d;

study of the $Ni_{0.5}Zn_{0.5}Fe_2O_4$@$BaTiO_3$ core-shell particles revealed that they are constructed by quasi-spherical nickel ferrite particles with an average diameter of 600 nm onto which have been attached needle-like $BaTiO_3$ nanograins with a length of 130 nm and an average diameter of 57 nm. The dielectric constant has values which are lower than those of the pristine perovskite phase. Moreover, the frequency dispersion of the dielectric constant increases with temperature and this increase is more pronounced at lower frequencies (ε=123 at 40 °C and 540 at 140 °C for a measurement frequency of 1 kHz). The dielectric losses were found to have values ranging between 0.2 and 0.4 for all frequencies at 40 °C and 1 at 140 °C for the same operating frequency. The experimental data suggested that, in terms of the dielectric loss the assembly of the magnetic grains and dielectric shells does not improve significantly the electric properties of the nanocomposites.

Mornet and coworkers proposed an interesting soft-solution route for the preparation of functional ferroelectric-ferro/ferrimagnetic nanocomposites [70]. In the first step commercial perovskite ($BaTiO_3$ and/or $Ba_{0.6}Sr_{0.4}TiO_3$) nanoparticles with different sizes (50-150 nm) were coated with a layer of SiO_2 and then their surface was modified by coupling aminosilane groups to induce positive charges (Figures 10a and b). This solution was treated with a ferrofluid containing superparamagnetic 7.5 nm γ-Fe_2O_3 nanoparticles coated with a silica layer (shell thickness ≈ 2 nm) whose surfaces were negatively charged upon addition of citric acid (Figure 10c). These two solutions were subsequently mixed together and ferroelectric-superparamagnetic ceramic nanostructures were obtained by electrostatic assembly between the aminated iron oxide nanograins and the perovskite nanoparticles (Figure 10e). The iron-oxide/perovskite particles can be assembled into 3D colloidal crystals via a sedimentation step (Figure 10d) followed by a drying and sintering process which yields a dense ferroelectric/superparamagnetic composite material and, at the same time, prevents interdiffusion and grain growth processes and preserves the individual building blocks distinct. The silica layer has proven to play a key role in the formation of ceramic nanocomposites because (i) it stabilizes the ferroelectric particles by playing a role of dielectric barrier between them, (ii) acts as a binding agent between the magnetic and dielectric phase and (iii) prevents the conductivity percolation among the iron oxide nanoparticles.

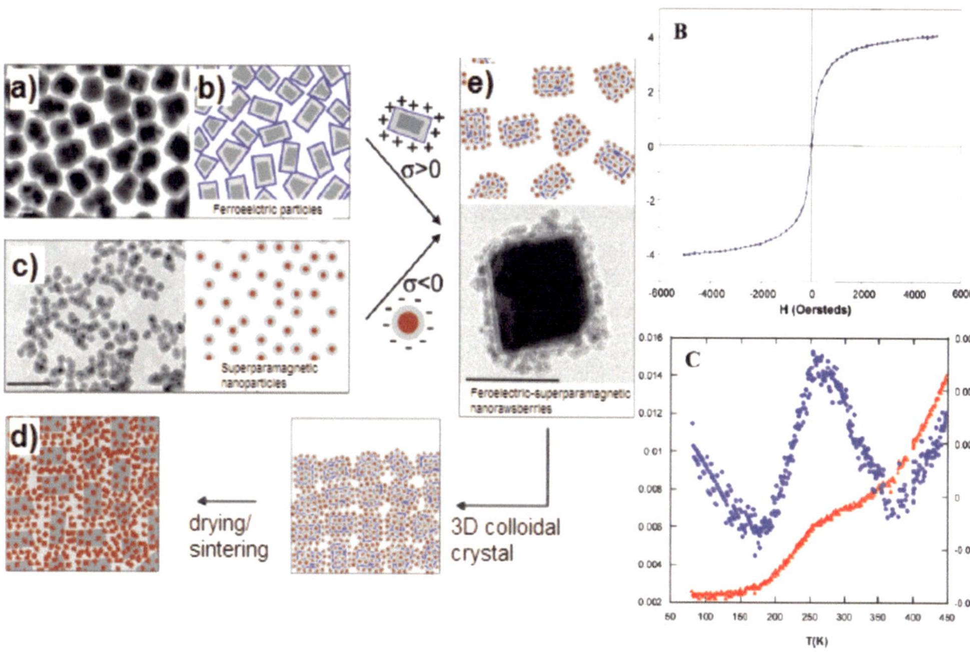

Fig. 10. Schematic of the successive steps used by Mornet and coworkers to fabricate perovskite-iron oxide core-shell ceramic nanocomposites, 3D colloidal crystals and dense powders

Interestingly, the γ-Fe_2O_3 did not convert to α-Fe_2O_3 (non-magnetic), as seen in Fig. 10b, which suggests that the silica layer protects them during the thermal annealing and stabilize the constituent phases and prevents their coalescence and interdiffusion. It has been found that the dielectric permittivity can be controlled by adjusting the thickness of the silica layer, as well as the size of the perovskite particles. For example, in Fig. 10c is displayed the temperature variation of the imaginary part of the dielectric permittivity which shows a maximum at 270 K, which is the transition temperature of $Ba_{0.6}Sr_{0.4}TiO_3$. The electron microscopy micrographs of the nanopowders show that the morphology of the ceramic nanoparticles is preserved upon sintering at high temperature (Figure 11a), whereas the resonance amplitude decreases progressively and disappears at temperatures up to 430 K, which crosses the ferroelectric-paraelectric temperature of the perovskite phase (T_c=405 K; Figure 11b).

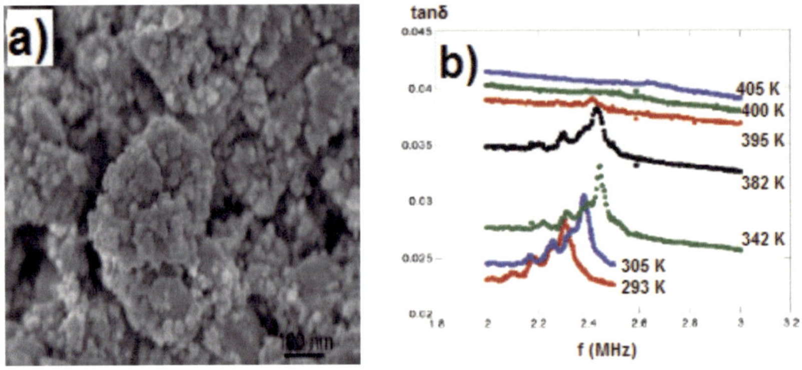

Fig. 11. (a) SEM image of the γ-Fe_2O_3@SiO_2/$BaTiO_3$@SiO_2 nanocomposites; (b) the frequency dependence of the piezoelectric amplitude of the nanopowders at different temperatures

Although the synthetic methodologies described above lead to the formation of core-shell ferroelectric/magnetic nanoparticles, the modification of their surfaces in such a way that they are soluble in polar solvents and, therefore, can be used in biomedical application is difficult. We describe in the following two different alternate strategies which can enable the post-synthesis functionalization of the nanoparticles. Two-phase ME nanoparticulate composites with a core-shell structure can be fabricated by a sequential bottom-up approach which combines a hydrolytic route with the liquid phase deposition (LPD). In the first step aqueous-based ferrofluids containing nearly monodisperse, water-soluble ferrite nanoparticles with different sizes and chemical compositions will be prepared by the complexation and controlled hydrolysis of metal transition ions in diethyleneglycol solutions at 250 °C. The formation of the metal oxide nanoparticles takes place through three successive steps depicted in Fig. 10 and leads to highly stable colloidal solutions in polar solvents, due to the passivation of the individual nanocrystals with polyol molecules [71,72]. The size of the magnetic nanocrystals can be controlled by adjusting the complexing power of polyol solvent (Figure 12).

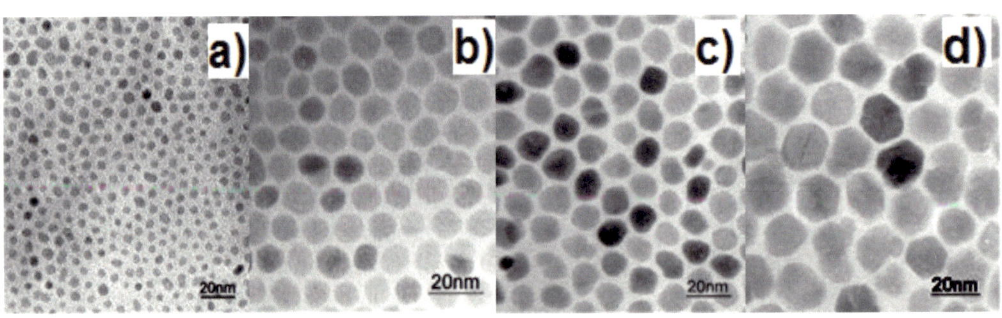

Fig. 12. (a) TEM images of monodisperse Fe_3O_4 nanoparticles with a diameter of (a) 6 nm; (b) 10 nm; (c) 14 nm and (d) 18 nm obtained in polyol/polyamine solutions

In the next step a uniform piezoelectric shell can be deposited on each individual magnetic nanoparticle by mixing the ferrofluid with a treatment solution in which metal-fluoro complexes are slowly hydrolyzed at temperatures below 50 °C. This method, called the liquid phase deposition (LPD) consists of the progressive replacement of the fluoride ions in the inner coordination sphere of the metal fluoro-complex by OH^- ions and/or water molecules with formation of the desired metal oxide. The hydrolysis rate can be controlled by temperature and/or by the consumption of the F^- ions by a fluoride scavenger such as H_3BO_3 which forms water soluble complexes with the fluoride ions and prevents the contamination of the as-deposited hydroxydes/oxyhydroxides by fluorine. We have successfully used this methodology for the deposition of both highly uniform transition metal ferrite and perovskite films and nanotubular structures with tunable chemical composition and controllable magnetic and ferroelectric properties [73]. As seen in Figure 13b the coating of the ferrite nanoparticles with the ferroelectric layer has a substantial effect on their magnetic properties: while the 14 nm $CoFe_2O_4$ nanoparticles are superparamagnetic at room temperature, $CoFe_2O_4@PbTiO_3$ core-shell nanoparticles are ferrimagnetic and present a coercivity of about 1200 Oe. This is the result of the modification of the contributions of the surface anisotropy and dipolar interactions between the adjacent nanoparticles on the average magnetic moment of the nanocomposite [74,75]. A wide variety of spinel ferrite nanoparticles can be coated with perovskite layers, such as $BaTiO_3$, $PbTiO_3$ and $Pb(Ti, Zr)O_3$ and their properties can be controlled by adjusting both the size of the magnetic core and the thickness of the perovskite shell.

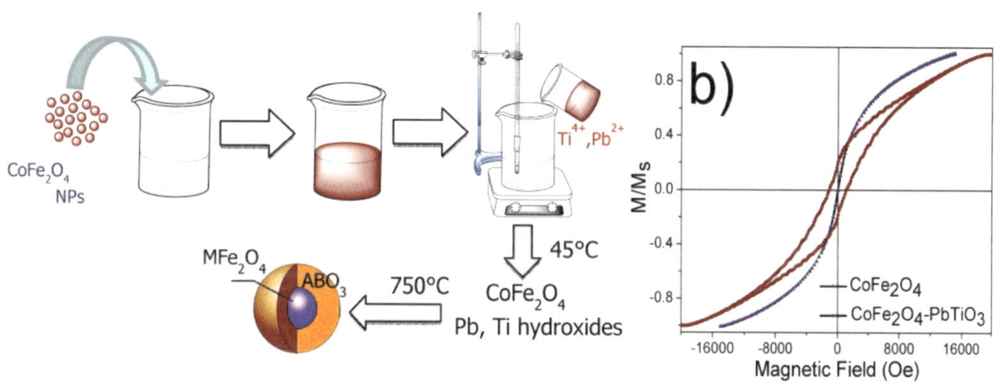

Fig. 13. (a) Schematic of the bottom-up synthetic strategy for the fabrication of ferrite-perovskite core-shell magnetoelectric nanostructures by liquid phase deposition; (b) Room temperature hysteresis loops of the 14 nm $CoFe_2O_4$ nanoparticles and $CoFe_2O_4@PbTiO_3$ core-shell nanoparticles

Moreover, upon treatment with a citric acid solution the ferroelectric-ferrimagnetic ceramic nanocomposites are passivated, thereby forming stable colloidal solutions in water and other polar solvents. Another synthetic methodology for the preparation of functional ferroelectric-ferro/ferrimagnetic nanocomposites consists of assembling pre-synthesized, monodisperse spinel and perovskite colloidal nanoparticles using stable oil-water microemulsion systems. Well-dispersed, size-uniform ferrite and perovskite nanoparticles can be mixed together in a stable micellar solution and confined in the oil microemulsion droplets [63]. The subsequent evaporation of the low-boiling solvent from the colloidal solution leads to a shrinkage of the individual droplets, decreasing of the distances between the nanoparticles which will stick together to form 3D spinel-perovskite colloidal spheres. This technique has been used for assembling either particles with the same chemical composition, such as Ag_2Se, $BaCrO_4$, CdS, Au, Fe_3O_4 [76-79] or dissimilar nanoparticles, such as Fe_3O_4 and Au [80] and CdSe/CdS, respectively [78]. The resulting nanoparticles can be dispersed in aqueous solutions since the capping agents passivating each individual nanoparticle interdigitate with those from the neighboring nanocrystals via hydrophobic Van-der-Waals interactions. Another advantage of this method is that the amount of ferrite phase in the nanocomposite can be varied in a wide range will be used for the optimization of the ME coupling coefficient.

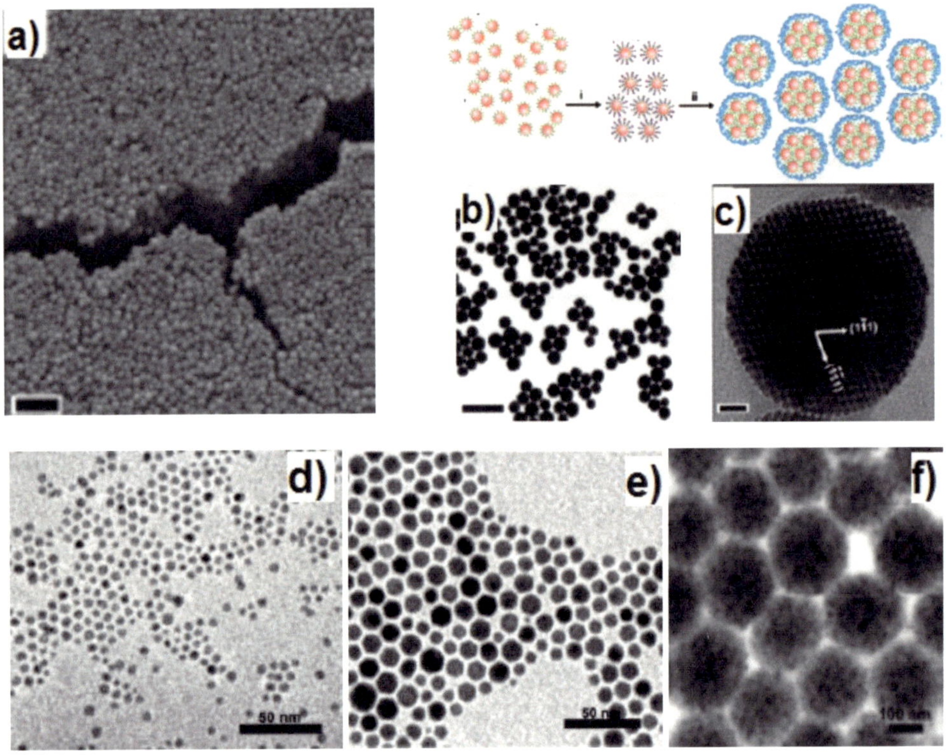

Fig. 14. (a) SEM image of multilayered assemblies of Fe_3O_4 superparticles with an average diameter of 190 nm; (b) and (c) TEM images of the 190 nm superparticles. The scale bars are: 1 μm (a); 500 nm (b) and 20 nm (c), respectively (Ref.63)

This method was used for the synthesis of colloidal superparticles, whereby the diameter of the colloidal spheres can be adjusted in a wide range, typically from 80 to 350 nm by controlling the size of the microemulsion droplets.

Fabrication of spinel-perovskite core-shell nanotubular structures

Another type of core-shell ferroelectric-ferromagnetic nanostructures with potential applications in the biomedical field is represented by coaxial nanotubular/nanorod structures. Xie and coworkers have recently reported on the fabrication of $CoFe_2O_4$-$PbZr_{0.52}Ti_{0.48}O_3$ core-shell nanofibers using a sol-gel process and coaxial electrospinning. [81] Two different solutions, containing the precursors for the ferrite and perovskite phases combined with polymethyl metacrylate (PMMA) and poly(vinyl) pyrollidone (PVP)w were electrospun by using a coaxial spinneret consisting of two commercial blunt needles with different lengths and diameters (Figures 15a and b). A voltage of 1.2 kV/cm was applied to the needle and the resulting nanofibers were heated at 750 °C in air for 2 h to induce the crystallization of the oxide phases and eliminate the organic components.

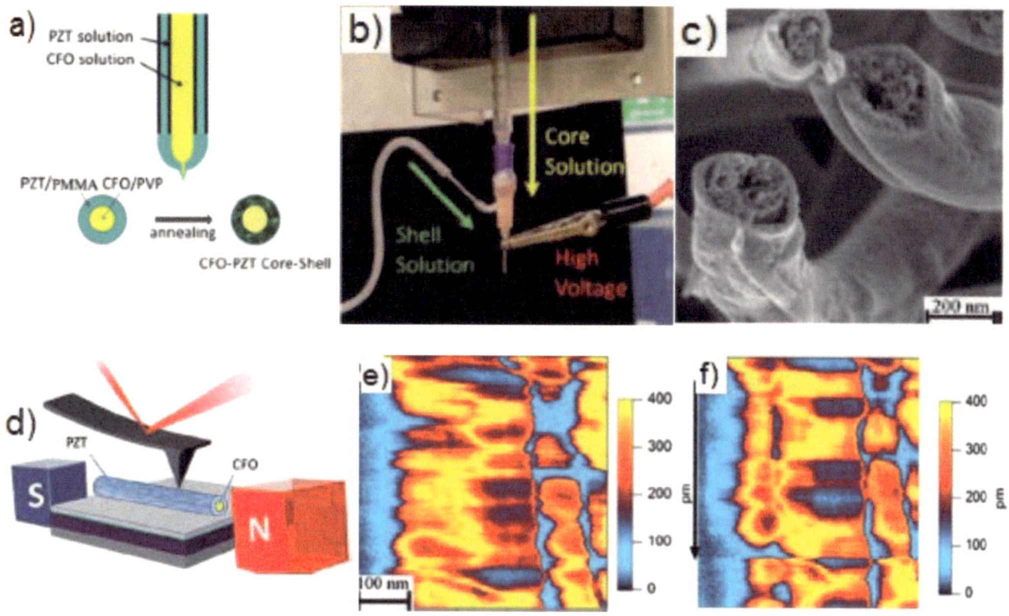

Fig. 15. (a) and (b) Schematic and a picture of the experimental setup used for the fabrication of core-shell nanofibers by electrospinning; (c) SEM image of a core-shell nanofibers; (d) Schematic of the PFM measurement of an individual ferrite-perovskite core-shell fiber; amplitude PFM contrast images of one composite fiber without (e) and in the presence of an external magnetic field (f)

The SEM image of one individual fiber (Fig. 15 c) shows that, indeed, a core-shell structure can be obtained by this approach; however, the magnetic core is porous compared to the perovskite shell, which looks much more compact. Such a discrepancy between the porosity of the core and shell components of the composite fiber can be detrimental for the ME coupling since the interfaces are not uniform and local chemical heterogeneities and gaps are present at the interphase boundaries. The existence of a local magnetoelectric coupling between the ferrite and perovskite phases was proven qualitatively by recording the amplitude PFM contrast images of one single fiber without and with a magnetic field. As seen in Figures 15e and f, the ferroelectric domain structure of the nanofibers is significantly altered when a magnetic field is applied, which suggests that an elastic coupling between the ferrite core and perovskite shell is induced by the magnetic field.

Nanotubular ferroelectric and magnetostrictive structures, as well as ME coaxial architectures can be fabricated by a template-assisted liquid phase deposition (LPD) method. This synthetic strategy is schematized in Figure 16. In the first step, ferroelectric nanotubes are deposited into the pores of alumina membranes (AAM) by the controlled hydrolysis of metal fluoro-complexes under mild conditions (Fig. 16a) similar to those used to create a perovskite shell around each individual water-soluble ferrite nanoparticle. The outside diameter of the tubes is determined by the size of the pores of the membrane and can be controlled in the range from 20 to 200 nm. In the second step the ferroelectric nanotubes immobilized within the pores of the membranes will be filled with a spinel ferrite leading to the formation of coaxial magnetoelectric ceramic nanostructures (Figure 16b).

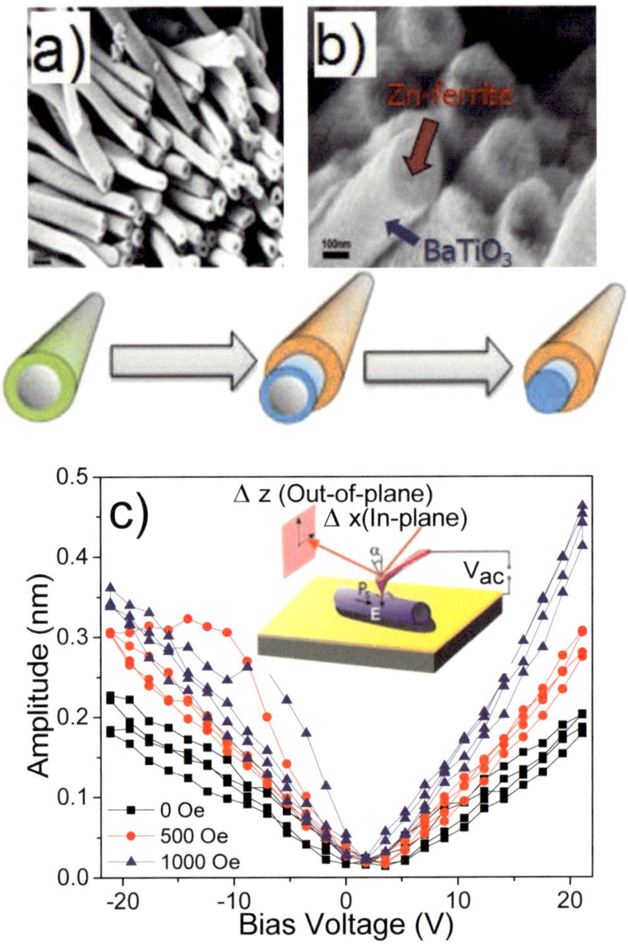

Fig. 16. (a) Side view FE-SEM image of $CoFe_2O_4$ (CFO) nanotubes with a diameter of 200 nm; (b) Side view FE-SEM image of $BaTiO_3$-$CoFe_2O_4$ coaxial nanowires obtained by filling the perovskite nanotubes with $CoFe_2O_4$; ; (c) Probing the ME response of an individual BTO-CFO coaxial nanowire by measuring its piezoelectric signal under different magnetic fields

Finally, the membranes are dissolved upon treatment in an alkaline solution, yielding free-standing spinel-perovskite coaxial 1-D nanostructures. Optimization of the ME properties of these structures will be carried out by either changing the chemical composition of the oxide nanotubes, or by varying the spinel/perovskite ratio which can be achieved by changing the wall thickness of the shell (a structural parameter which is hard to control in the sol-gel approaches conventionally used in the synthesis of oxide nanotubes). The magnetoelectric coupling between the oxide phases in these 1-D nanocomposites was proven by piezoresponse force microscopy (PFM) in the presence of a magnetic field. As seen in Figure 16c, there is a notable change in the amplitude of the PFM signal and the slope of the experimental curves when a magnetic field is applied in the plane of the coaxial nanostructure, indicating that the piezoelectric deformation of the nanocomposite is influenced by the magnetic field. The value of the ME coupling coefficient calculated from the slope of the butterfly- loops is α_E=1.08 V/cm Oe, which indicates a strong elastic coupling between the constituent phases of the nanocomposite.

Future applications

One of the areas where control of channel gating may be of particular importance is generation of action potentials in neurons. When neurons are not being stimulated and not conducting impulses, their membrane is polarized. This so called resting potential is of the order of 70-90 mV for typical cells and it is a measure of an imbalance of sodium and potassium ions across the membrane resulting from the interplay between several mechanisms of ion transport coexisting in the membrane. The active mechanisms (such as sodium-potassium pumps) require energy form ATP phosphorylation to "pump" ions against the concentration gradient, while passive mechanisms (such as ion channels) flux the ions down their gradients. As a result, there is a higher concentration of sodium on the outside than the inside and a higher concentration of potassium on the inside than the outside. The resting potential will be maintained until the membrane is disturbed or stimulated. If the stimulus is sufficiently strong, an action potential will occur.

An action potential is a very rapid change in membrane potential that occurs when a nerve cell membrane is stimulated. During the time course of the action potential, the membrane potential goes from the resting value (typically -70 mV) to some positive value (typically about +30 mV) in a very short period of time, of the order of a few ms. Action potentials are triggered by stimuli that can alter the potential difference by opening some sodium channels in the membrane. The stimulus can be for instance the binding of neurotransmitters to ligand gated ion channels or the membrane can be directly stimulated electrically from a glass pipette through a patch clamp. Once the potential difference reaches a threshold voltage (typically 5 - 15 mV less negative than the resting potential), it starts a chain of events (opening and closing of potassium channels) that leads to the action potential. This phenomenon has been first described mathematically in the celebrated Hodgkin-Huxley model [82].

As we mentioned action potentials can be triggered chemically (neurotransmitter binding) or electrically, as a result of membrane depolarization caused by electric impulses delivered locally through an electrode or globally by electric shock. The drawback for the electrode stimulation is that it is highly invasive and therefore of a very limited applicability for therapy, while the electric shock is global and makes targeted stimulation very difficult. This is, however, the basis for neuroscience research as well as electrostimulation therapy, for instance for pain. Using multiferroic nanoparticles seems very promising alternative to these techniques. The method we describe can be used for membrane stimulation and it eliminates the main drawbacks of hitherto used methods: it is noninvasive and localized. It can be used for localized stimulation and triggering of action potentials in studies of neural connectivity, and it can potentially become an effective pain treatment.

Emerging research on multiferroic materials is primarily seeking applications in information technology and wireless communication [7,8]. This work seems to be first to indicate great potential of these new materials in biology and medicine. Stimulation of the nervous system by means of defibrillators, electroshock, or other electro-physical therapy devices for resuscitation, depression therapy, or pain treatment utilizes high voltages and affects large portions of a patient's body. On the other hand, locally applicable electro-physical devices such as pace makers or contemporary acupuncture equipment require invasive methods to be implemented (surgery or puncture).

The method proposed in this article takes advantage of nanotechnology to stimulate functions on the cellular level by remotely controlling voltage generated by nanosized objects placed either inside or outside the cells in their close vicinity. Although this article targets ion channels to control the ion transport across membranes of mammalian cells, the multiferroic particles have potential for new applications in other fields of biology and medicine. For instance, the signals generated by nanoparticles have similar characteristics to those of the electric signals propagating in the neural cells. Therefore, it can be expected that the electric fields of the nanoparticles can interfere with the signals received by the neural cells. The magnitude and the shape of the signals generated by the nanoparticles can be controlled by the time characteristics of the external magnetic field. This creates an opportunity to remotely stimulate functions of neural cells for pain treatment or for enhancing response of damaged cells.

Another unexplored area of research is the effect of electric fields generated by the multiferroic nanoparticles inside the cell on the organelles and DNA. Electric field may have either stimulating or damaging effect on cells, depending on its strength and frequency. Strong static fields may result in electrolysis of the cytoplasm, electroporation etc., and high frequency electromagnetic fields can cause local hyperthermia. Current research and existing physical-therapy devices use global electric fields and the response of organelles to fields delivered locally is unknown. It is likely that strong enough electric fields inside cells can significantly perturb the functions of their organelles. This opens a possibility to use multiferroic nanoparticles for cancer treatment by a method different from chemotherapy or hyperthermia. Interestingly, the multiferroic particles may have a dual function depending on the frequency of the applied magnetic field. The magnetic component of the particles can be used at high frequencies (in the radio- or microwave range) for hyperthermia whereas it can generate electric signals for stimulation of membranes or organelles in the acoustic (kHz) frequency range. At this point it is important to notice that the values of magnetoelectric voltage coefficient (1V/cmOe) used in our estimates refer to non-resonant performance of the composite nanoparticles. This coefficient can be enhanced by a factor of 100 by applying magnetic field pulses with higher frequencies (in MHz or GHz) which correspond to electromechanical resonance. Although ion channels may not respond to these high frequency fields, the strength of the electric fields produced by the nanoparticles may be sufficient to cause irreversible electroporation of the cell membranes and consequently their damage. This effect can provide alternative tool for cancer treatment and be used separately (for short series of pulses) or jointly with hyperthermia (for longer series). Also, static magnetic field can be used to drag the particles to certain organs for highly targeted applications. More intriguing problem is to recognize the effects of weaker fields. There have been reports indicating some effect on cell proliferation and growth, ionic transport, and neural signaling [83-85]. The method we proposed opens unique opportunities in this area. Composite multiferroic particles are ideal objects to produce electric fields on submicrometer scale which have never been tested. We anticipate great potential for applications of mutiferroic nanoparticles in several fields of biomedicine, such as cancer research, neurology, brain functions, pain treatment, and we strongly believe that the necessary technology exists and these applications will soon be put to test.

2. Acknowledgments

The authors acknowledge financial support through DARPA grant HR0011-09-1-0047.

3. References

[1] W. Andrä, U. Häfeli, R. Hergt and R. Misri, "Applications of magnetic particles ion medicine and biology", in: Handbook of Magnetism and Advanced Magnetic Materials, ed. H. Kronmüller, S. Parkin Vol. 4 (Novel Materials) pp.2536-2568, John Wiley & Sons 2007

[2] J.M.D. Coey, in: Magnetism and Magnetic Materials, Cambridge University Press, USA 2010, pp. 555-565

[3] P. Moroz, S.K. Jones and B.N. Gray, Magnetically mediated hyperthermia, current status and future directions, *International Journal of Hyperthermia* 18 (2002) 267-284

[4] H. Huang, S. Delikanli, H. Zeng, D.M. Ferkey and A. Pralle, "Remote control of ion channels and neurons through magnetic-field heating of nanoparticles", *Nature Nanotechnology* 5 (2010) 602-606

[5] J. Dobson, "Nanomagnetic actuation: remote control of cells", *Nature Nanotechnology* 3 (2008) 139-143

[6] L.W. Martin, S. P. Crane, Y-H. Chu, M. B. Holcomb, M. Gajek, M. Huijben, C-H. Yang, N. Balke and R. Ramesh Multiferroics and magnetoelectrics: thin films and nanostructures *J. Phys.: Condens. Matter* 20 (2008) 434220

[7] C. W. Nan, M. I. Bichurin, S. Dong, D. Viehland and G. Srinivasan, "Multiferroic magnetoelectric composites: Historical perspective, status, and future direction", *J. Appl. Phys.* 103 (2008) 031101, pp.1-35

[8] M. Fiebig, "Revival of the magnetoelectric effect", *J. Phys. D: Appl. Phys.* 38 (2005) R123–R152

[9] W. Eerenstein, N. D. Mathur and J. F. Scott, "Multiferroic and magnetoelectric materials", *Nature* 442 (2006) 759-765

[10] C. Foged, B. Brodin, S. Frokjaer and A. Sundblad, "Particle size and surface charge affect particle uptake by human dendritic cells in an in vitro model, *Int. J. Pharmaceutics* 298 (2005) 315-322

[11] C. N. R. Rao and C. R. Serraoab," New routes to multiferroics", *J. Mater. Chem.* 17 (2007) 4931–4938

[12] M. I. Bichurin, V. M. Petrov, S. V. Averkin, and E. Liverts, "Present status of theoretical modeling the magnetoelectric effect in magnetostrictive-piezoelectric nanostructures. Part I: Low frequency and electromechanical resonance ranges" *J. Appl. Phys.* 107 (2010) 053904 pp.1-11

[13] M. I. Bichurin, V. M. Petrov, S. V. Averkin, and E. Liverts," Present status of theoretical modeling the magnetoelectric effect in magnetostrictive-piezoelectric nanostructures. Part II: Magnetic and magnetoacoustic resonance ranges" *J. Appl. Phys.* 107 (2010) 053905 pp.1-18

[14] R. Liu, Y. Zhao, R. Huang, Y. Zhao, and H. Zhou, "Multiferroic ferrite/perovskite oxide core/shell nanostructures" , *J. Mater. Chem.* 20 (2010) 10665-10670

[15] K. Raydongia, A. Nag, A. Sundaresan, and C. N. R. Rao, "Multiferroic and magnetoelectric properties of core-shell $CoFe_2O_4@BaTiO_3$ nanocomposites", *Appl. Phys. Lett.* 97 (2010) 062904 pp. 1-3

[16] K. C. Verma, S. S. Bhatt, M. Ram, N. S. Negi,and R. K. Kotnala,"Multiferroic and relaxor properties of Pb0.7Sr0.3[Fe 2/3 Ce1/3)(0.012)Ti-0.988]O-3 and Pb0.7Sr0.3[(Fe2/3La1/3) (0.012)Ti-0.098]O-3 nanoparticles", *Materials Chem. and Phys.* 124(2-3) (2010) 1188-1192

[17] C.-C. Chang, L. Zhao, and M.-K. Wu, "Magnetoelectric study in SiO_2-coated Fe_3O_4 nanoparticle compacts", *J. Appl. Phys.* 108 (2010) 094105 pp. 1-5

[18] R. P. Maiti, S. Basu, S. Bhattacharya, and D. Charkravorty, "Multiferroic behavior in silicate glass nanocomposite having a core-shell microstructure", *J. Non-Crystalline Sol.* 355(45-47) (2009) 2254-2259

[19] V. Corral-Flores, D. Buano-Baques, R.F. Ziolo, Synthesis and characterization of novel $CoFe_2O_4$-$BaTiO_3$ multiferroic core shell-type nanostructures. *Acta Materialia* 58 (3) (2010) 764-769

[20] M. Liu, H. Imrane, Y. Chen, T. Goodrich, Z. Cai, K. Ziemer, J. Y. Huang, and N. X. Sun, "Synthesis of ordered arrays of multiferroic $NiFe_2O_4$-$Pb(Zr_{0.52}Ti_{0.48})O_3$ core-shell nanowires", *Appl. Phys. Lett.* 90 (2007) 152501 pp. 1-3

[21] V. M. Petrov, M. I. Bichurin, and G. Srinivasan, "Electromechanical resonance in ferrite-piezoelectric nanopillars, nanowires, nanobilayers, and magnetoelectric interactions", *J. Appl. Phys.* 107 (2010) 073908, pp 1-6

[22] P. Fulay, in: Electronic, Magnetic and Optical Materials, CRC Press, Taylor and Francis Group,LLC, USA, 2010

[23] R.C. Smith, in: Smart Material Systems: Model Development, Society for Industrial and Applied Mathematics, USA 2005

[24] B. Hille, Ionic Channels of Excitable Membranes. Sinauer Associates Inc., Sunderland Massachussetts 1992.

[25] F.M. Ashcroft, Ion Channels and disease. Academic Press, San Diego, London 2000.

[26] F. Bezanilla, E. Peroso, and E. Stefani.. Gating of Shaker K^+ channels: II. The components of gating currents and a model of channel activation. *Biophys. J.* 66 (1994) 1011-1021

[27] W.N. Zagotta, T. Hoshi, and R.W. Aldrich. Potassium channel gating: III. Evaluation of kinetic models for activation. *J. Gen. Physiol.* 103 (1994) 312-362

[28] M.M. Millonas and D.R. Chialvo. Control of voltage-dependent biomolecules via non-equilibrium kinetic focusing. *Phys. Rev. Lett.* 76 (1996) 550-553

[29] H. Qian, From discrete protein kinetics to continuous Brownian dynamics: a new perspective. *Protein. Sci.* 11 (2002) 1-5

[30] D. Sigg, H. Qian, and F. Bezanilla. Kramer's diffusion theory applied to gating kinetics of voltage-dependent ion channels. *Biophys. J.* 76 (1999) 782-803

[31] D. Sigg and F. Bezanilla. A physical model of potassium channel activation: from energy landscape to model kinetics. *Biophys. J.* 84 (2003) 3703-3716

[32] B. Sakmann and E. Neher. Single-channel Recording. Plenum Press, New York - London 1995.

[33] Y. Jiang, et al. X-ray structure of a voltage-dependent K^+ channel. *Nature* 423 (2003) 33-41

[34] Y. Jiang, V. Ruta, J. Chen, A. Lee, and R. MacKinnon. The principle of gating charge movement in a voltage-dependent K^+ channel. *Nature* 423 (2003) 42-48

[35] A. Cha, G. E. Snyder, P. R. Selvin, and F. Bezanilla. Atomic scale movement of the voltage-sensing region in a potassium channel measured via spectroscopy. *Nature* 402 (1999) 809-813

[36] M.C. Menconi, M. Pellegrini, M. Pellegrino, and D. Petracci. Periodic forcing of single ion channel: dynamic aspects of the open-closed switching. *Eur. Biophys. J.* 27 (1998) 299

[37] A. Kargol, A. Hosein-Sooklal, L. Constantin, and M. Przestalski. Application of oscillating potentials to the Shaker potassium channel. *Gen. Physiol. Biophys.* 23 (2004) 53-75

[38] A. Kargol, B. Smith, and M.M. Millonas. Applications of nonequilibrium response spectroscopy to the study of channel gating. Experimental design and optimization. *J. Theor. Biol.* 218 (2002) 239-258

[39] A. Kargol and K. Kabza. Test of nonequilibrium kinetic focusing of voltage-gated ion channels. *Phys. Biol.* 5 (2008) 026003

[40] M.M. Millonas and D.A. Hanck. Nonequilibrium response spectroscopy of voltage-sensitive ion channel gating. *Biophys. J.* 74 (1998) 210229

[41] C.R. Doering, W. Horsthemke, and J. Riordan. Nonequilibrium fluctuation-induced transport. *Phys. Rev. Lett.* 72 (1994) 2984

[42] A. Fulinski, Noise-simulated active transport in biological cell membranes. *Phys. Lett. A* 193 (1994) 267

[43] W. Horsthemke and R. Lefever. Noise-induced transitions in electrically excitable membranes. *Biophys. J.* 35 (1981) 415

[44] K. Lee and W. Sung. Effects of nonequilibrium fluctuations on ionic transport through biomembranes. *Phys. Rev. E* 60 (1999) 4681

[45] B. Liu, R.D. Astumian, and T.Y. Tsong. Activation of Na^+ and K^+ pumping modes of (Na,K)-ATPase by an oscillating electric field. *J. Biol. Chem.* 265 (1990) 7260

[46] D.C. Camerino, T. Domenico, and J.-F. Desaphy. Ion channel pharmacology. *Neurotherapeutics* 4 (2007) 184-198

[47] S.F. Cleary, "In vitro studies of the effects of nonthermal radiofrequency and microwave radiation", in: J.H. Bernhardt, R. Matthes, M. Repacholi (eds.) Non-thermal effects of RF electromagnetic fields. Proc. Intl. Seminar of the Biological effects of non-thermal pulse and amplitude modulated RF electromagnetic fields and related health hazards", Munich-Neuherberg, 20-22 Nov. 1996, 119-130

[48] R.P. Liburdy "Biological interactions of cellular systems with time-varying magnetic fields", *Ann. N.Y. Acad. Sci.*, 649 (1992) 74-95

[49] M.H. Repacholi, "Low-level exposure to radiofrequency electromagnetic fields: health effects and research needs", *Bioelectromagnetics* 19 (1998) 1-19

[50] M.H. Repacholi, B. Greenebaum "Interaction of static and extremely low frequency electric and magnetic fields with living systems: health effects and research needs", *Bioelectromagnetics* 20 (1999) 133-160

[51] A.D. Rosen "Effect of a 125 mT static magnetic field on the kinetics of voltage activated Na^+ channels in GH3 cells", *Bioelectromagnet.* 24 (2003) 517-523

[52] J.E. Tattersall, I.R. Scott, S.J. Wood, J.J. Nettell, M.K. Bevir, Z. Wang, N.P. Somarisi, X. Chen, "Effects of low intensity radiofrequency electromagnetic fields on electrical activity in rat hippocampal slices", *Brain Res.* 904 (2001) 43-53

[53] R.A. Deyo, N.E. Walsh, D.C. Martin, L.S. Schoenfeld, and S. Ramamurthy. A controlled trial of transcutaneous electrical nerve stimulation (TENS) and exercise for chronic low back pain. *N. Engl. J. Med.* 322 (1990) 1627-1634

[54] J.S. Perlmutter and J.W. Mink. Deep brain stimulation. *Ann. Rev. Neurosci.* 29 (2006) 229-257

[55] A.N. Morozovska, M.D. Glinchuk, and E.A. Eliseev, Phase transitions induced by confinement of ferroic nanoparticles. *Phys. Rev. B* 76 (2007) 014102

[56] M.T. Buscaglia, V. Buscaglia, L. Curecheriu, P. Postolache, L. Mitoseriu, A.C. Ianculescu, B.S. Vasile, Z. Zhe, Z. and P. Nanni, Fe2O3@BaTiO3 Core-Shell Particles as Reactive Precursors for the Preparation of Multifunctional Composites Containing Different Magnetic Phases. *Chem. Mater.* 22 (2010) 4740-4748, doi:10.1021/cm1011982

[57] L.Q. Weng, Y.D. Fu, S.H. Song, J.N. Tang, J.Q. Li, Synthesis of lead zirconate titanate-cobalt ferrite magnetoelectric particulate composites via an ethylenediaminetetraacetic acid-citrate gel process. *Scripta Mater.* 56 (2007) 465-468, doi:10.1016/j.scriptamat.2006.11.032

[58] G. Srinivasan, E.T. Rasmussen, B.J. Levin, and R. Hayes, Magnetoelectric effects in bilayers and multilayers of magnetostrictive and piezoelectric perovskite oxides. *Phys. Rev. B* 65 (2002) doi:13440210.1103/PhysRevB.65.134402

[59] G. Srinivasan, E.T. Rasmussen, J. Gallegos, R. Srinivasan, Y.I. Bokhan, V.M. Laletin, Magnetoelectric bilayer and multilayer structures of magnetostrictive and piezoelectric oxides. *Phys. Rev. B* 64 (2001) 214408, doi:214408

[60] G. Srinivasan, C.P. DeVreugd, C.S. Flattery, V.M. Laletsin, and N. Paddubnaya, Magnetoelectric interactions in hot-pressed nickel zinc ferrite and lead zirconante titanate composites. *Appl. Phys. Lett.* 85 (2004) 2550-2552, doi:10.1063/1.1795365

[61] C.W. Nan, G. Liu, Y.H. Lin, Influence of interfacial bonding on giant magnetoelectric response of multiferroic laminated composites of $Tb_{1-x}Dy_xFe_2$ and $PbZr_xTi_{1-x}O_3$. *Appl. Phys. Lett.* 83 (2003) 4366-4368, doi:10.1063/1.1630157

[62] R. Sun, B. Fang, L. Zhou, Q. Zhang, X. Zhao, H. Luo, Structure and magnetoelectric property of low-temperature sintering $(Ni_{0.8}Zn_{0.1}Cu_{0.1})Fe_2O_4$/[0.58PNN-0.02PZN-0.05PNW-0.35PT] composites. *Curr. Appl. Phys.* 11 (2011) 37-42, doi:DOI: 10.1016/j.cap.2010.06.015

[63] N. Zhang, W. Ke, T. Schneider, G. Srinivasan, Dependence of the magnetoelectric coupling in NZFO-PZT laminate composites on ferrite compactness. *J. Phys. -Cond. Matter* 18 (2006) 11013-11019, doi:10.1088/0953-8984/18/48/029

[64] L. Mitoseriu, V. Buscaglia, Intrinsic/extrinsic interplay contributions to the functional properties of ferroelectric-magnetic composites. *Phase Trans.* 79 (2006) 1095-1121, doi:10.1080/01411590601067284

[65] R.Z. Liu, Y.Z. Zhao, R.X. Huang, Y.J. Zhao, H.P. Zhou, Multiferroic ferrite/perovskite oxide core/shell nanostructures. *J. Mater. Chem.* 20 (2010) 10665-10670, doi:10.1039/c0jm02602f

[66] L.P. Curecheriu, M.T. Buscaglia, V. Buscaglia, L. Mitoseriu, P. Postolache, A. Ianculescu, P. Nanni, Functional properties of $BaTiO_3/Ni_{0.5}Zn_{0.5}Fe_2O_4$ magnetoelectric ceramics prepared from powders with core-shell structure. AIP Vol. 107 (2010)

[67] M.T. Buscaglia, C. Harnagea, M. Dapiaggi, V. Buscaglia, A. Pignolet, P. Nanni, Ferroelectric $BaTiO_3$ Nanowires by a Topochemical Solid-State Reaction. *Chem. Mater.* 21 (2009) 5058-5065, doi:10.1021/cm9015047

[68] M.T. Buscaglia, V. Buscaglia, R. Alessio, Coating of $BaCO_3$ crystals with TiO_2: Versatile approach to the synthesis of $BaTiO_3$ tetragonal nanoparticles. *Chem. Mater.* 19 (2007) 711-718, doi:10.1021/cm061823b

[69] A. Lotnyk, S. Senz, D. Hesse, Formation of $BaTiO_3$ thin films from (110) TiO_2 rutile single crystals and $BaCO_3$ by solid state reactions. *Solid State Ionics* 177 (2006) 429-436, doi:10.1016/j.ssi.2005.12.027

[70] S. Mornet, C. Elissalde, O. Bidault, F. Weill, E. Sellier, O. Nguyen, M. Maglione, Ferroelectric-based nanocomposites: Toward multifunctional materials. *Chem. Mater.* 19 (2007) 987-992, doi:10.1021/cm0616735

[71] D. Caruntu, Y. Remond, N.H. Chou, M.J. Jun, G. Caruntu, J.B. He, G. Goloverda, C. O'Connor, V. Kolesnichenko, Reactivity of 3d transition metal cations in diethylene glycol solutions. Synthesis of transition metal ferrites with the structure of discrete nanoparticles complexed with long-chain carboxylate anions. *Inorg. Chem.* 41 (2002) 6137-6146

[72] D. Caruntu, G. Caruntu, C.J. O'Connor, Magnetic properties of variable-sized Fe_3O_4 nanoparticles synthesized from non-aqueous homogeneous solutions of polyols. *J. Phys. D-Appl. Phys.* 40 (2007) 5801-5809, doi:10.1088/0022-3727/40/19/001

[73] A. Yourdkhani, A.K. Perez, C. Lin, G. Caruntu, Magnetoelectric Perovskite-Spinel Bilayered Nanocomposites Synthesized by Liquid-Phase Deposition. *Chem. Mater.* 22 (2010) 6075-6084, doi:10.1021/cm1014866

[74] C.R. Vestal, Z.J. Zhang, Synthesis and Magnetic Characterization of Mn and Co Spinel Ferrite-Silica Nanoparticles with Tunable Magnetic Core. *Nano Lett.* 3 (2003) 1739-1743, doi:10.1021/nl034816k

[75] J.L. Dormann, D. Fiorani, E. Tronc, On the models for interparticle interactions in nanoparticle assemblies: comparison with experimental results. *J. Magn. Magn. Mater.* 202 (1999) 251-267, doi:10.1016/s0304-8853(98)00627-1

[76] J.Q. Zhuang, H.M. Wu, Y.A. Yang, Y.C. Cao, Supercrystalline colloidal particles from artificial atoms. *J. Am. Chem. Soc.* 129 (2007) 14166, doi:10.1021/ja076494i

[77] F. Bai, D.S. Wang, Z.Y. Huo, W. Chen, L.P. Liu, X. Liang, C. Chen, X. Wang, Q. Peng, Y.D. Li, A versatile bottom-up assembly approach to colloidal spheres from nanocrystals. *Angewandte Chemie-International Edition* 46 (2007) 6650-6653, doi:10.1002/anie.200701355

[78] J.Q. Zhuang, A.D. Shaller, J. Lynch, H.M. Wu, O. Chen, A.D.Q. Li, Y.C. Cao, Cylindrical Superparticles from Semiconductor Nanorods. *J. Am. Chem. Soc.* 131 (2009) 6084, doi:10.1021/ja9015183

[79] P.H. Qiu, C. Jensen, N. Charity, R. Towner, C.B. Mao, Oil Phase Evaporation-Induced Self-Assembly of Hydrophobic Nanoparticles into Spherical Clusters with Controlled Surface Chemistry in an Oil-in-Water Dispersion and Comparison of Behaviors of Individual and Clustered Iron Oxide Nanoparticles. *J. Am. Chem. Soc.* 132 (2010) 17724-17732, doi:10.1021/ja102138a

[80] X. Zhang, J.S. Han, T.J. Yao, J. Wu, H. Zhang, X.D. Zhang, B. Yang, Binary superparticles from preformed Fe_3O_4 and Au nanoparticles. *Cryst. Eng. Comm.* 13 (2011) 5674-5676, doi:10.1039/c1ce05664f

[81] S. Xie, F. Ma, Y. Liu and J. Li, Multiferroic CoFe2O4-Pb(Zr0.52Ti0.48)O3 core-shell nanofibers and their magnetoelectric coupling, *Nanoscale,* 3 (2011) 3152-3158

[82] A. Hodgkin and A.F. Huxley. A quantitative description of membrane current and its application to conduction and excitation in nerve. *J. Physiol.* 116 (1952) 507-544

[83] J. McCann, F. Dietrich, and C. Rafferty. The genotoxic potential and magnetic fields: an update. *Mutation Research* 411 (1998) 45-86

[84] M.A. Macri, S. Di Luzio, and S. Di Luzio. Biological effects of electromagnetic fields. *Int. J. of Immunopathology and Pharmacology,* 15(2) (2002) 95-105

[85] L. Huang, L. Dong, Y. Chen, H. Qi, and D. Xiao. Effects if sinusoidal magnetic field observed on cell proliferation, ion concentration, and osmolarity in two human cancer cell lines. *Electromagnetic Biology and Medicine,* 25(2) (2006) 113-126

5

Micro-Fabrication of Planar Inductors for High Frequency DC-DC Power Converters

Elias Haddad, Christian Martin, Bruno Allard,
Maher Soueidan and Charles Joubert
*Université de Lyon, Université Lyon 1,
CNRS UMR5005 AMPERE,
France*

1. Introduction

The inductors are essential elements for radiofrequency (RF) integrated circuits. A large number of communication devices functioning in RF such as mobile phones or wireless ethernet require transceivers, filters and power amplifiers in which inductors are critical components. Recently, the push toward miniaturisation of electronic components enabled to embark more and more portable equipments and accessories of high energy consumption. Embedded systems used in these devices are facing energy shortage that leads designers to spread power electronic converters to achieve dynamic voltage and frequency scaling [Zhao]. Wherever it is possible, the linear low-drop converter is replaced by inductive DC-DC converters and/or capacitive DC-DC converters in order to improve the whole efficiency of the system. Since then the demand on power converters for portable electronic devices has attracted great interest [Sugawara]. Efficiency and footprint (or volume) are the main design criteria to ensure respectively a large operating range and a smaller device. These applications use switching mode power supply (SMPS) or inductive DC-DC converter which typically require the following characteristics: 1 W, V_{IN}=3.6 V, V_{OUT}=1 V, I_{OUT}=1 A. Passive components fill a significant part of the chip area occupied by power converters, even when the components are optimized for minimum area. Thus, for this power range, System-In-Package (Fig. 1) is more appropriate than monolithic integration (also named System-On-Chip).

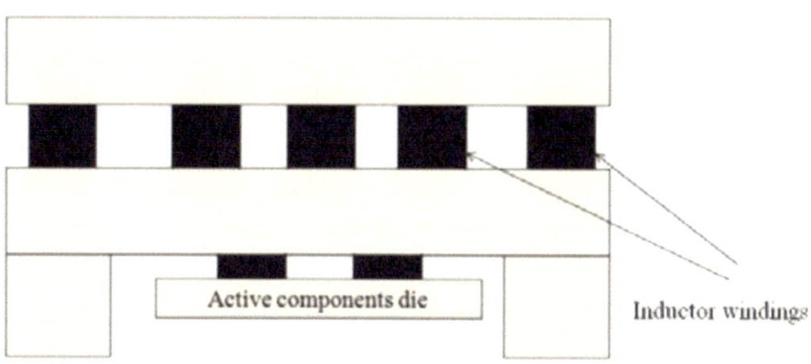

Fig. 1. Hybrid integration of System-In-Package

However, System-In-Package applications are still facing miniaturization issues due to passive components size [Musunuri]. Therefore, by increasing the switching frequency of DC-DC converters in the 10 MHz - 100 MHz frequency range [Chesneau], the size of filter passive components is dramatically reduced. Thus, the area devoted to passive components can be reduced below the 10 mm^2 range. In this condition, and for a typical 1 W DC-DC converter, the inductance value targeted for the output filter passive component is about 25 nH.

Since the system will be placed in a low-profile package, a planar inductor structure is preferred. Moreover, planar devices offer several advantages like better thermal management and higher power density.

Although research on planar inductors is concentrated on integrating air core inductors on silicon wafers [Wang], the application of ferrite magnetic substrates to planar inductors enables to increase the inductance value without increasing the stray capacitance between the coils and the ground plane. In our application, surface area is the key point of designing the inductor since the aim is to integrate the inductor on the top of a SMPS die in a surface area of 3mm^2.

This paper deals with a design methodology starting with Finite Element Method (FEM) simulation based on Flux2D simulator. The design aspect and boundary conditions will be presented and a specific figure of merit (MHz/mΩ.mm^2) will be proposed to evaluate our inductor performance targeting the DC-DC converter application. Next, the fabrication process using electroplating technique will be detailed. Finally, micro-fabricated inductors will be presented and their measured characteristics will be shown in the last section.

2. Context

Passive (inductor and capacitor) components have a major role in switching mode power supply since they ensure filtering function. However, they can occupy up to 30 per cent of the volume of the system. In such converter, current and voltage waveforms are far from sinusoidal. For example in a buck converter, voltage and current waveforms applied on the output inductor are square and triangular respectively. In this application, the output inductor is used to decrease the output current ripple defined as follows:

$$\Delta I_L = \frac{\alpha(1-\alpha)\cdot V_E}{L\cdot F} \tag{1}$$

where (L) is the inductance value, (F) represents the switching frequency, (V_E) is the input voltage, and (α) represents the duty cycle. Therefore, by increasing the frequency, the current ripple can be kept constant with a lower inductance value. For 1W - 1A DC-DC converter and 100 MHz switching frequency, an inductance value can be estimated at 30 nH. This inductor can be manufactured using coreless technology but it would require a large footprint area. Hence, the use of a magnetic material will allow fabricating the inductor with the same inductance value on a smaller surface area.

Since the inductor will be located as near as possible of the power and command circuit, electromagnetic interferences (EMI) must be limited using a magnetic circuit. Fig. 2 describes a qualitative magnetic field distribution in 3 cases: (a) coreless inductor (without magnetic material), (b) single layer (bottom side) and (c) double layer of ferrite magnetic

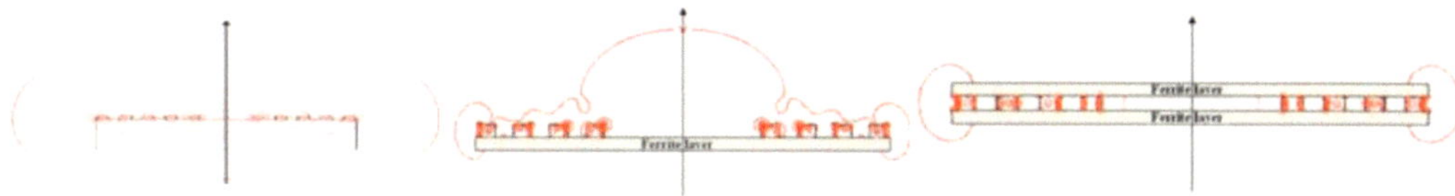

(a) (b) (c)

Fig. 2. Qualitative magnetic field distribution in: (a) coreless inductor (without magnetic material); (b) single layer (bottom side) and (c) double layer of ferrite magnetic material

material. We can observe in Fig.2 (b) and (c) that the magnetic field is confined within the magnetic material and thus electromagnetic disturbances are reduced.

Considering a low current ripple, which depends on the inductance value computed at the switching frequency, the current flow in the inductor could be considered equal to its average value. That is why direct current is the main contributor to losses in the windings and thus specific geometrical parameters for the winding must be optimized to reduce Joule effect losses at low frequency.

Inductance values at 100 MHz and DC resistance are computed using finite element method. Details are presented in the section below.

3. Design methodology

3.1 Finite Element Method (FEM) simulation

The design and analysis of the electromagnetic behaviour of magnetic components require finite element method software since analytical formulas are not available for the inductance and magnetic field calculation.

Based on finite element method, three-dimensional (3D) structures like the square planar inductor described in Fig. 3 require lots of memory, and computation time becomes impractical since high density meshing is required. Since commonly guidelines indicate that density meshing depends on skin effect in all materials, a frequency increase imply both a higher density mesh and computation time increase. This technique cannot be used in the design process.

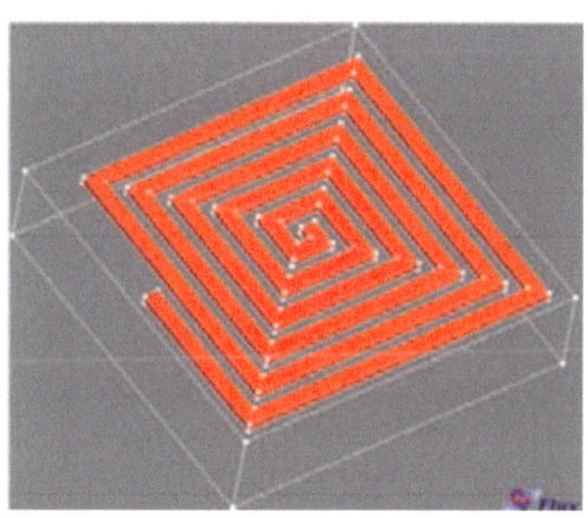

Fig. 3. 3D representation of a square planar inductor

Despite the three-dimensional form for such structure, we can assume a geometrical approximation to simplify our study. This approach consists in modifying the 3D structure by changing rectangular conductors to circular and concentric circles. Assuming this approximation, an axial symmetry appears as shown in Fig. 4.

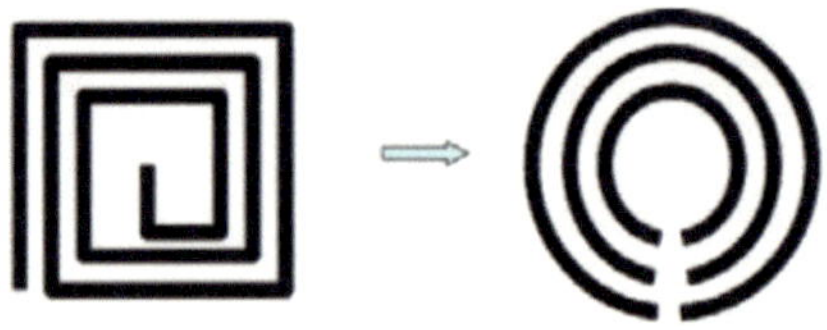

Fig. 4. Description of the geometrical approximation

Assuming these approximations, the planar inductor will be modelled in two-dimensional space as depicted in Fig. 5, which will allow to speed-up simulation. Therefore, the impact of geometrical parameters can be evaluated.

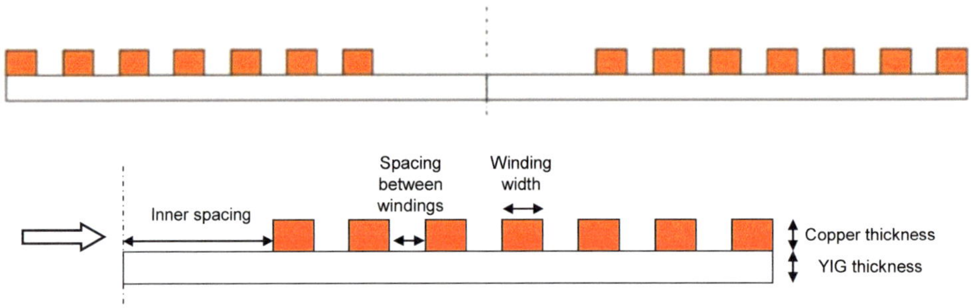

Fig. 5. Simplified 2D inductor representation

Fig. 5 describes the model dimensions and geometrical parameters. The number of turns, the turn width, the spacing between turns, the inner spacing, the magnetic substrate thickness (in this article we have used Yttrium Iron Garnet (YIG) as magnetic substrate) and copper thickness are taken into account. Simulations are performed with Flux2D® software. Regarding the inductor description, only the conductor cross-sections and magnetic substrate thickness will be described on the inductor half section. Boundary conditions applied for the magnetic field are tangential and capacitive behaviour will be neglected for frequency below 1 GHz [Yue].

An equivalent R-L circuit can be computed as depicted in Fig. 6. The impacts of both the geometry and frequency changes are observed on the magnetic field variation. Resistance and inductance values are calculated from active (P) and reactive (Q) power respectively as follow:

$$R = \frac{P}{I_{RMS}^{2}} \qquad L = \frac{Q}{\omega \cdot I_{RMS}^{2}} \tag{2}$$

where R is the inductor resistance, and I_{RMS} is the RMS current value, and ω is the angular frequency.

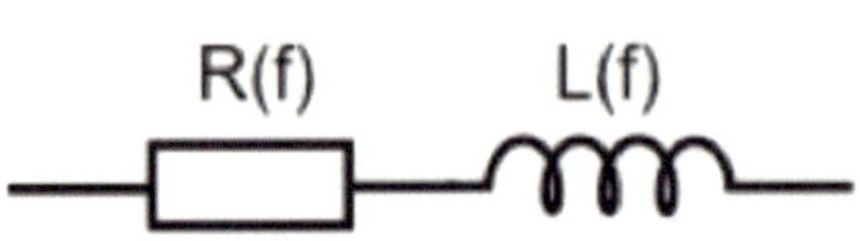

Fig. 6. Equivalent R-L circuit of the inductor

To evaluate the influence of geometrical parameters, we have considered a piecewise linear B-H curve for the magnetic material. In the linear region, a constant relative permeability is set to 25 while in the saturation region the flux density is below 0,25 T and relative permeability is equal to 1. Magnetic saturation and relative permeability are determined from characterization as described in section 3.2. Copper is also considered as a pure material of resistivity $1{,}7.10^{-8}$ Ω.m. Skin effect is taken into consideration as it contributes to increasing series resistance of the inductor at high frequency.

The choice of the frequency is important to compute resistance and inductance values. In many applications, the quality factor (Q) is commonly used to evaluate the AC performance of a structure. In such application, current and voltage waveforms are sinusoidal and the frequency of both signal is used according to equ (2).

$$Q = \frac{L\omega}{R} \tag{3}$$

As presented in the section 2, current and voltage waveforms are far from sinusoidal. Since the inductor is designed for reducing current ripple, the inductor excitation current waveform has a considerable DC component. Thus the inductor DC resistance is the main contributor to losses. On the other hand, the inductance value at switching frequency (100 MHz) will be computed to evaluate the current ripple.

In our application, a specific Merit Factor (equ (4)) taking into account DC resistance (R), inductance value (L) and footprint (S) is proposed to compare different models [Martin] and to evaluate a good power inductor performance.

$$MF = \frac{L_{100MHz}(nH)}{R_{DC}(m\Omega)\cdot S\left(mm^2\right)} \tag{4}$$

The following guidelines have been deduced from this parametric study:

- YIG doubles inductance value compared to an air core inductor with the same winding and reduces electromagnetic interferences. Moreover, simulations have also shown that increasing YIG thickness above 200 µm has only a slight effect on the inductance value.
- Copper thickness has a very small effect on inductance, but it contributes to a lower DC resistance.
- Increasing the space between the turns lowers the interwinding magnetic coupling, and thus decreases the inductance value.
- Widening the conductors reduces the inductance value, but it has an overall positive effect on Merit factor due to R_{DC} decreasing more significantly.
- Inductance value increases with the number of turns, but Merit Factor remains constant. This can be explained by the fact that improvement of MF is slowed down by the increase of the series resistance.
- Narrowing the inner spacing implies increasing the negative magnetic coupling and hence lowering the inductance value.

Simulations allowed evaluating our design with the optimal geometry contributing to the highest Merit Factor with taking into consideration the limitations of the technology parameters. Subsequently, we designed a planar spiral inductor of 3 mm² surface area (maximum surface area allowed in the converter).The inductor has four turns with a turn width of 75 µm, a 75 µm spacing between the turns, an inner diameter of 200 µm and a conductor thickness of 50 µm. The inductance value targeted is 30 to 40 nH with a series resistance as low as possible. The copper thickness and width are constrained to the maximum resolution of the dry film photoresist. The magnetic material which also serves as a mechanical support has a thickness of 500 µm. Though, the inductor design exhausted the theoretical limits.

3.2 Magnetic material characterization

Two specific test benches have been set up in order to characterize the properties of magnetic materials. Both tests require a toroidal shaped material in order to deduce magnetic properties from electrical quantities with geometric characteristics of the core. Material characteristics (relative permeability μ_R and magnetic saturation B_{SAT}) obtained are used in the FEM simulations.

3.2.1 Hysteresis graph

B-H curves are measured using a hysteresis graph curve recorder developed in our lab using a transformer approach. The bench used is shown in Fig. 7. Inner and outer diameters of the sample are quite equal in order to assume a homogeneous field in the core. Hysteresis graph is used to measure B-H cycles (minor and major loops) for frequency up to 20 kHz.

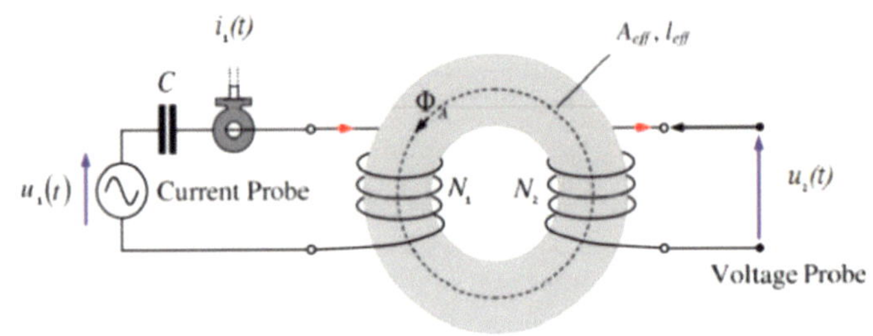

Fig. 7. Principle of B-H measurement

Magnetic field is imposed by the primary winding and is computed from the current. Flux density is obtained using induced voltage at the secondary winding. Starting from Ampere law presented in equ (5), magnetic field can be deduced from the current in the primary winding as described in equ (6). Flux density can be computed from equ (7). This method is used at low frequencies. The maximum frequency is estimated thanks to the equation of the skin depth.

$$\oint_{l_{eff}} \vec{H} \cdot \vec{dl} = \sum_{i=1}^{n} N_i \cdot i_i \tag{5}$$

$$H(t) = \frac{N_1}{l_{eff}} \cdot i_1(t) \tag{6}$$

$$B(t) = \frac{1}{A_{eff} \cdot N_2} \cdot \int u_2(t) \cdot dt \tag{7}$$

N_1 and N_2 represent the number of turns in primary and secondary side.

l_{eff} and A_{eff} are geometrical characteristics, the average length and section respectively of the toroidal sample.

Several ferrite materials have been tested and compared to datasheet. Since magnetic properties of ferrite material are not affected by machining, ring shape samples of YIG material have been machined from a ferrite plate. Fig. 8 shows B-H curve measured at 25 °C ambient temperature and for 1 kHz sinusoidal field excitation. Saturation induction is equal to 0.2 T and can be observed for a magnetic excitation beyond 200 A/m.

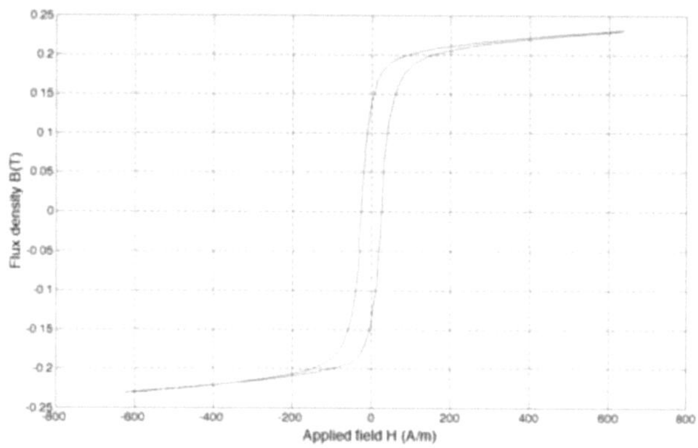

Fig. 8. Example of B-H curve

Permeability of the material can be deduced from the previous B-H cycle by using equ (8). However, this characterization method is not well adapted for high frequency. Complex permeability will be determined from impedance measurement presented in section 3.2.2.

$$\mu_{r_h} = \frac{1}{\mu_0} \cdot \left. \frac{\partial B}{\partial H} \right|_{H=h} \tag{8}$$

3.2.2 Impedance measurement

Permeability in the alternating-current magnetic field is defined as complex relative permeability ($\mu_{complex}$). The real part of the complex relative permeability (μ') represents the amount of energy stored in the magnetic material from alternating-current magnetic field. On the other hand, the imaginary part (μ'') indicates energy loss to the alternating-current magnetic field. Complex relative permeability can be computed from impedance measurement (equ 9).

$$\mu_{complex} = -j \cdot \frac{\underline{Z} \cdot l_{eff}}{2 \cdot \pi \cdot f \cdot \mu_0 \cdot S} = \mu' - j \cdot \mu'' \tag{9}$$

Impedance analyzer Agilent 4294A associated with the dedicated adapter for magnetic material characterization (Agilent 16454A) are used and permeability measurement can be performed in a wider frequency range (40 Hz - 110 MHz) (Fig. 9). The measurement respects the setup described in the manual guide and recommended by the manufacturer.

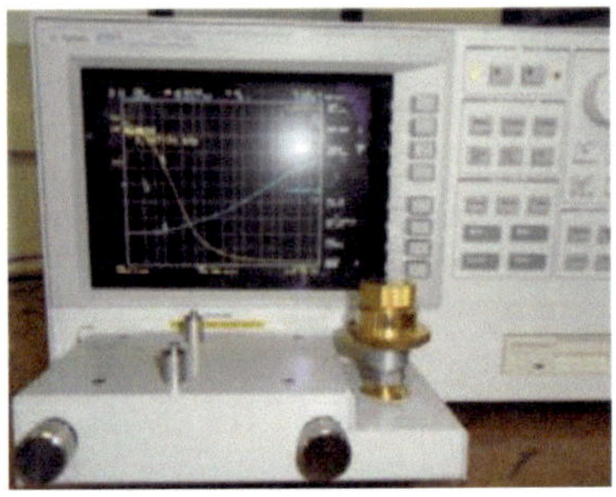

Fig. 9. Impedance meter Agilent 4294A and magnetic module

Characterizations have been performed on a toroidal shaped sample of YIG material in the 100 kHz - 110 MHz frequency range. Complex impedance (module and phase) is shown in Fig. 10. From this measurement, complex relative permeability has been computed using equ (9) and exhibited in Fig. 11. A behaviour change can be observed from 3 MHz with the change of the slope. This change can be explained by the permeability variation as described in Fig. 11.

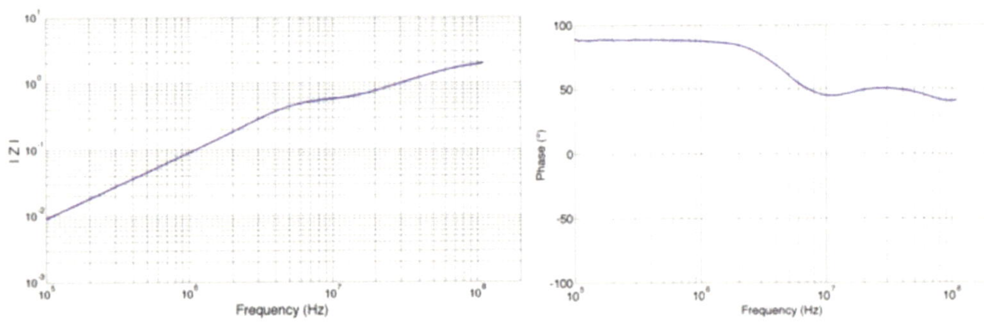

Fig. 10. Module (left) and phase (right) of the impedance

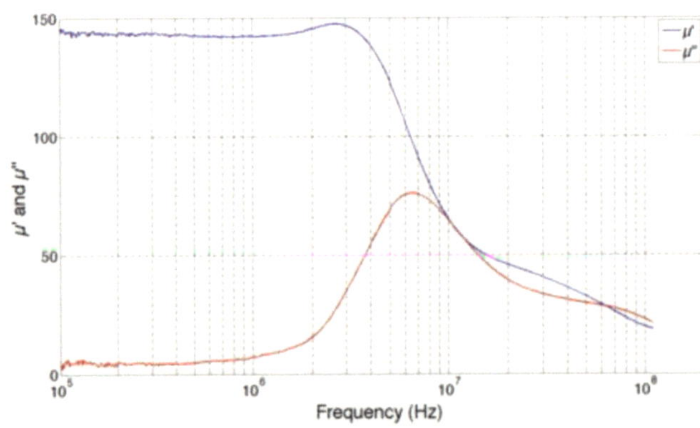

Fig. 11. Complex relative permeability (μ′ and μ′′)

Results detailed in Fig. 11 have good adequacy with the literature [Kedous] [Siblini]. Up to 3 MHz the real part of the permeability is quite constant and equal to 140 and decreases for higher frequencies. At high frequencies, the magnetic field is non-uniform and diffuses in the material which implies a permeability decrease. This phenomenon depends on the angular frequency (ω), material resistivity (ρ) and permeability (μ). It can be estimated with the expression of the skin depth (δ) (equ. 10). This decrease in permeability reaches 25 for 100 MHz.

$$\delta = \sqrt{\frac{2 \cdot \rho}{\omega \cdot \mu}} \tag{10}$$

This test bench allowed the computation of the permeability in a wide frequency range. The low excitation level permits to identify μ' (the real part of the permeability) in order to fill in the simulation parameters.

4. Technological process

At 100 MHz switching frequency, the inductance value required is small and the integration of the inductor on the SIP using micro-fabrication techniques may be practical. Although significant previous work has been done on integrated micro-fabricated inductors for power conversion applications [Fukuda, Prabhakaran, Gao], much of this work has focused on the frequency range of less than 10 MHz. To be practical, the micro-inductor technology must be implemented in a cost effective manner which normally depends on the size of the component and the number of steps used in the fabrication. Thus, the electroplating technique, a low temperature and a simple technique is used to deposit thick copper conductors and hence allowing to minimize DC resistance with high current handling (1 A). In comparison to [Orlando], a higher inductance to DC resistance ratio was achieved in a relatively large area and with a relatively more complex process. Other techniques can be used but they present several technological disadvantages (complexity; high temperature process, cost). Sputtering technique was used in [Nakazawa] to deposit a thin layer of magnetic material. This technique is not cost effective since it allows low deposition rate (1-2 microns per hour) and is highly dependent on temperature. Sputtering technique can also be used to deposit the conductor lines but also presents the same inconvenient. Although copper can be deposited by using chemical vapor deposition (CVD) [Pan] or electroless [Jiang, Schacham-Dtamand] but electroplating remains the main technique for depositing thick copper conductors because it Provides a higher deposition rate than other methods, good ductility and high adhesion at low temperature in a simple and low cost manner [Bunshah].

In this section, we detail the fabrication process. In order to electroplate the copper windings, a metallic thin layer serving a double purpose as a conductive layer and with a relatively high adhesion to the magnetic substrate is deposited. Ti 500 Å /Cu 1500 Å seed layer is chosen to be deposited on the magnetic substrate using evaporation technique (Fig. 15). This choice allows minimizing the stress at the interface of the seed layer/electroplated copper due in particular to the difference in the CTE of the materials. The magnetic material used is a solid substrate with a thickness of 500 μm.

Dry film photoresist is then laminated and patterned using UV photolithography (Fig. 12, Fig. 16)). Dry film photoresist is an alternative solution for epoxy-based resin (SU-8,..) since it requires fewer number of technological steps to be processed.

Copper windings are deposited using electroplating technique (Fig. 17). This process is cost effective since it is simple to implement at ambient temperature (25 °C) and it allows us to

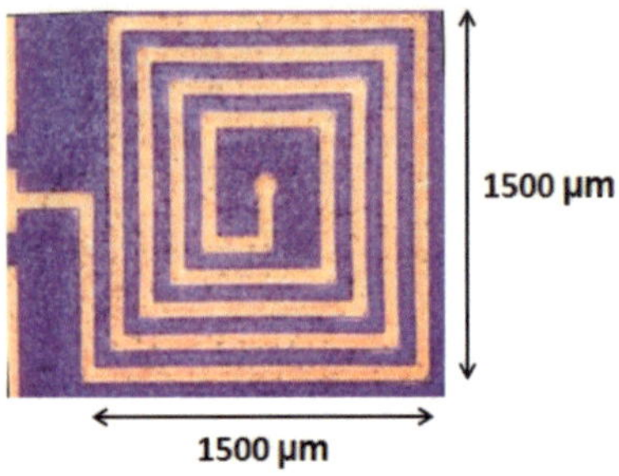

Fig. 12. Patterned inductor windings

reach a high thickness of copper in a reasonable time (copper deposition rate around 15 µm/h). Copper deposition is carried out in a horizontal cell (Fig. 13). Current densities used are in the 10-20 mA/cm^2 range (calculated for the active area on the sample). As a result, deposit height is uniform across the sample and electroplated copper shows a good crystalline quality.

After electroplating the copper windings, the remaining dry film is removed and the seed layer is wet etched to isolate the coil turns (Fig.14, Fig.17).

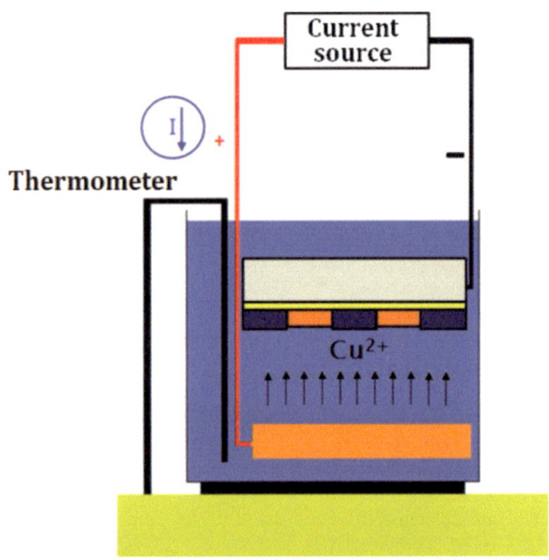

Fig. 13. Copper electroplating bench

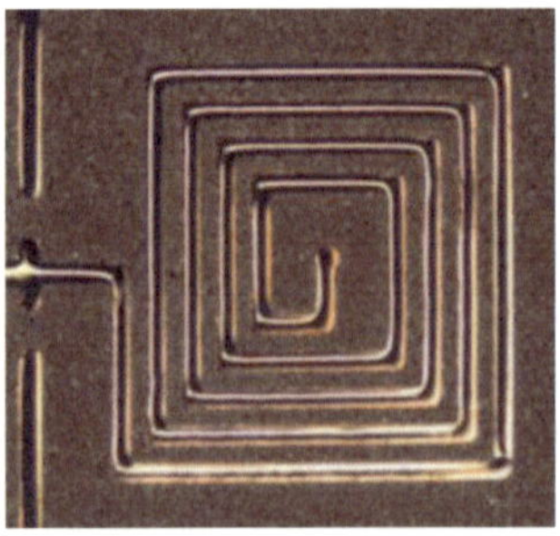

Fig. 14. Isolated inductor windings after dry film and seed layer removal

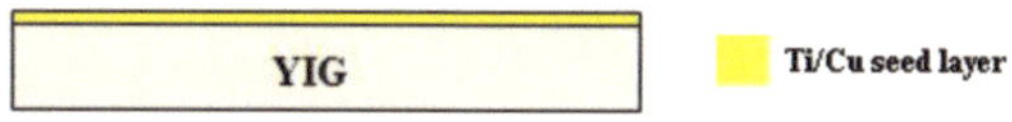

Fig. 15. Deposition of the Ti/Cu seed layer

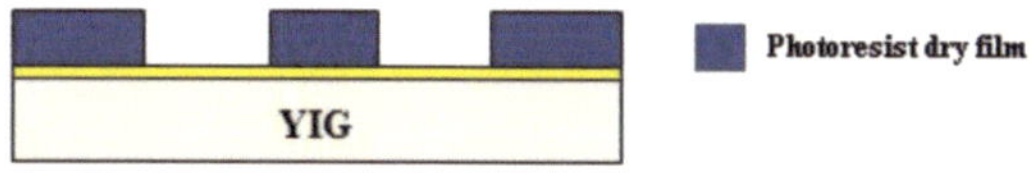

Fig. 16. Inductor windings patterning

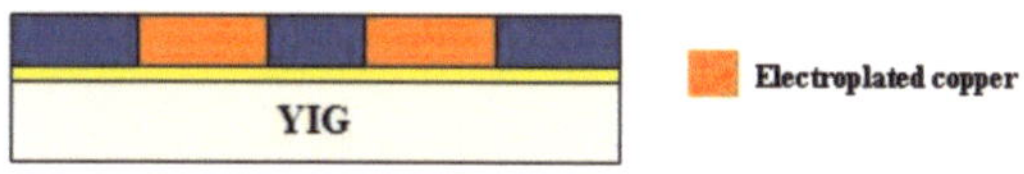

Fig. 17. Copper electroplating to fill the molds

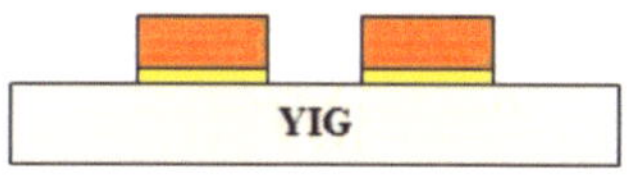

Fig. 18. Dry film and seed layer removal

5. Experimental results

Fig 19 shows SEM images of the fabricated micro-inductor fabricated. Inductors have a high aspect ratio : 75 µm in width and 50 µm in copper thickness sufficient to hold a 1 A current flow.

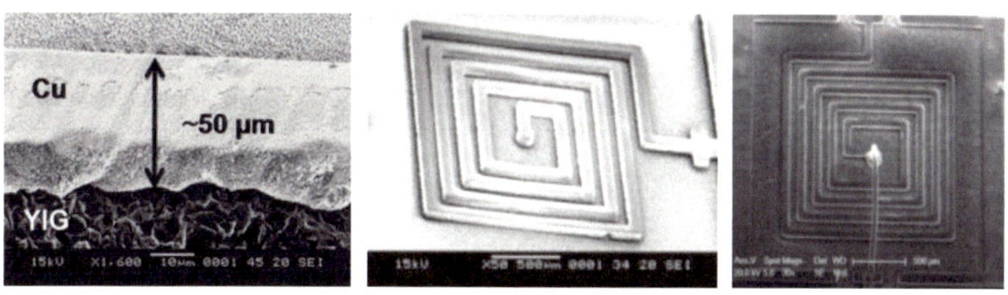

Fig. 19. SEM images of the micro-fabricated inductor

In order to perform electrical characterization, bonding has been realized (fig. 19 right) to connect the inner terminal to the external bump contact of the inductor. Device measurements have been carried out using a Vector Network Analyzer and Ground-Signal-Ground (GSG) probes from 10 MHz to 1GHz. Two-port scattering parameters (S-parameters) have been measured and then changed to admittance parameters (Y-parameters). Inductance L of the spiral inductor is subsequently computed from the resultant Y_{11} parameters after the bonding and the de-embedding procedure. Its expression is given by:

$$L = \frac{1}{\mathrm{Im}ag(Y_{11})} \cdot \frac{1}{2 \cdot \pi \cdot f} \tag{11}$$

Fig. 20 shows the simulated and measured inductance from 10 MHz to 1 GHz. Referring to equ.1, increasing the frequency with keeping the inductance value constant results in a lower current ripple and hence a satisfying voltage ripple. But since, the inductor will be integrated in a DC-DC converter with 100 MHz switching frequency, we are interested in the 100-200 MHz frequency range. Moreover, we measured an inductance value of 36 nH at 100 MHz. The perturbations at low frequency (10-50 MHz) are due to the fact that the vector network analyzer is expected to measure the s-parameters at higher frequencies. There is a 15% difference between the measured and simulated inductance value at 100 MHz because of the approximations done to perform 2D simulations. DC resistance was also measured using a Source Measuring Unit and we obtained a value of 20 mΩ (Fig. 5). Merit factor corresponding to our application is equal to 0,6 nH/mΩ.mm². Compared to [O'donnell], we have reached a higher Merit Factor for the inductor with the same number of turns and with a smaller area. In [Orlando], a higher inductance to DC resistance ratio was achieved but in a relatively large area of 5.6 x 5.6 mm² and with a relatively complex process. Thus our inductor design combined with a high permeability magnetic material allowed us to achieve good results compared to literature.

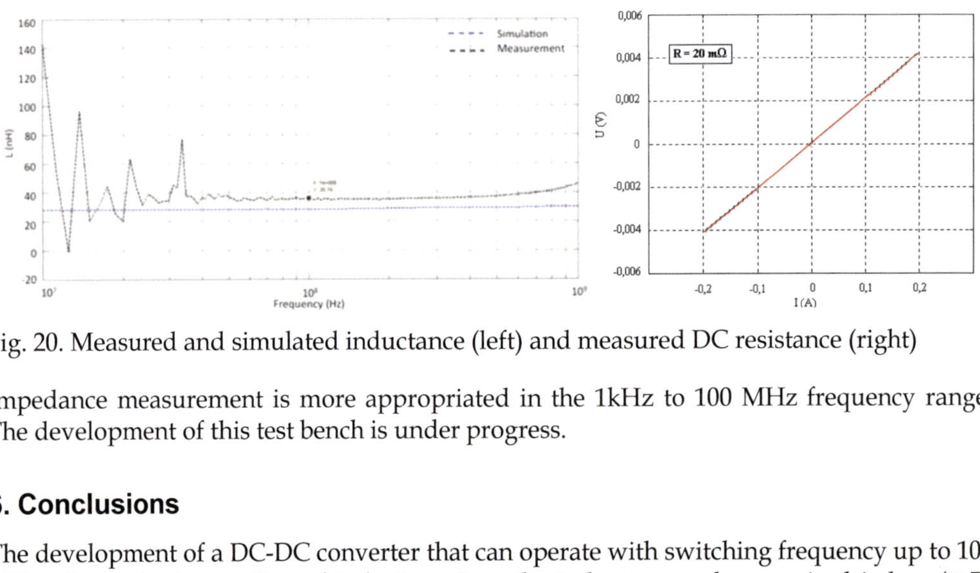

Fig. 20. Measured and simulated inductance (left) and measured DC resistance (right)

Impedance measurement is more appropriated in the 1kHz to 100 MHz frequency range. The development of this test bench is under progress.

6. Conclusions

The development of a DC-DC converter that can operate with switching frequency up to 100 MHz allows the integration of inductors since the inductance value required is low (≤ 50 nH). This implies a lower footprint and a lower profile compatible with hybrid integration for nomad applications.

This chapter presents the development of planar micro inductor using a ferrite magnetic material. A design methodology based on finite element method and dedicated software have been presented. According to geometrical approximations, a comparative study has been led to evaluate the influence of geometrical parameters. A design has been defined according to the inductor's specifications.

Magnetic properties of the magnetic material have been extracted from the hysteresis graph and impedance measurements in order to identify B-H loop and permeability, respectively. Based on these results, FEM simulations have been carried out and results were discussed. Merit Factor was presented respecting critical characteristic of integrated power inductor.

Planar spiral inductors were realized by electroplating technique. This technique and the fabrication process have been detailed. Advantages and limits have been shown. The inductors characterization have been carried out from 10 MHz to 1 GHz. Finally, simulation and characterization results have been compared with good adequacy.

7. References

J. Zhao, X. Dong, Y. Xie, "An energy-efficient 3D CMP design with fine-grained voltage scaling", Design, Automation & Test in Europe Conference & Exhibition, 14-18 March 2011.

S. Sugawara, A. Nakamori, Z. Hayashi, M. Edo, H. Nakazawa, Y. Katayama, M. Gekinozu, K. Matsuzaki, A. Matsuda, E. Yonezawa, K. Kuroki: "Characteristic of a monolithic dc–dc converter utilizing a thin-film inductor", Proc. IPEC, Tokyo, Japan, 2000.

S. Musunuri and P. L. Chapman, "Optimization issues for fully-integrated CMOS dc-dc converters," Proc. Conf. Record IEEE Industrial Applications Society Annu. Conf., 2002, pp. 2405–2410.

D. Chesneau, F. Hasbani, "Benefits and Constraints of SMPS Integration in Wireless Multi Media Terminals", 1st IEEE International Workshop on Power Supply on Chip (PwrSoC)", Cork, Sept. 21-24, 2008.

N. Wang, T.O'Donnell, S. Roy, P. McCloskey, S.C. O'Mathuna, "Micro-inductors integrated on Silicon for Power supply on chip", Journal of Magnetism & Magnetic Materials, 2007, vol. 316, n. 2, pp. 233-237.

C. P. Yue and S. S. Wong, "Physical modeling of spiral inductors on silicon", IEEE Trans. Electron Devices, vol. 47, no. 3, pp. 560–568, Mar. 2000.

C. Martin, B. Allard, D. Tournier, M. Soueidan, J.-J. Rousseau, D. Allessem, L. Menager, V. Bley, J. -Y. Lembeye, "Planar inductors for high frequency DC-DC converters using microwave magnetic material", IEEE ECCE, 2009, pp. 1890-1894.

A. Kedous-Lebouc, "Materiaux magnétiques en génie électriques 2", edition Lavoisier 2006, ISBN 2-7462-1461-X.

A.Siblini, I.Khalil, JP.Chatelon, JJ.Rousseau, "Determination of initial magnetic permeability of YIG thin films using the current sheet method", Advanced Materials Research, 2011, Vol. 324, pp. 290-293.

S. Prabhakaran, Y. Sun, P. Dhagat, W. Li and C. R. Sullivan, "Microfabricated V-Groove power inductors for high-current low-voltage fast-transient DC-DC converters", IEEE 36thPower Electronics Specialists Conference, 2005, PESC '05, pp.1513–1519.

Y. Fukuda, T. Inoue, T. Mizoguchi, S. Yatabe and Y. Tachi, "Planar inductor with ferrite layers for DC-DC converter", IEEE Transactions on Magnetics, Vol. 39, No. 4, July 2003, pp. 2057-2061

X. Gao, Y. Zhou, Wen Ding, Ying Cao, Chong Lei, Ji An Chen,and Xiao Lin Zhao, "Fabrication of ultralow-profile micromachined inductor with magnetic core material", IEEE Transactions On Magnetics, Vol. 41, No. 12, December 2005, pp. 4397-4400.

B. Orlando, R. Hida, R. Cuchet, M. Audoin, B. Viala, D. Pellissier-Tanon,X. Gagnard, and P. Ancey, "Low-resistance integrated toroidal inductor for power management", IEEE Transactions On Magnetics, Vol. 42, No. 10, October 2006, pp. 3374-3376.

H. Nakazawa, M. Edo, Y. Katayama, M. Gekinozu, S. Sugahara, Z.Hayashi, K. kuroki, E. Yonezawa, and K. Matsuzaki, "Micro-DC/DC converter that integrates planar inductor on power IC", IEEE Transaction on Magnetics, Vol. 36, No. 5, September 2000, pp. 3518-3520.

T. Pan, A. Baldi, E. Davies-Venn, R. Drayton, and B. Ziaie, "Fabrication and modeling of silicon-embedded high-Q inductors", Journal of Micromechanics and Microengineering, Vol 15, 2005, pp. 849-854.

H. Jiang, Y. Wang, J. Yeh, and N. C. Tien, "On-Chip spiral inductors suspended over deep copper-lined cavities", IEEE Transactions on Microwave Theory and Techniques, Vol. 48, No. 12, December 2000.

Y. Shacham-Dtamand , V. M. Dubin, "Copper electroless deposition technology for ultra-large-scale integration (ULSI) metallization", Microelectronic Engineering 33 (1997) 47-58.

R. F. Bunshah, "Handbook of deposition technologies for films and coatings", ISBN: 0-8155-1337-2.

T. O'Donnell, N. Wang, R. Meere, F. Rhen, S. Roy, D. O'Sullivan, C. O'Mathuna, "Microfabricated inductors for 20 MHz DC-DC converters", APEC 2008. Twenty-Third Annual IEEE, February 2008, pp. 689-693.

6

Fe-Al Alloys' Magnetism

F. Plazaola[1], E. Apiñaniz[3], D. Martin Rodriguez[4], E. Legarra[1] and J. S. Garitaonandia[2]

[1]*Elektrika eta Elektronika Saila, Euskal Herriko Unibertsitatea UPV/EHU, Bilbao,*
[2]*Fisika Aplikatua II Saila, Euskal Herriko Unibertsitatea UPV/EHU, Bilbao,*
[3]*Fisika Aplikatua I Saila, Euskal Herriko Unibertsitatea UPV/EHU, Bilbao,*
[4]*Jülich Centre for Neutron Science and Institute for Complex Systems, Forschungszentrum Jülich GmbH, Jülich,*
[1,2,3]*Spain*
[4]*Germany*

1. Introduction

Intermetallic compounds and ordered intermetallic structures have attracted great interest for both scientific reasons and their possible technological applications [1, 2]. These materials are widely used as high temperature structural materials, functional materials for scientific applications, as diffusion barriers, and as contacts and interconnections in microelectronics.

In this chapter we will study the magnetism of Fe-Al intermetallic alloys. Interest in these alloys grew after 1930 when their excellent oxidation resistance was discovered. Iron aluminides are intermetallics, which apart from the good oxidation resistance offer excellent sulphidation resistance and potentially lower cost compared to many other high temperature materials. Additionally, they have densities that are about 30 % lower than commercial high temperature structural materials, such as stainless steel or Ni based superalloys. It has also been found that Fe-Al alloys with different magnetic and physical properties can be obtained by varying the composition and their heat treatments [3]. However, the limited ductility at room temperature and the decrease in strength above 600 °C are still drawbacks that limit their exploitation for structural applications.

The mechanical and magnetic properties of the Fe-Al intermetallic alloys strongly depend on the deviation from the stoichiometry, and the addition of a ternary alloy component can improve the ductility at room temperature and the strength at high temperature.

All the properties mentioned make these alloys of great technological importance; that is the reason why they have been widely studied for the last decades. In addition, many first principle investigations on the electronic and magnetic structure have been conducted to provide a microscopic understanding of the chemical bonding, the formation of clusters, the surfaces and the phase stability.

As far as magnetism is concerned, one of the most prominent features of Fe-rich Fe-Al alloys is that any structural change is directly reflected by their magnetic behavior. That is to say, any slight mechanical deformation produced on the ordered alloy causes an abrupt increase of the ferromagnetic signal [4-12]; in fact, this alloy presents an unusual ease to undergo order-disorder transitions. Indeed, with a simple crushing, a solid solution structure is induced and the cell parameter can increase by even as much as 1% in some of the alloys. This change is reflected immediately by the magnetic behavior of the material. For instance, for certain compositions ordered samples present a great paramagnetic contribution, but they become strongly ferromagnetic when crushed. Moreover, small changes in Fe content of the alloy (less than 1 at. %) induce large changes in its magnetic behavior [13, 14]. In order to explain these behaviors the intimate relationship between microstructure and magnetism has to be taken into account; indeed, this is linked to the fact that the iron rich side of FeAl phase diagram (see fig. 1) presents three main phases: ordered D03 and B2, and disordered A2. For this reason, Fe-rich FeAl intermetallic alloys are considered as a "test field" to test theories and hypothesis of fundamental magnetism. Figure 2 shows the three structures mentioned before. A2 structure is a solid solution, and therefore the Fe and Al atoms are distributed at random in the crystallographic positions of a bcc structure. B2 structure is a CsCl type structure, with a stoichiometric composition of $Fe_{50}Al_{50}$, where the Fe atoms site is the vertex (position A) and Al atoms sit in the center (position B) of the cube. The D03 structure consists of four interpenetrating face centered cubic sublattices. For the stoichiometric composition ($Fe_{75}Al_{25}$) Fe atoms occupy A, B and C non-equivalent positions and Al atoms occupy D positions.

Apart from the properties mentioned above, these alloys show an important effect that it is worth mentioning: magnetostriction, i.e. change of the sample dimensions in response to an applied field. This property makes Fe-Al alloys interesting because of their potential use as low-cost sensor devices. Room temperature magnetostriction measurements of Fe-Al alloys indicated a five-fold rise in magnetostriction with Al additions up to 30% Al [16, 17]. These works, performed on single crystals, concluded that there was a large temperature dependence of the magnetostriction of the materials and they added that the stabilization of the disordered bcc structure was a fundamental component in the increase of the magnetostriction of the materials. On the other hand, higher magnetostriction values were found for rapidly quenched ribbons [18, 19, 20] and the microstructure and the room-temperature magnetostriction of polycrystalline FeAl alloys have also been studied [21]. These last works conclude that the magnetostriction of this system is very dependent on the heat treatment (and therefore, the structure), the temperature and the composition.

Another important magnetic property that will be studied in this chapter is the spin-glass and re-entrant spin-glass (or mictomagnetism) phenomena found at low temperatures for certain alloys. One of the direct consequences is the anomalous magnetization-increase that the magnetization curves show at low temperatures with the increase of the temperature [5, 22], in a certain range of Fe concentrations. Spin-glass systems are characterized by an absence of magnetic order below the denominated freezing temperature (T_f) as a consequence of weakening of magnetic exchange mechanisms among magnetic moments.

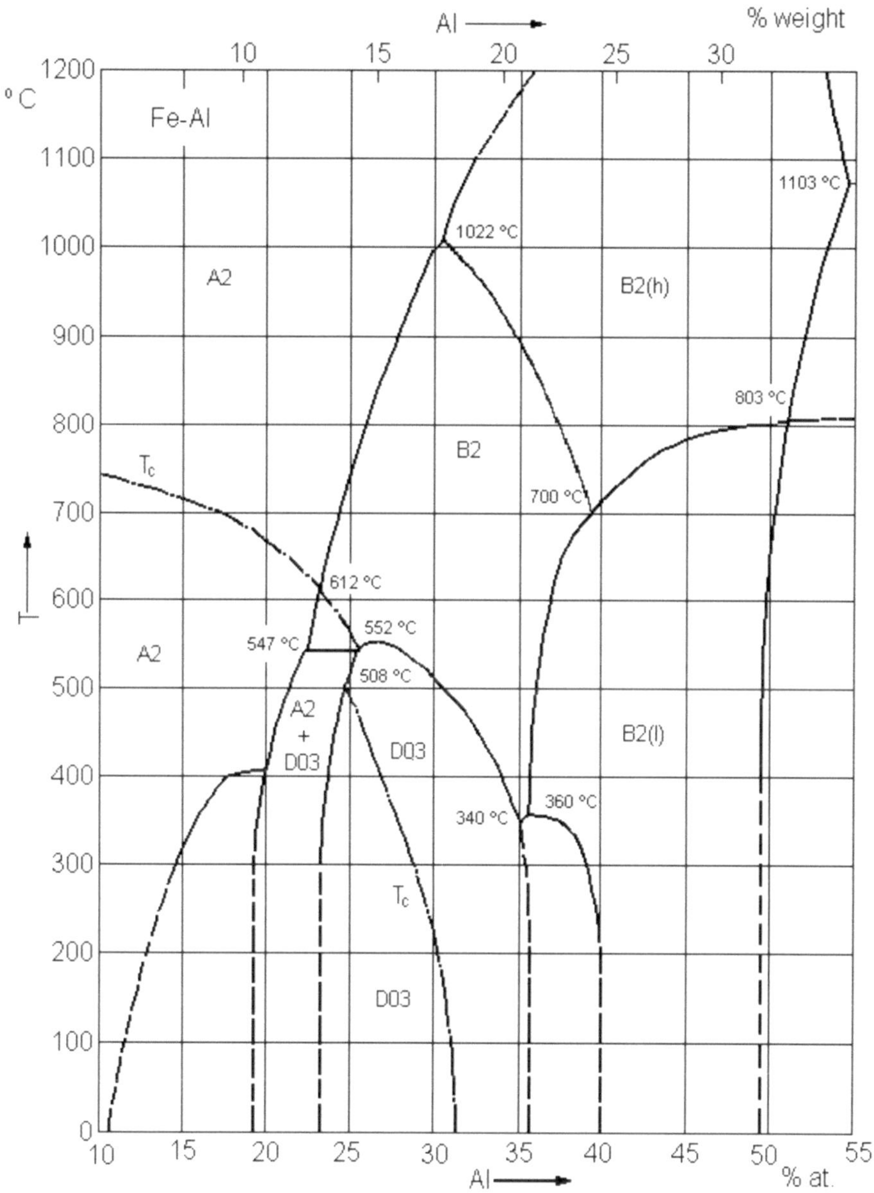

Fig. 1. Phase diagram of the Fe rich side Fe-Al system [15].

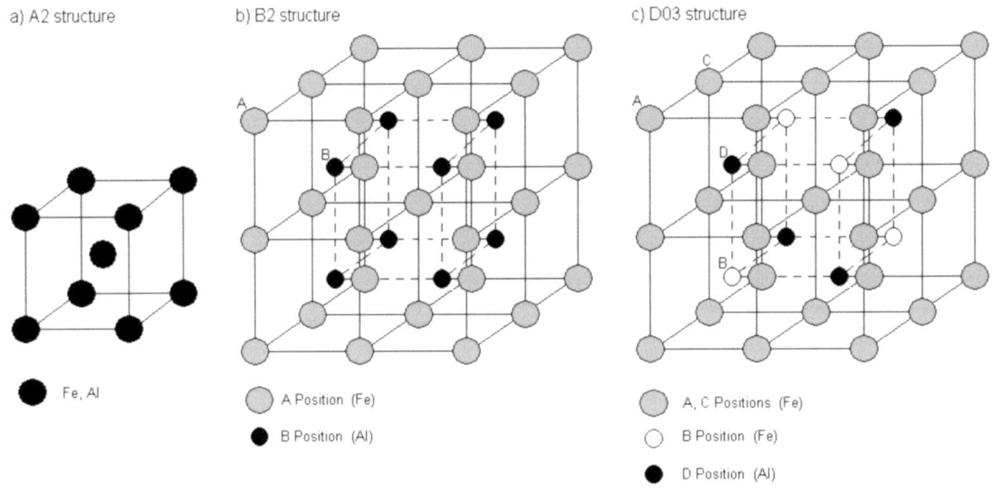

Fig. 2. Main crystallographic structures of the Fe rich side phase diagram: a) disordered A2 structure, b) B2 structure and c) D03 structure.

The Fe-Al system has a 3d transition element. It is widely known that one of the most important topics of magnetism is the study of the magnetic interactions between 3d magnetic elements and the magnetic effects and properties presented in the magnetic materials that they form. There are two main reasons that make this topic very important: On the one hand, the 3d based magnetic materials are used in many industrial applications; any study that can help in understanding the origin of the magnetic properties of this kind of materials could be used to improve them or to produce materials adapted or optimized for each application. On the other hand, these kinds of studies are fundamental from the point of view of basic magnetism. Most of the theories that try to explain the physical mechanism of the magnetic exchange interactions between 3d elements are qualitative. Usually the source of these theories is the Bethe-Slater curve [23], which explains the different magnetic behaviors of the materials in terms of differences in the exchange interactions between the magnetic elements. The verification of this curve and the theory hidden behind is one of the main aims in the field of magnetism. Fe-Al alloys are suitable to study the role of the structure on the magnetism of the materials, because they represent a simple model with well known ordered structures for studying the basic properties. In addition, it also presents a wide variety of magnetic behaviors as mentioned above (ferromagnetic, paramagnetic, spin-glass, re-entrant spin-glass), which is a clue to the existence of different kinds of exchange interactions in this alloy system.

The origin of magnetism of these alloys has been investigated by different techniques, as will be shown in the following sections. We have divided the experimental study of the magnetism of Fe-Al alloys in three subsections. In the first one the unusual magnetic properties of the ordered alloys are described. In the next one the contributions to the magnetic signal increase with disorder and the disordering process are presented and in the last one the reordering process is discussed. It has to be mentioned that the most complex magnetism in FeAl binary alloys is around $Fe_{70}Al_{30}$ composition; therefore, in this chapter we will pay plenty of attention to such a composition.

Finally, we will end up our work by showing the results of theoretical calculations performed for FeAl alloys for different structures and compositions.

2. Ordered alloys

FeAl alloys are a well-suited system for the study of the properties of magnetic materials and in particular, for study of the role of the structure on the magnetic character of those materials [4, 6, 8, 10, 24-26]. Due to the existence of only one magnetic atom and the structural simplicity of a binary system, the theoretical results can be easily related to the magnetic properties [27-30].

The room temperature magnetic moment of ordered iron aluminides ($Fe_{1-x}Al_x$) decreases slowly with increasing Al content up to x=0.2, which is consistent with dilution models. With further dilution the magnetic moment decreases more rapidly, becoming zero for alloys with $x \geq 0.325$ of Al [31]. Martin Rodriguez et al. [32] and Schmool et al. [33] showed by Mössbauer spectra measured at room temperature how the ferromagnetic network breaks into magnetic clusters with the addition of Al at room temperature above x=0.275 and the magnetic hyperfine field becomes zero at room temperature for x≥0.325. X-ray diffraction patterns show traces of D03 structured domains for x<0.325. However, for 0.325≤x≤0.5 the ordered alloys present only B2 structure, showing evidences that both, a magnetic transition and a structural transition, occur around x=0.325 at RT, which suggests a strong correlation between the intermetallic order and the magnetic behavior in this alloy system [32].

According to the molecular field model, saturation magnetization is expected to decrease with increasing temperature. In contrast, in Fe-Al alloys the opposite tendency has been observed in samples with Al concentration in the range 0.275<x<0.325. In this section we will focus on discussing that unusual increase of the magnetic signal that occurs for samples with about 68-72 at.% Fe content. Figure 3 shows some selected M(H) curves obtained at several temperatures, for the $Fe_{70}Al_{30}$ ordered sample. Although the magnetization increase rate depends strongly on the applied field, figure 3 indicates that the maximum increase of the magnetization curve is located somewhere between 150 K and 200 K. Arrott and Sato [5, 34] observed this effect in 1959 for the first time. The structural characterization published two years before [4] showed that the structure of the $Fe_{70}Al_{30}$ ordered sample was Fe_3Al type (D03 structure). Based on that crystallographic structure and on the evolution with temperature of the hysteresis curves, the authors proposed a model where nearest-neighbors Fe-Fe ferromagnetic exchange competed with indirect Fe-Al-Fe antiferromagnetic super-exchange. However, later neutron diffraction results [35] proved that, below the Curie temperature, the alloy always presented a ferromagnetic character, and the model was abandoned. Since those results, several papers have been published, suggesting the

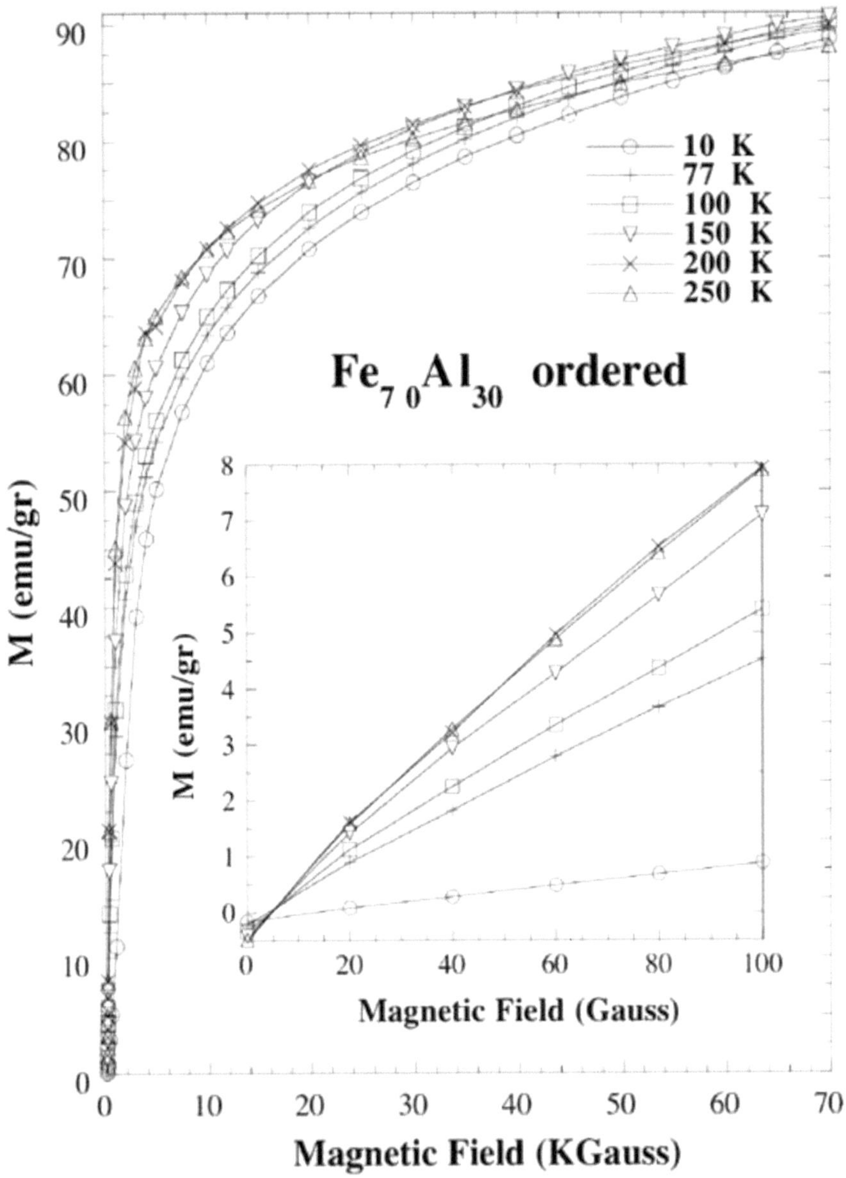

Fig. 3. Magnetization curves, M(H), of the $Fe_{70}Al_{30}$ alloy obtained at different temperatures. Before measuring every M(H) curve the sample was demagnetized. The inset shows the detail of the M(H) curves for applied fields below 0.01 T.

existence of a phenomenon known as mictomagnetism [10, 36, 37]. The mictomagnetism consisted in a collective freezing of the spin re-orientations at certain temperatures without long magnetic order. In the case of the FeAl alloys, this last fact implied that the behavior of the magnetism was local. Nowadays, we know that almost all mictomagnetic processes are associated with spin-glass or reentrant spin-glass processes. The transition to a spin-glass state is clearly detected by a comparison of the evolution of the magnetic signal with the temperature when the system has been zero-field-cooled (ZFC) and field-cooled (FC) from a paramagnetic regime. In the first case the material is cooled without applied field from temperatures higher than the Curie one (T_c), which makes the moments keep their random orientations at lower temperatures. The ZFC curve is obtained starting from such a situation, applying a small external field while increasing the temperature, which adds energy to the system and allows the magnetic moments to start orientating gradually into the direction of the applied field. However, in the FC case the system is cooled under an external magnetic field, which aligns the moments in the direction of the applied field once the temperature decreases below T_c. The difference in the initial state (moments already oriented or disoriented) is reflected in the different behaviors with the temperature of the ZFC-FC curves up to the freezing temperature (T_f), when the magnetic exchange among moments is strong enough to establish a long range order and to maintain intrinsically the moments oriented.

As it can be observed in Figure 4, $Fe_{70}Al_{30}$ ordered sample enters in a spin-glass like regime below T_f ~ 90 K. This transition is also consequence of the predominance of the local mechanisms governing the magnetism of this alloy.

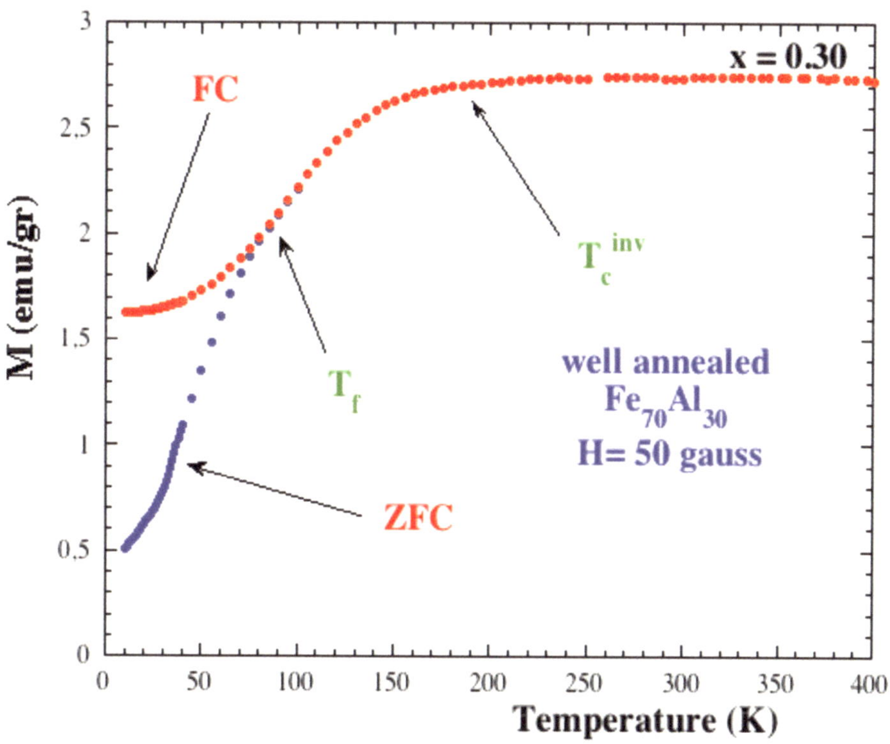

Fig. 4. ZFC and FC curves of the $Fe_{70}Al_{30}$ alloy. The FC curve was obtained under an applied field of 50 gauss. T_c^{inv} and T_f are transition temperatures (see text).

Figure 4 shows the ZFC-FC magnetization curves obtained at 50 gauss applied field for the studied sample. The T_f and T_c^{inv} temperatures indicate magnetic transition points. The transition points represent the limits of three different zones. The first zone ranges from the lowest temperatures up to T_f ~ 90 K. This zone is concerned with different evolutions of magnetization with temperature for field cooled or zero field cooled samples. In the second zone, which ranges from T_f temperature up to T_c^{inv} labeled temperature (~180 K), the magnetization increases with temperature. In the last zone (from T_c^{inv} temperature up) the magnetic signal remains constant.

Other peculiarity of the magnetization curves (figure 3) is the pronounced slope they present in the high field zone. The signal does not reach saturation even at 70 kgauss applied field. This fact is a clear indication of the existence of paramagnetic type behavior in the sample. Magnetic neutron scattering results as well as detailed magnetic measurements showed that the mictomagnetism is accompanied by a superparamagnetic character of FeAl alloys [37-40]. The superparamagnetism is the consequence of complex magnetic structure, composed by dynamical ferromagnetic clusters accommodated in a paramagnetic crystalline matrix. The discussion is centered on the evolution with temperature of the paramagnetic contribution, and on its role in the magnetic interaction among clusters [38, 41-43]. The magnetic picture of the $Fe_{70}Al_{30}$ ordered sample is the result of the coexistence of two different magnetic phases that follow different behaviors with temperature and, so, with different magnetic relations between them.

Mössbauer spectroscopy is a very useful technique for studying the electronic structure of solids such as chemical bonding or magnetism as it allows the detection of variations in the nuclear energy levels due to the electromagnetic coupling between nuclear and electronic charges. These variations are known as hyperfine interactions and they may shift energy levels or lift their degeneracy. In the presence of a magnetic field the interaction between the nuclear spin moments with the magnetic field removes all the degeneracy of the energy levels resulting in the splitting of energy levels. For iron atoms this magnetic splitting will result in a sextet. Thus, Mössbauer spectroscopy is able to distinguish magnetic and non-magnetic phases. For cubic structures, as the ones studied in this chapter, paramagnetic structures are fitted with singlets whereas ferromagnetic structures are fitted with sextets. In order to fit the spectra it is very important to know well the environment of the magnetic atoms. In a crystalline solid the iron atoms can be situated in different non-equivalent positions and therefore, their environments change. In this case, a different spectrum will be obtained from each non-equivalent iron atom and the resulting spectra will be the sum of the independent subspectra obtained for each non-equivalent iron atom. This may cause difficulties in separating the subspectra. When it is not possible to separate properly several subspectra, Mössbauer spectra can be fitted by a hyperfine field distribution, where non magnetic contribution is the central part around 0 T. The shape of the Mössbauer spectra supports the picture of a magnetic structure composed by ferromagnetic clusters surrounded by a paramagnetic phase. Figures 5.1 and 5.2 show the Mössbauer spectra obtained at several temperatures along with the corresponding hyperfine field distributions. The spectrum obtained at 77 K is composed mainly by a paramagnetic-like central contribution (peak around 0 mm/s velocity). Besides that contribution, the spectrum presents also certain not very formed ferromagnetic contribution joined to the central one. The ferromagnetic contribution evolves with temperature presenting soft out lines without marked transitions, and its contribution to the spectra shows up clearer and clearer as temperature increases. At the temperature of 200 K, two shoulders at both sides of the central paramagnetic peak around the 1mm/s and -1mm/s velocities are clearly appreciated. The shoulders become more evident as temperature increases and remain in all the spectra obtained at the higher temperatures. The distributions present two clear components.

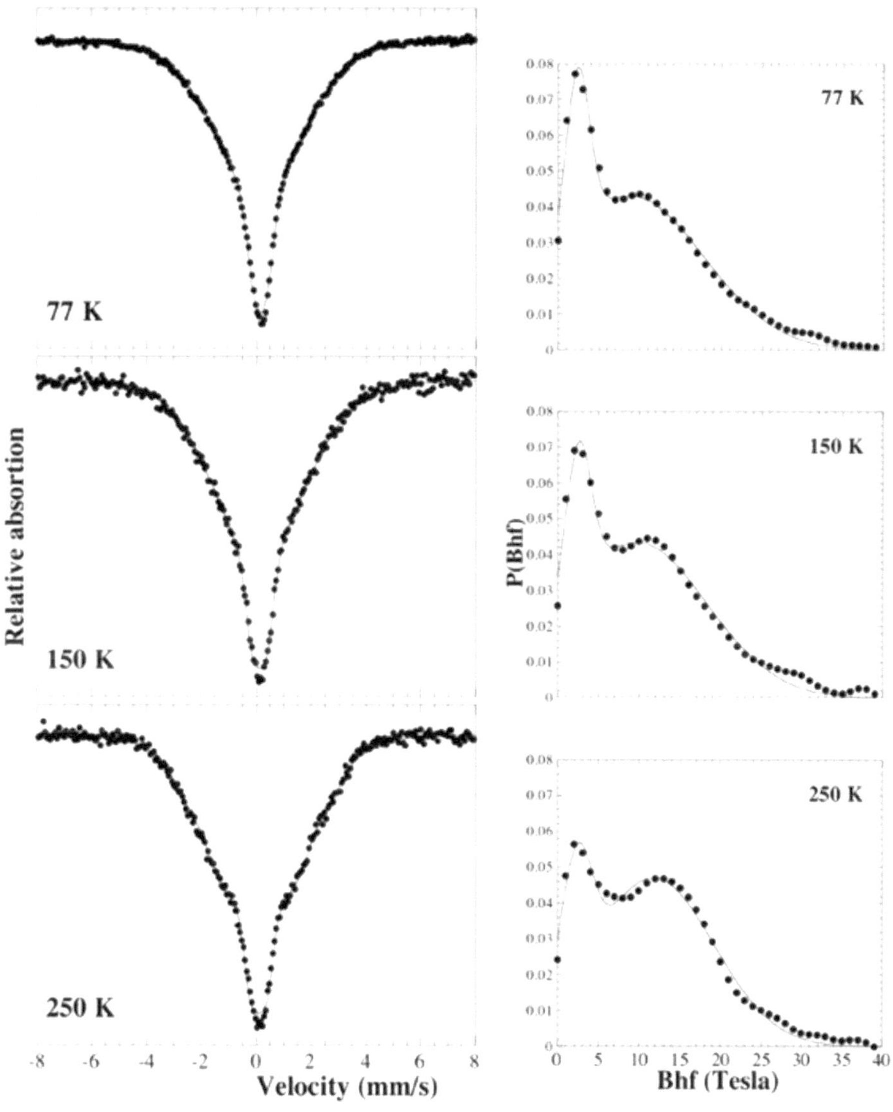

Fig. 5.1. Mössbauer spectra of the $Fe_{70}Al_{30}$ alloy measured at 77 K, 150 K and 250 K. The curves on the right show the corresponding hyperfine field distributions.

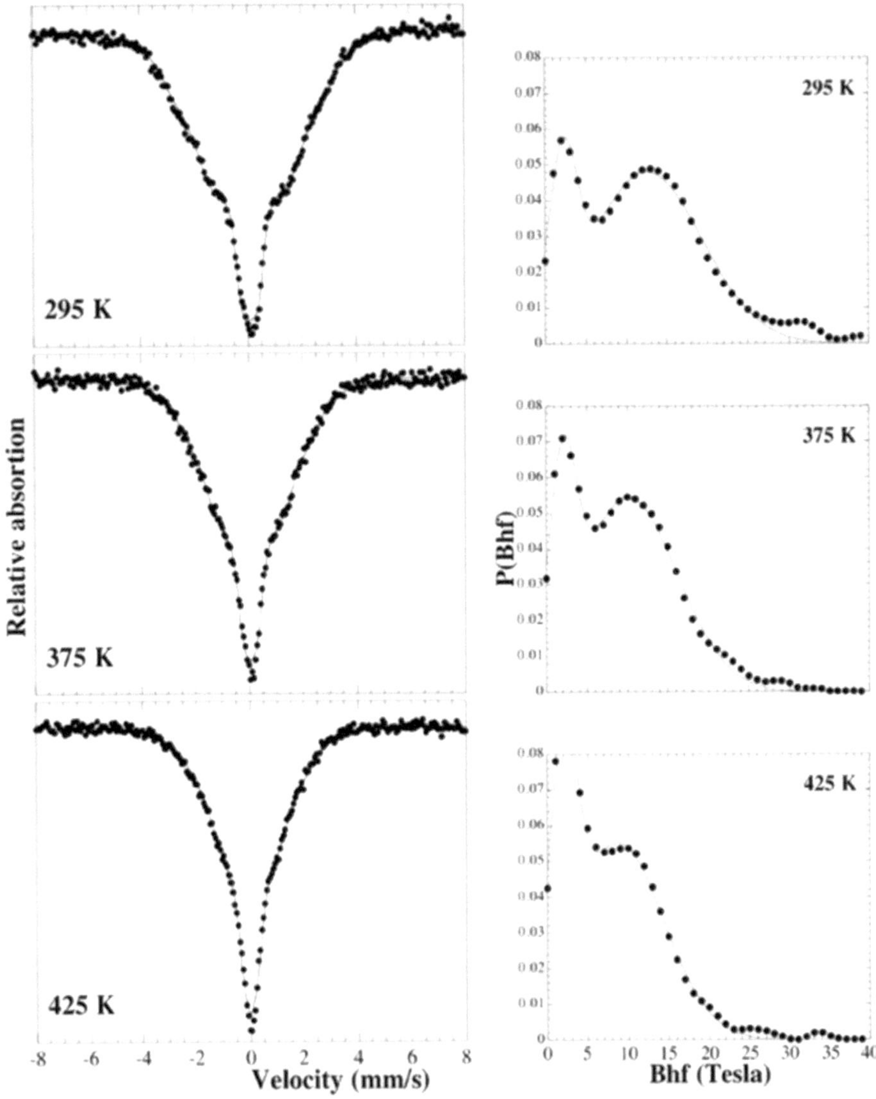

Fig. 5.2. Mössbauer spectra of the $Fe_{70}Al_{30}$ alloy measured at 295 K, 375 K and 425 K. The curves on the right show the corresponding hyperfine field distributions.

One of them is located around two Tesla and it is associated with the paramagnetic-like contribution to the spectra. The second component is situated at higher fields and gives a description of the broad ferromagnetic contribution. Both components evolve with temperature. From figure 5.1, an increase of the high field component with temperature, at the expense of the low field component, is clearly observed and, at the same time, the maximum of the high field component shifts towards higher values. The scenario offered by figure 5.2 is just the opposite. In this temperature range, the component due to the ferromagnetic contribution decreases with temperature and the maximum returns to lower fields. A fitting of the hyperfine field distributions by means of two gaussians; one for each component provides different and distinct information about the changes of both contributions to the Mössbauer spectra. Figure 6 shows the evolution of the intensity of the ferromagnetic contribution. The intensity of this component increases up to ~180 K; from that temperature up, however, decreases monotonically. Figure 6 describes the relative changes with temperature of the quantity of ferromagnetic Fe atoms in the sample. Thus, the increase means that the number of ferromagnetic atoms increases with temperature up to ~180 K, that is, up to the same temperature where the T_c^{inv} transition was observed in the ZFC-FC curves and within the temperature range where the maximum of the magnetic signal was observed in the M(H) curves of figure 3. From that temperature up, the intensity decreases. This can be interpreted as an evolution of the sample towards a more paramagnetic state. Mössbauer spectra confirm the results observed by Cable et al. [44]. Using a neutron diffuse scattering technique, they concluded that ferromagnetic clusters existed in the sample all over the regions observed in the ZFC-FC curve (figure 4) and the size of those ferromagnetic clusters was continuously increasing with temperature.

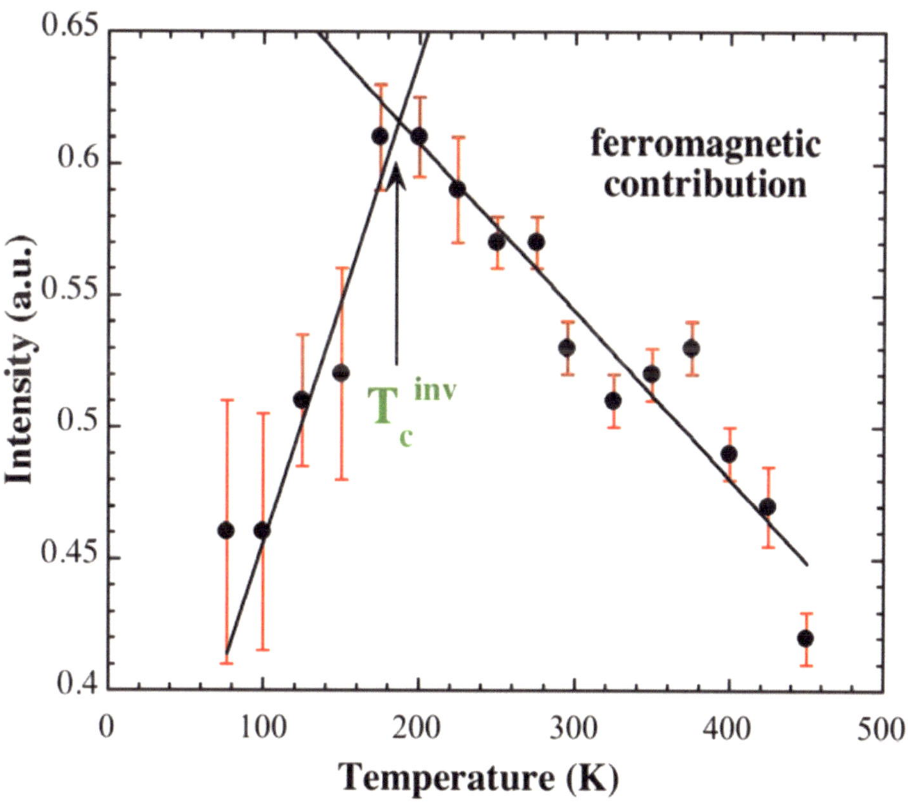

Fig. 6. Evolution with the temperature of the resonant area corresponding to the ferromagnetic contribution, as obtained from the fitting of the hyperfine field distributions of the Mössbauer spectra.

The most widely accepted model to explain the local magnetic character was firstly proposed by Srinivasan et al [45], and afterwards updated by Besnus et al. [8], and by Cable et al. [44]. Based on magnetic and neutron diffraction results, they proposed that the magnetism is related to the local magnetic environment. They distinguished two kinds of Fe atoms, magnetic Fe atoms (those with four or more than four iron nearest neighbors) and paramagnetic Fe atoms (those with less than four Fe nearest neighbors), and their model basically consisted in the assignment of an adequate magnetic moment to each magnetic Fe atom. This model was very successful because it was able to explain qualitatively and quantitatively the effects such as the decrease of the magnetization signal with Al content in $Fe_{100-x}Al_x$ alloys, the coexistence of the ferromagnetic and paramagnetic contributions in the Mössbauer spectra and the presence of ferromagnetic clusters in the sample. According to the model, the ferromagnetic clusters are the consequence of regions composed by Fe atoms with four or more than four Fe nearest neighbors, that is, regions formed by magnetic Fe atoms. In this sense, the evolution of the ZFC-FC curves of $Fe_{70}Al_{30}$ alloy and the magnetic transitions observed in those curves (see figure 4) were explained as temperature dependent interactions between the paramagnetic phase and the ferromagnetic clusters due to the existence of random fields [46, 47].

The increase of the volume of the ferromagnetic phase and a probable increase of the quantity of magnetically interconnected clusters is behind the strong decrease of the slope value (see figure 7) observed between ~50 K and ~ 180 K, which is interpreted as a turning of the sample towards a more ferromagnetic state. The value of the slope changes again towards a constant value at T_c^{inv} temperature, which would mean that at that temperature, the sample has acquired its maximum ferromagnetic character.

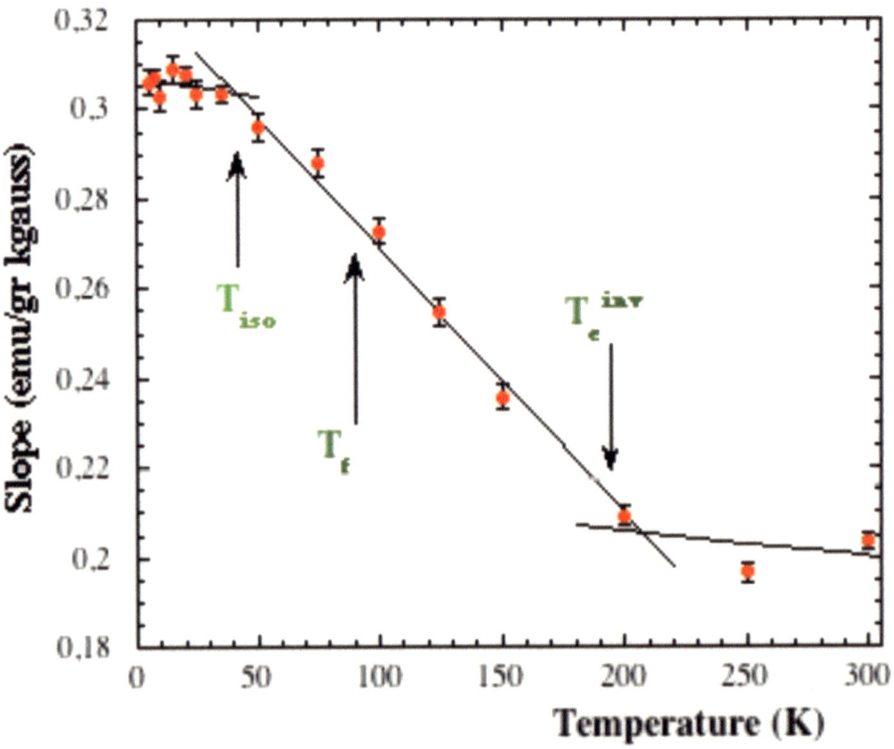

Fig. 7. Slope of the M(H) curves at high fields (between 4 and 7 T) versus temperature. T_c^{inv} and T_f are transition temperatures.

At T_c^{inv} the resonant area corresponding to the ferromagnetic component is 0.61, high enough for the clusters to be in physical contact, creating a ferromagnetic network in the sample by cluster interconnection. From that point on, posterior increase of cluster size might not imply an increase of the macroscopic ferromagnetic behavior of the sample and the value of the slope of the magnetization curves at high field remains constant.

Studies on magnetic fluctuations with temperature of the $Fe_{70}Al_{30}$ ordered sample performed by neutron depolarization experiments show the same trend [41]. The average neutron polarization Po decreased as the temperature increased and from T_c^{inv} temperature up it remained constant with a value close to zero. This change is explained as caused by the transition of the sample from spin-glass (cluster) state to the ferromagnetic state. Similarly, from inelastic neutron scattering data, published by Motoya et al. [38], it is observed that ferromagnetic spin-waves were formed above T_c^{inv} temperature. The authors interpreted that fact as an increase of the coupling degree between the spins of the paramagnetic region and the ferromagnetic network. Therefore, all these data suggest that T_c^{inv} transition is a consequence of the growth of the cluster size, but that the process related with the growth does not finish at that temperature.

The reentrant spin-glass state is defined as the entrance of the system to a spin-glass like state from ferromagnetism. In the $Fe_{70}Al_{30}$ ordered sample, this state is formed below T_f ~90 K temperature and it has usually been explained as the consequence of the total magnetic isolation of ferromagnetic clusters. The evolution of the ZFC curve with temperature (see Fig. 4) indicates that a magnetic disconnection of the clusters occurs, but also that such a disconnection is only effective while the applied field is zero.

3. Mechanically disordered alloys

3.1 Volume expansion/Chemical disorder contributions

As mentioned above, the room temperature magnetic moment of ordered iron aluminides ($Fe_{1-x}Al_x$) decreases slowly with increasing Al content, consistently with dilution models, up to x=0.2. With further dilution the magnetic moment decreases more rapidly, becoming zero for alloys with $x \geq 0.35$ of Al [25]. However, disordered $Fe_{1-x}Al_x$ alloys are ferromagnetic at room temperature even for alloys with $x>0.35$ [6, 14, 26, 48-56]. Thus, paramagnetic to ferromagnetic transition linked to an order-disorder transition can be observed after mechanical deformation. Experimentally, the influence of structural disorder on the magnetic properties has been evidenced in FeAl, in different types of microstructures such as cold worked single crystals [49, 26], quenched or cold worked polycrystalline materials [6, 8, 51], or ball-milled and mechanically alloyed systems [14, 26, 31, 48, 50, 52, 53, 55, 56]. From a theoretical point of view, the magnetism of diluted and disordered transition metal (TM) intermetallic alloys has been traditionally explained by the local environment model [10, 57]. In this model the magnetic moment of a given TM atom depends on the number of nearest-neighbor TM atoms: (i) either the TM atoms have their full moment when surrounded by a given minimum number of TM neighbors and zero otherwise [6, 8, 57] or (ii) the moment progressively decreases with reducing the number of TM nearest neighbors below a critical number [10, 48].

Using this simple model, the effect of Al substitution and disorder in FeAl can be qualitatively explained [6, 8, 10, 48, 57]. However, usually no quantitative agreement can be reached [6, 8, 10, 34, 48, 57-60].

It is noteworthy to take into account that the disordered state in FeAl alloys is accompanied by an increase of volume in the deformed state [14, 48, 50, 54-56]. Taking into account that variations in the distance between Fe atoms have profound effects on the magnetism [61-63], it was actually argued that the origin of the magnetic interactions in disordered FeAl intermetallics may not arise solely from nearest-neighbors magnetism (i.e., local environment model) but also from changes in the band structure of the material induced by lattice parameter variation (Δa_o) [50, 14]. Actually, in the band structure calculations of disordered FeAl alloys an expansion of the lattice parameter is also found [27, 28, 64-67]. Moreover, there are clear indications from band structure calculations performed in $Fe_{50}Al_{50}$ and $Fe_{75}Al_{25}$, that Δa_o could play a role in the magnetic moment of disordered FeAl intermetallics [68, 69]. Nonetheless, experimentally, the problem remained on how to separate disorder effects from Δa_o effects. One possibility was to reduce a_o by deformation without altering the disorder. For that purpose x-ray magnetic circular dichroism (XMCD) was studied in ball-milled $Fe_{60}Al_{40}$ alloy under applied pressure, with the aim to separate the effects of disorder from those of lattice expansion on the magnetic properties [70]. The normalized XMCD integrated intensity (i.e. magnetic moment) and the normalized saturation magnetization do not practically change up to 1.4 GPa applied pressure (which corresponds to lattice parameter values of around 0.2905 nm; that is $\Delta a_o/a_o$~ 0.3%) (see Fig. 8). However, as the pressure is increased (the lattice parameter decreases) a magnetic phase transition is observed, leading to a rapid decrease of the normalized integrated XMCD

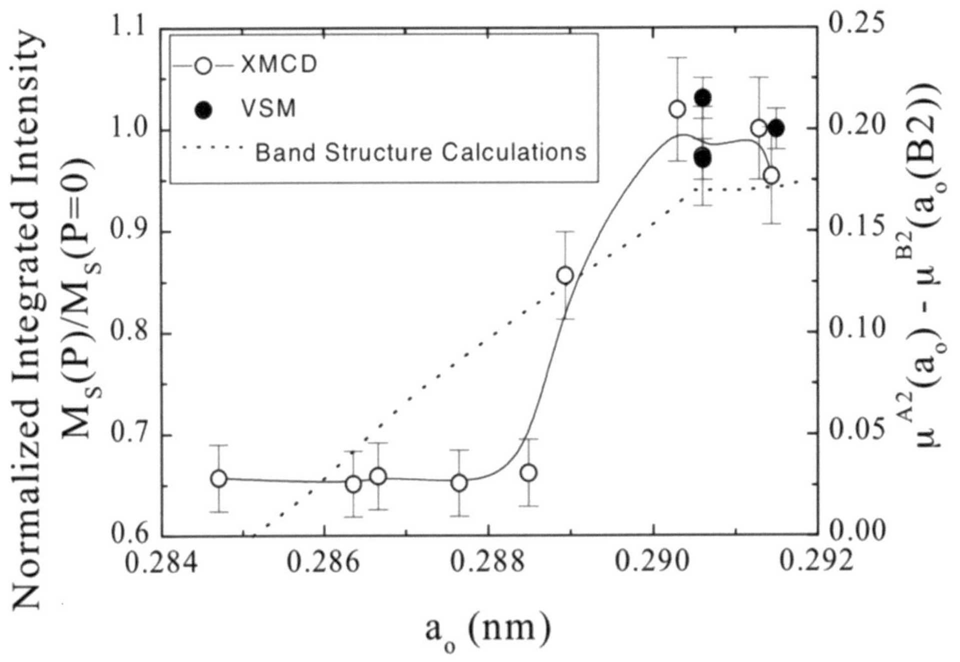

Fig. 8. Normalized XMCD integrated intensity (open symbols) and normalized saturation magnetization (filled symbols) vs the lattice parameter, a_o, for a ball milled $Fe_{60}Al_{40}$ alloy. The dashed line shows the evolution of the theoretically calculated difference between the magnetic moment of the disordered alloy, μ^{A2}, and the equilibrium magnetic moment of the ordered alloy, $\mu^{B2}[a_o(B2)]$, as a function of the lattice parameter, a_o, for a $Fe_{62}Al_{38}$ alloy. The continuous lines are guides to the eye. Figure taken from reference [70].

intensity (i.e. magnetic moment) reaching a value that does not change with further pressure. This indicates that when the lattice parameter a_o reaches approximately the one of the ordered sample, then up to 35±5% of the magnetic moment of the sample vanishes. This sharp magnetic transition indicates the existence of a moment-volume instability, which is not related to any structural phase transition, because XRD measurements do not show any phase transition in the studied pressure range. The band structure calculation results shown in figure 8 give a contribution of volume change (Δa_o) to the total magnetic moment of about 45±10%; which is intermediate between the ones calculated in $Fe_{50}Al_{50}$ and $Fe_{75}Al_{25}$ [68]. Therefore, experimental and theoretical results demonstrate that the magnetism in this kind of system arises from both the atomic disorder and the disorder-induced lattice expansion. This is in contrast to previous studies where only near-neighbor effects were considered to explain the magnetic behavior of similar alloys. In the case of disordered $Fe_{60}Al_{40}$, experimentally, the contribution of disorder and lattice expansion account for 65% and 35% of the magnetism of the alloy, respectively.

3.2 Disordering process

Once having studied the volume effect on the magnetic properties of disordered alloys let's tackle the study of the disordering process as a whole. The $Fe_{70}Al_{30}$ alloy was chosen to systematically study and characterize the evolution of different surroundings of Fe atoms with mechanical deformation (milling time), during the order-disorder transition. This alloy presents a weak magnetism at room temperature (see section 2) [71].

XRD measurements performed at room temperature on the $Fe_{70}Al_{30}$ alloy show that with ball milling a complete transition from the ordered alloy to the disordered one is obtained. The area of the (100) superstructure peak decreases with milling time, up to 5 h (see Fig. 9), when it disappears, and then only the disordered A2 phase can be distinguished. Fig. 9 shows a progressive lattice parameter increase that reaches a maximum after six milling hours. The maximum lattice parameter increase in the order-disorder transition amounts to 0.7%, which is in good agreement with the volume increase obtained after deformation in alloys in the range 27.5-35 at% Al [32].

DTA measurements show an exothermic peak at about 200 °C for samples milled for more than 1 h (see inset of Fig. 10). Figure 10 shows the evolution of the enthalpy, obtained from the integration of the exothermic peak, with milling time. The enthalpy increases up to 6 h of milling when it saturates. Taking into account the previous XRD measurements that show a complete transition from the ordered phase to the A2 disordered one for milling times larger than 5 h, the calorimetric peak is attributed to structural reordering of the sample. That is to say, the larger the area of the exothermic peak the larger the disorder of the ball milled sample. At 6 milling hours the enthalpy saturates; therefore, it can be concluded that after 6 milling hours a complete disorder is obtained (see Fig. 10). It is worth mentioning that in the 1-hour milled sample no peak can be distinguished at 200 °C.

Therefore, the calorimetric data, consistent with XRD data, indicate that the disordering process in the transition between the ordered alloys to the completely disordered one is a progressive one. With milling time the A2 phase starts forming (growing enthalpies) and it goes on increasing till the entire sample gets the disordered A2 structure.

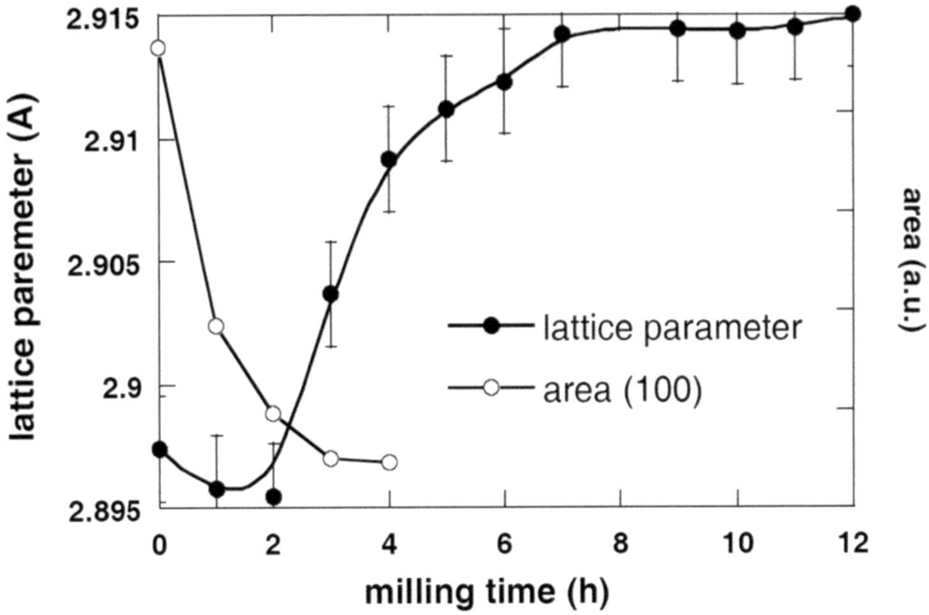

Fig. 9. (100) superstructure peak (empty circles) and lattice parameter (full circles) evolution with milling time. The solid lines are guides for the eye. Figure taken from reference [71].

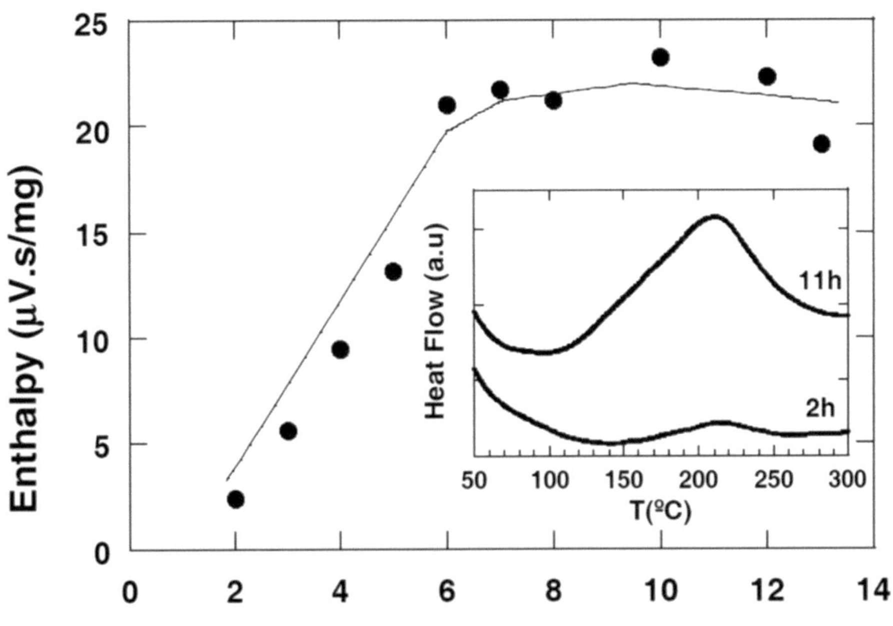

Fig. 10. Evolution with milling time of the enthalpy of the exothermic peak obtained around 200 C in the DTA (the solid line is a guide for the eye). The inset shows the calorimetric data obtained for samples milled for 2 and 11 h. Figure taken from reference [71].

Fig. 11 shows magnetization curves for different milling times [72]. The figure shows an abrupt magnetic property change from four to five milling hours, which it is also observed in Fig. 12 [72]. The magnetization value obtained at applied fields of 7 T measured at 10 K (see Fig. 12) indicates clearly that before complete disorder is obtained in the samples (X-rays indicate that this happens after five milling hours) magnetic saturation is not reached at 7 T. Once complete disorder is obtained, (above 5 h of milling time) magnetic saturation is reached at around 4 T. In addition, Fig. 12 shows clearly the increase of magnetization with disorder. The shape of the magnetization curves for few milling hours (less than 5 h) is similar to the one of the annealed sample; this indicates that there is still order in those samples. It is interesting to indicate that even though the XRD and calorimetric data indicate a monotonous increase of the disorder with the milling time, the magnetic data (see Figs. 11 and 12) show an abrupt change between 4 and 5 h of the milling time.

Figure 13 shows the Mössbauer spectra measured at room temperature for samples milled for different times. The annealed sample presents a large paramagnetic contribution superposed on a wide non-defined magnetic one. After the first milling hour, the paramagnetic contribution decreases and broadens and the wide magnetic contribution is yet to be defined. After 3 h of milling (even after 2 h) the magnetic contribution starts to be defined and a wide sextet-like contribution starts to appear superposed on a paramagnetic contribution. Therefore, in order to fit the spectra of Fig. 13 left a hyperfine field distribution, P(Bhf), was used. However, it is not until the samples are milled for 4 or more hours that the spectra show a clear sextet (see Fig. 13 right). Therefore, the fitting of the spectra of Fig. 13 right (4 or more milling hours) has been performed with a singlet and discrete sextets. Fig 14 shows, in the case of samples milled for less than 4 h the P(Bhf) obtained fitting the spectra. In the case of samples milled for 4 h or more, besides the discrete sextet fittings, in order to compare them with the ones performed for alloys milled during shorter time, we have made also fittings with hyperfine distributions in the spectra up to 5 milling hours (see figure 14).

In the annealed sample the P(Bhf) shows two main peaks at low fields (around 2 and 10 T) and a very small one at 28 T. Evidently, the peak located at 2 T must correspond to the paramagnetic contribution of the spectrum. After the first milling hour the pattern is similar but the P(Bhf) shows a clear evolution, although the first two peaks are still present, the peak at 2 T decreases quite significantly and a wide bump around 28 T appears (it is worth mentioning that in the calorimetric measurements no meaningful difference between the annealed and 1 h milled sample was obtained). The evolution continues monotonically with the milling time and already after 3 h of milling the main contribution to the spectrum comes from the bump centered at 28 T. The P(Bhf) of the spectra milled up to 3 h can be fitted using three Gaussians. However, three Gaussians are not enough to fit the P(Bhf) of the spectra with more milling hours, which is another indication of the magnetic order increase with milling time; and indicates that the proper fitting of those spectra must be made discretely.

As the value of each sextet depends on the environment of each iron atom, the discrete fitting of the spectra with 4, 5, 6 and 11 milling h (see Fig. 14) has been performed using six sextets and one singlet. Each sextet corresponds to Fe atoms that have a fixed number of Fe nearest neighbors in a bcc structure, like the disordered A2 one. Figure 14 shows that the peaks of the P(Bhf) built out of the discrete fit (each peak corresponds to a Gaussian that has

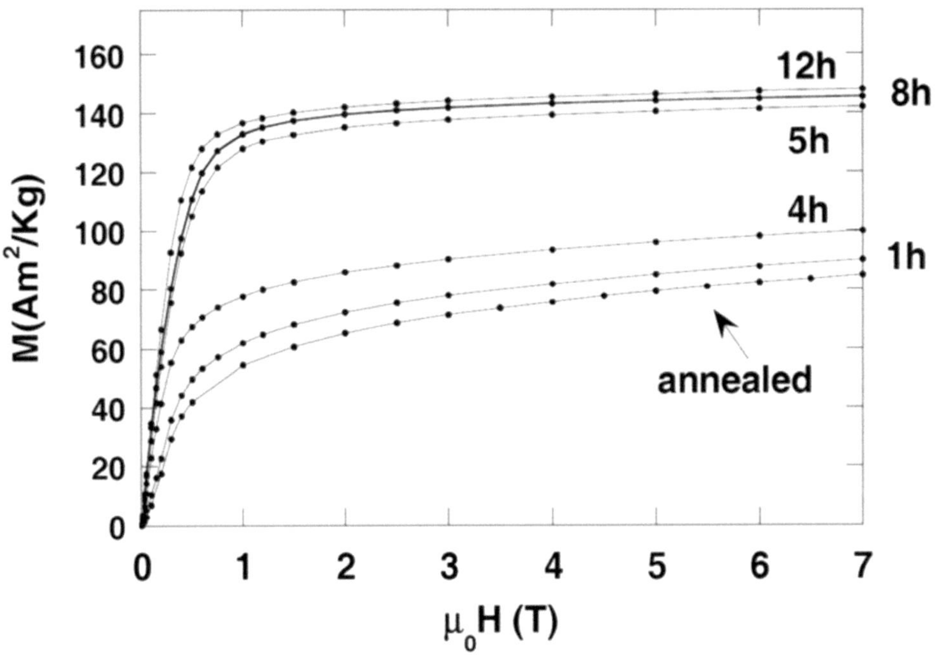

Fig. 11. Magnetization curves at 10 K for samples milled for different hours. Figure taken from reference [72].

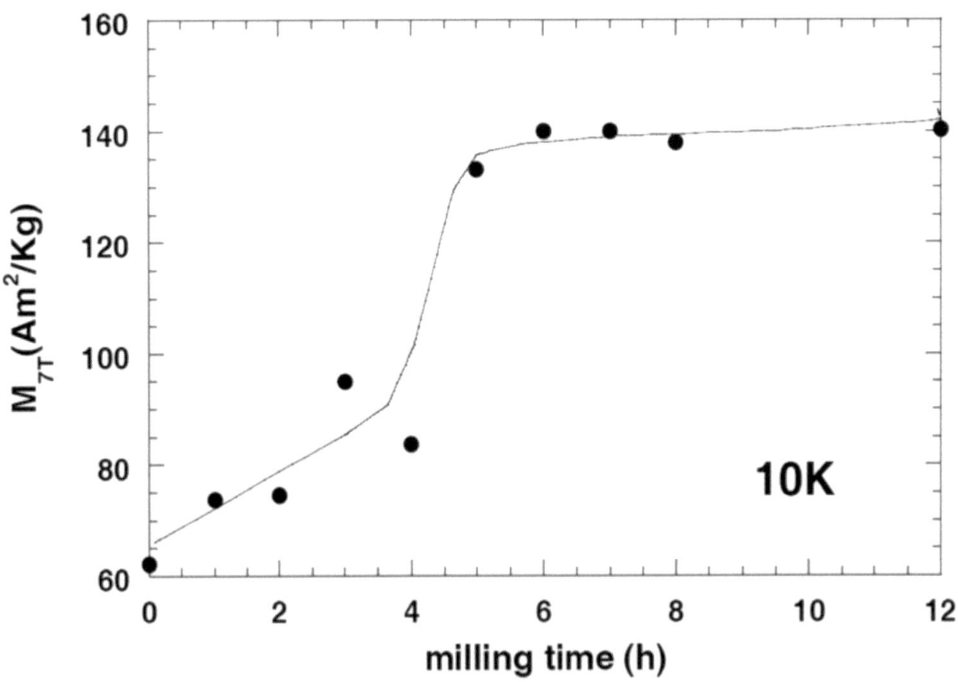

Fig. 12. Magnetization of $Fe_{70}Al_{30}$ alloy measured at 10 K and 7 T versus milling time.

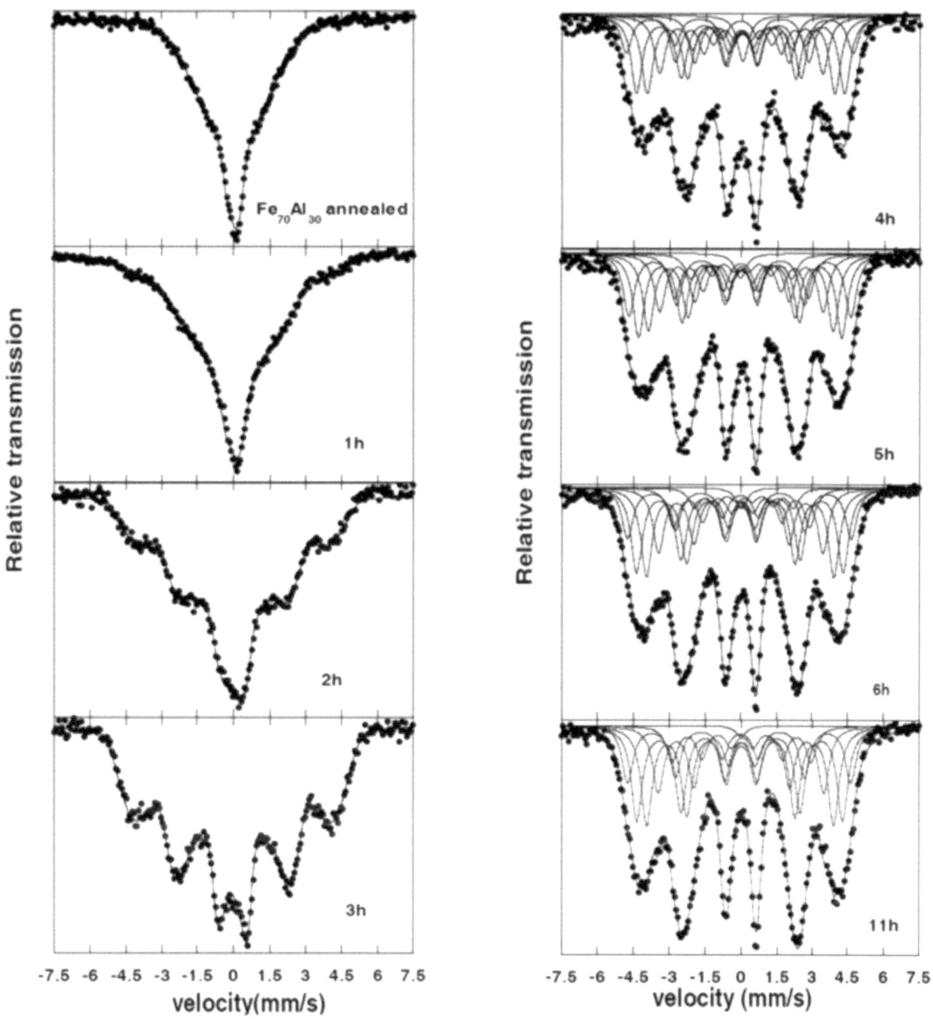

Fig. 13. Mössbauer spectra measured at RT in $Fe_{70}Al_{30}$ alloy after different milling time.

been built taking into account the width and area of each subspectra obtained from the discrete fits) correspond very well to the values inside the second and third bumps of the P(Bhf) distribution. Besides, Fig. 14 shows that the area of the peaks (corresponding to sextets), with Bhf lower (larger) than 20 T, decreases (increases) with milling time. Indeed, at long milling hours the peak around 10 T disappears. Therefore, the main magnetic contribution to the spectra in the annealed and short time milled samples disappears completely in the completely disordered state. On the other hand, the small magnetic contribution centered around 28 T, present in the annealed sample, seems to be the seed of the main contributions obtained in the completely disordered samples. Moreover, the area of the sextet used in the fitting of a completely disordered alloy (11 h of milling time) agrees quite well with the shape of the binomial distribution corresponding to the A2 structure of $Fe_{70}Al_{30}$[72].

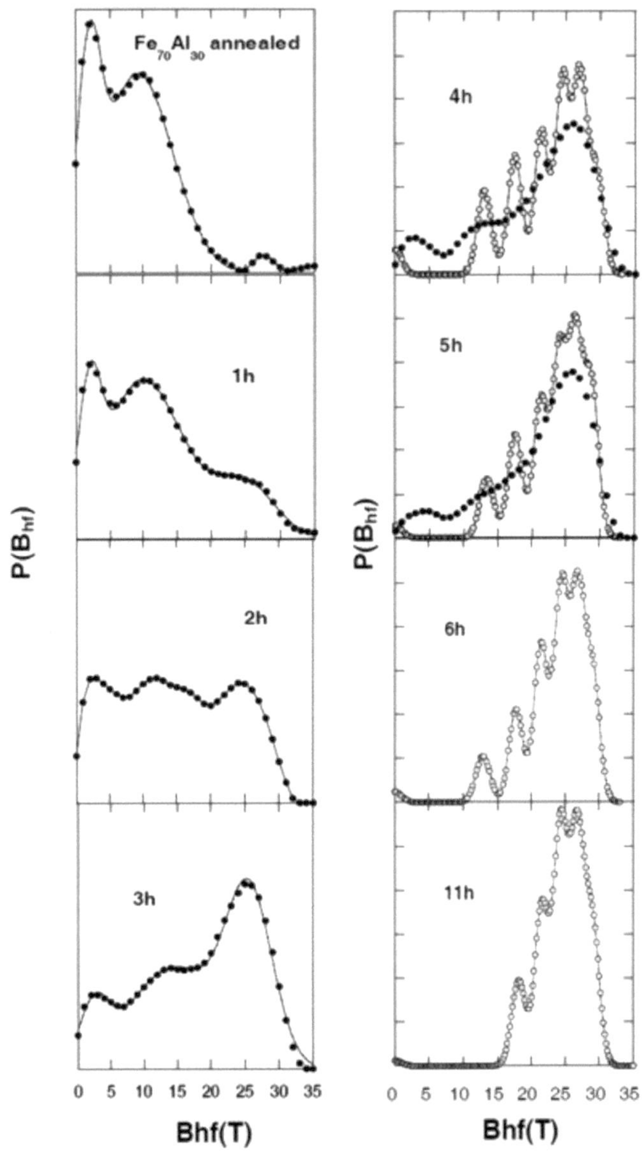

Fig. 14. (Full circles) P(Bhf) of the hyperfine field distribution obtained from the fit of Mössbauer spectra of Figure 11. (Empty circles) simulated P(Bhf) from the discrete sextets fit of the spectra (see text).

Summarizing, the different techniques show that the milling is a dynamical process. The ball milling causes the structural order-disorder transition in the sample and a lattice parameter increase that amounts to 0.7%. At the same time magnetic order is induced in the alloy.

The experiments performed indicate that the order-disorder transition is a monotonous process from the structural point of view that (in our conditions) this transition is accomplished completely after 6 milling hours. However, the magnetization shows an abrupt change at low temperature before complete disorder takes place. The Mössbauer spectra suggest that the main contribution of the weak magnetism of the ordered sample disappears and that the small contribution is the seed of the enhancement of the magnetism in the order-disorder transition. Moreover, the area of the sextet used in the fitting of the completely disordered alloy agrees with the shape of the binomial distribution corresponding to the A2 structure of $Fe_{70}Al_{30}$.

4. Reordering of disordered alloys

In the former subsection the effect of mechanical deformation on the magnetic properties of Fe-Al alloys has been shown. The overall effect of the deformation on the sample was the induction of a disordered state that follows the binomial distribution, corresponding to the A2 disordered structure, and volume expansion. The amount and type of deformation can determine the change on the magnetic signal of the sample, even causing a transition between paramagnetic and ferromagnetic behaviors (or from paramagnetic to ferromagnetic state). However, this transition is metastable and the former state can be recovered just by heating the sample. The processes involved are called *recovery and recrystallization processes* [73]. As the deformed state has locally many defects and each kind of defect has its own thermal activation temperature, this process is not monotonous and it usually occurs in several stages. When appropriate temperature is reached for some type of defect, the mobility of the defects is activated and they migrate to the surface where they disappear. It is evident that the smallest defects, like point defects, need smaller energy to migrate; therefore, these are the type of defects that are going to be removed first when the sample is heated. Larger defects, like planar defects (antiphase boundaries (APB)), will need higher energy to be moved, and their activation temperature will be higher. The study of these processes and its influence on the magnetic properties of the studied alloys can be important to identify which defects are important in the change on the magnetic properties, and more particularly, in the case of Fe-Al alloys.

The studies found in literature show that in the case of the Fe-Al binary alloy system two stages have been found for alloys in the B2 phase field independent of the type of deformation applied to the samples: cold rolled [26, 74], ball milled [14, 50, 54] or crushed (ours). The first stage takes place around 400-500 K. During this stage a peak has been observed in calorimetric measurements [26, 54, 71], and neutron diffraction patterns [75] show a nucleation of new small ordered domains with B2 structure accompanied by a decrease of the lattice parameter. The second stage takes place between 600-700 K when the defects introduced during deformation disappear completely. In this stage, the B2 domains present in the alloys start to grow until all the strains in the sample are released by annealing [75].

In the case of alloys in the D03 field of the phase diagram, the recovery process occurs through the creation of transient B2 phase. In fact, the two stages are the same as in the case of D03 [76]: nucleation of small B2 domains in the first stage and growth of the existing B2 domains in the second one. However, in this case growth of the existing D03 domains takes place immediately after the second stage a domain.

In both cases (B2 and D03), the recovery processes follow the phase transformations described in the classical work by Allen and Cahn [77].

The influence of these processes is very strong due to the recovery of the intermetallic order. Fig. 15 shows the magnetization as a function of temperature for different mechanically deformed Fe-Al alloys with an applied field of 0.15 T. In the case of alloys in the B2 field of the phase diagram (35 at. % Al), the magnetization drops to zero in a single stage, and this occurs around 400-500K. This large drop of magnetization takes place simultaneously with the first stage of the recovery process, described above [54, 75, 76]. After this, the contribution of the remaining deformations in the alloy is negligible.

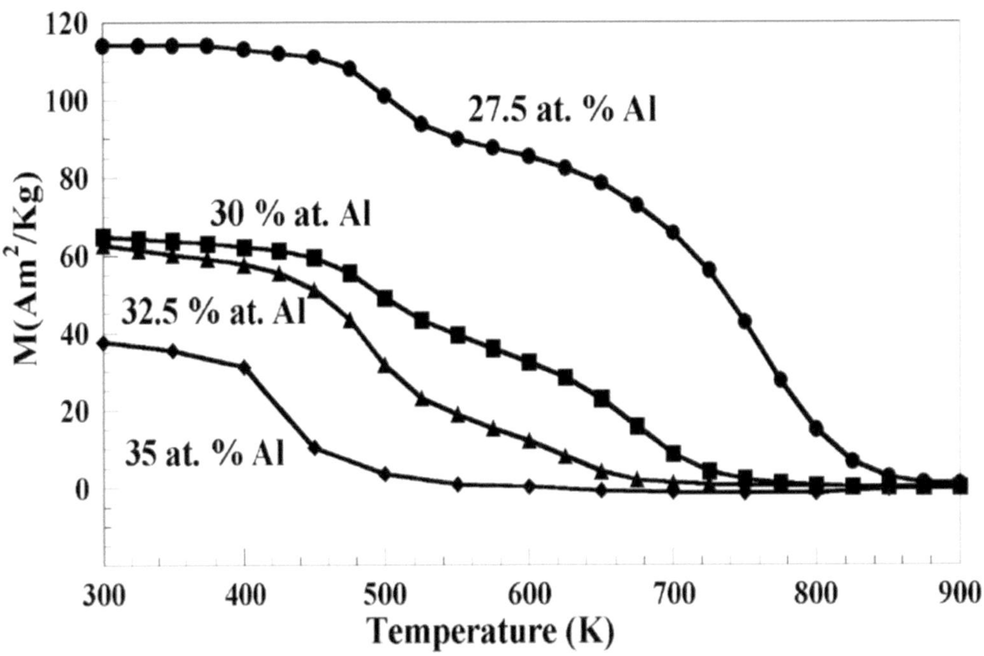

Fig. 15. Magnetization as a function of temperature of mechanically deformed alloys with an applied field of 0.15 T. The deformation of the alloys is explained in refs. [75] and [76].

Mössbauer spectroscopy measurements also show the same image, as can be found in figure 16. This figure confirms that after the first stage there is no magnetic contribution [75, 76]. In the case of alloys with D03 structure the magnetic evolution is different. There is a marked drop in magnetization around 400-500 K, however this time the magnetizaton does not drop to zero. Mössbauer spectroscopy measurements also show magnetic contributions after this stage in the recovery process. The magnetization in alloys with D03 structure will drop to zero around 800 K.

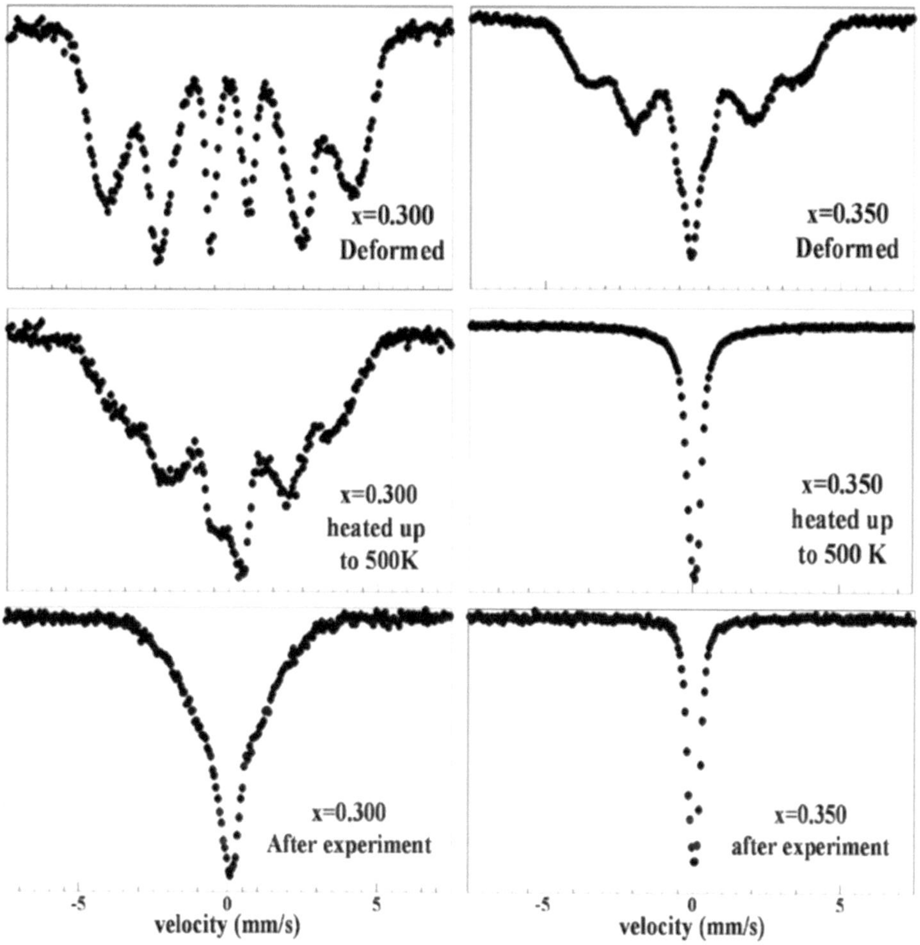

Fig. 16. Mössbauer spectra of alloys with 30 (left) and 35 (right) at. % Al in different situations: (top) mechanically deformed, (middle) heated up to 500 K (i. e., after the first stage in the recovery process), (bottom) heated up to 900 K (i. e., after the second stage in the recovery process). From references [75] and [76].

According to the process described above, the first stage in the recovery process of alloys with B2 and/or D03 can be attributed to removal of some kind of defect that causes an increase in volume of the lattice. Furthermore, this kind of defect also causes a strong increase in the magnetic signal. In literature this stage has been attributed to the removal of point defects [54, 76] like vacancies, and this is plausible, as the energy necessary to move this kind of defects is low. On the other hand, studies performed on cold rolled alloys [10, 74, 77] attribute this stage to APB tube [79, 80] removal. However, these two explanations are compatible since the APB tube removal takes place by means of vacancy migration [26]. The second stage in the recovery process, on the other hand, must be due to removal of larger defects. In literature, this has been attributed to removal of planar defects [54] and more specifically to superdislocations [26]. It is also important to remark that according to the work on cyclically deformed alloys performed by Yasuda et al. [81], the main contribution to the magnetic signal comes from APB tubes and superdislocations' contribution is negligible. In the case of alloys with D03 structure, there is still a strong magnetic signal remaining in the alloys after the first stage. As stated above, the superdislocations on B2 domains have no significant magnetic contribution, and therefore this remaining magnetic signal must come from the existing D03 domains. However, these D03 domains must be deformed because the Curie temperature of the alloy with 30 at.% Al is around 500 K and at that temperature the signal is still strong.

All the processes described allow us to study the effect of the Fe local environment and the Fe-Fe interatomic distance on the magnetic properties of Fe-Al alloys. Up to now, this influence for Fe-Al alloy system was focused on the magnetic moment, but taking into account the spin-glass properties, found in this alloy system (see section 2), there must exist an influence on the competition between ferromagnetic and antiferromagnetic exchange interactions. This influence on the exchange interactions has been used to explain the Invar effect on Fe-Ni alloys [82] and has been proven in Fe-R (R=rare earth) alloy systems [83].

This can explain the fact that for B2 alloys, the strong drop of magnetization in the first stage of recovery takes place due to the decrease in volume, because after this stage, the intermetallic order is not completely recovered and its magnetic contribution is negligible. This indicates that in the case of alloys in the B2 phase field of the phase diagram, the change in volume contributes in a higher degree than the change in Fe local environment. However, in the D03 alloys, the magnetic contributions do not disappear after the first stage. The only difference in this case is the presence of domains with D03 structure. In this case, as the D03 structure has more Fe-rich local environments, the changes in Fe local environment due to the deformation will lead to Fe-richer Fe-local environments. Therefore, both changes in Fe-Fe interatomic distance and in Fe local environment have an important influence.

In summary, the studies of the recovery process in deformed Fe-Al alloys demonstrate that both the Fe local environment and the Fe-Fe interatomic distance can be responsible for the origin of magnetism in Fe-Al alloys.

5. Theoretical calculations

Because of the great difficulties in understanding these compounds and the interesting properties they show, during the last years the band calculation has been used to study this system.

The electronic structure of magnetic transition-metal (TM) aluminides with stoichiometric composition has already been studied many times by various methods and in different approximations. In the earliest calculations within the Korringa-Kohn-Rostoker (KKR) [84] and modified KKR methods [85-87] the problem of filling up of the transition metal (TM) d-bands by Al p-electrons was discussed in detail and the charge transfer from Al to a TM site was shown. The trends in the chemical bonding and the phase stability of transition metal aluminides with equiatomic composition have been studied with the full-potential linearised augmented plane-wave (FLAPW) method [88]. A review of electronic structure calculation results, along with band structures, densities of states and Fermi surfaces of many TM aluminides can be found in ref. [89]. Another study using the full-potential linearized augmented Slater-type orbital method [90] reports the formation energies and equilibrium volume of many 3d aluminides. In addition, cohesive, electronic and magnetic properties of the transition metal aluminides have been calculated using the Tight Binding Linear Muffin Tin Orbital (TB-LMTO) method [91] where it was found that FeAl retains its magnetic moment. These findings coincide with the results of earlier Linear Muffin Tin Orbital (LMTO) calculations for NiAl and FeAl intermetallic compounds [92]. In all these calculations for the $Fe_{50}Al_{50}$ stoichiometric composition a magnetic moment was found, but Mohn et al. [93] found a non-magnetic ground state for this alloy and composition using corrected Local Density Approximation (LDA+U).

On the other hand, there are relatively few calculations aimed at the study of the influence of defects on the electronic and magnetic structure of TM aluminides. The LMTO method has been applied to study the electronic structure of antisite (AS) defects in FeAl where point defects were modeled by suitably chosen supercells [94]. Finally, the Linear Muffin Tin Orbital Coherent Potential Approximation (LMTO-CPA) technique has been used to discuss the order-disorder transition in FeAl alloys [29]. The supercell approach has been used in order to study the antiphase boundary in NiAl and FeAl [95], as well as point defects in these aluminides [96]. The onset of magnetism in Fe-Al system as a function of the defect structure was studied using the CPA within the KKR method for the disordered case and the TB-LMTO for the intermetallic compound [27], where they found appearance of large local magnetic moments associated with the Fe antisite defect.

In our work we mainly performed calculations based on TB-LMTO in order to study these alloys, and the results we obtained are useful in giving an idea of the general trend of the magnetism in these alloys. Nowadays, the Vienna Ab-Initio Simulation Package (VASP) with a wide choice of potentials is commonly used and this is why we decided to perform some tests with this method in order to see what results could be obtained. This latter method is more time consuming, which is a disadvantage when making supercell calculations; however, it allows introducing vacancies as well as ion relaxation in a simple way and it could be useful for the study of these alloys.

In the previous sections we have studied experimentally the cause of the large change in the magnetic properties with structural transition and the increase of the magnetic signal with the temperature observed in ordered $Fe_{70}Al_{30}$ sample. Theoretical calculations are very useful in order to understand the mechanism that leads to these changes of magnetic properties.

We study ordered (A2, B2, D03, B32) and disordered (B2) structures present in the Fe-Al phase diagram for different Fe compositions. Moreover, the influence of disorder on the magnetic properties of the Fe-Al system is also studied. Details on these calculations are given in ref. [28].

We calculated the lattice parameter, the total energy, the cohesive energy, the density of states (DOS), the magnetic moment and other interesting data that will be shown in this work. We performed the calculations both for non-polarized and spin-polarized cases. First, we got the convergence for non-polarized calculations and then we divided the moments and potentials into two, one belonging to each spin, and we made the calculations again for these new values. It is worth mentioning that, as shown in figure 17, in most of the structures studied the spin-polarized cases showed a lower energy, and therefore it can be said that this system is magnetic for most of the studied cases.

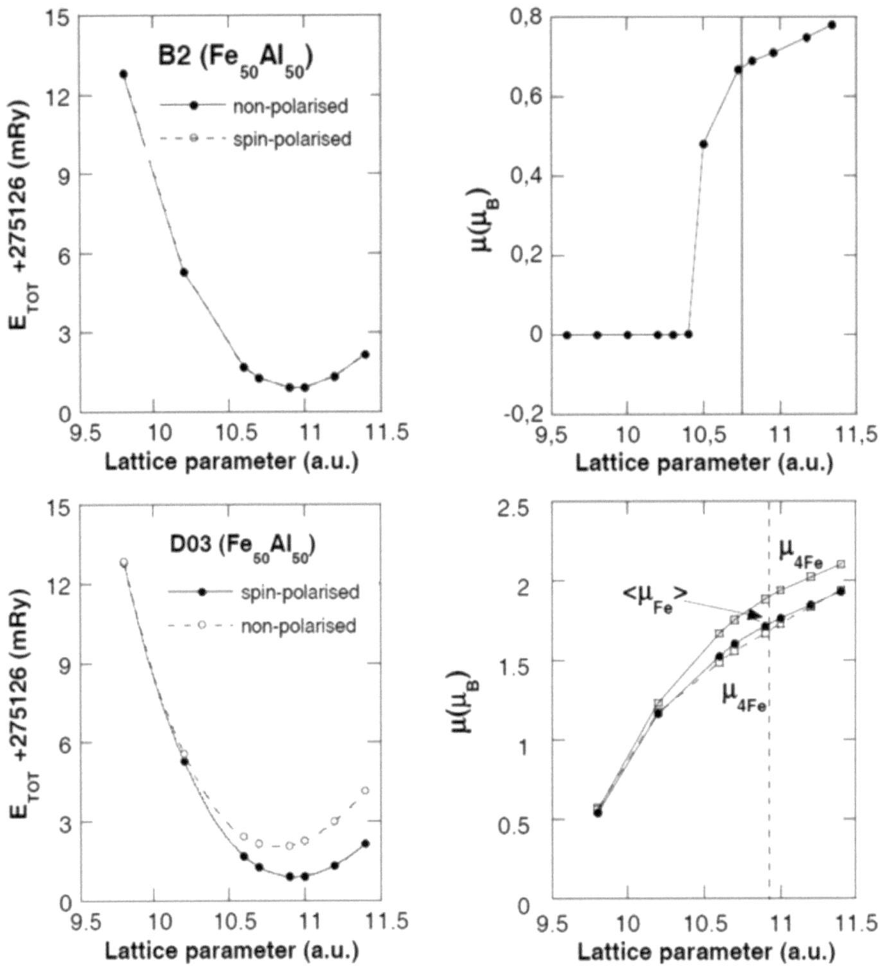

Fig. 17. (a) Energy and (b) magnetic moment evolution with lattice parameter. The experimental lattice parameter for $Fe_{50}Al_{50}$ alloy is almost 1% larger than the one calculated theoretically in this work, in agreement with the general experience that local density approximation (LDA) underestimates the lattice parameter (see Table II). Figure taken from reference [28].

5.1. Ordered structures

Three different ordered structures have been studied B2, D03 and B32 for different compositions. We first analyzed the $Fe_{50}Al_{50}$ thoroughly and used these conclusions to explain the results of the other compositions. Also, the study helps in understanding the influence of the nearest neighborhood of Fe sites on its magnetic properties in different structures.

Table I shows the results obtained and table II presents a comparison between the experimental values of $Fe_{75}Al_{25}$ and $Fe_{50}Al_{50}$ compositions with the theoretical ones obtained by different authors for the structures and compositions shown in the phase diagram.

The calculations performed for the B2 structure and $Fe_{50}Al_{50}$ composition are in very good agreement with the results that appear in the literature [27, 90, 91, 97].

Composition	Structure	a (a.u.)	B (Gpa)	$\mu_{Fe}(\mu_B)$	E_{coh}(eV)	E_f (eV)	E_{min}(eV)
$Fe_{50}Al_{50}$	**B2**	10.74 (10.77)	200	0.64	-5.91	0.462	-3743.302 (-3743.301)
	D03	10.93 (10.82)	173	1.71	-5.66	-0.204	-3743.258 (-3743.240)
	B32	10.91 (10.82)	44	1.67	-5.73	-0.286	-3743.270 (-3743.244)
$Fe_{75}Al_{25}$	**B2**	10.6 (10.50)	192	1.07	-6.25	-0.245	-4086.982 (-4086.966)
	D03	10.65 (10.51)	176	1.86	-7.62	-0.252	-4086.984 (-4086.969)
	B32	10.71	190	1.97	-6.18	-0.171	-4086.913
$Fe_{81.25}Al_{18.75}$	D03	10.63 (10.47)	197	1.90	-6.31	-0.177	-4153.892 (-4153.871)
$Fe_{87.5}Al_{12.5}$	**B32**	10.62 (10.43)	208	2.02	-6.38	-0.180	-4215.130 (-4215.101)

Table I. Summary of the results obtained for the ordered structure, where a is the lattice parameter, B the bulk modulus, μ_{Fe} the mean magnetic moment per Fe atom, E_{cohe} the cohesive energy for the spin-polarized calculations, E_f the formation energy for the spin-polarized calculations, E_{mim} is the minimum energy obtained. The numbers in brackets correspond to the energy minimum for the non-polarized calculations.

It is found that $Fe_{50}Al_{50}$ retains a magnetic moment of $0.64\mu_B$, even though the phase diagram of this system shows that this alloy is not magnetic at room temperature [12]. The energy for the spin-polarized calculations has a lower value than for the non-polarized ones; however, it is within the error (~0.01eV) that can be obtained with this method; therefore, taking the results of these calculations into account we cannot conclude whether this structure is magnetic or not. It is worth mentioning that this is not the first time that theoretical calculations predict a magnetic moment for this structure and composition; most of the literature shows the same theoretical results that go against experiments in this $Fe_{50}Al_{50}$ [27, 30, 97, 98]. Moruzzi and Marcus [98] seeing that the energy difference was so small used another criteria to determine whether this alloy for this structure and composition was magnetic or not. They calculated the bulk modulus and compared it with the experimental results and they got to the conclusion that this alloy was ferromagnetic. Kulikov et al. [27] noticed that since the energy difference between the ground states was so small one could conclude that the formation of a spin-glass state was possible in a Fe-Al system. Bogner [30] explained this discrepancy between experiment and calculations by stating that the high density of defects, found in real systems, destroys the atomic periodicity, and this lack of periodicity could cause a decrease of the magnetic moment. In contrast to these results Mohn at al. [93] used LDA+U approximation and got a non-magnetic $Fe_{50}Al_{50}$ stoichiometric alloy.

	$Fe_{50}Al_{50}$		$Fe_{75}Al_{25}$	
a_{exp} (a.u.)	[98][12][99][100][87]	5.309–5.495	[101] [99]	10.945 10.926
a_{theo} (a.u.)	[90][91][27][98][101]	5.330–5.398	[99]	10.8
present work (LDA)	5.37		10.65	
present work (P-W)	5.49		11.00	
B_{exp} (Gpa)	[100] [27]	152 150		
B_{theo} (Gpa)	190– 205 [91][102][98][101]			
present work (LDA)	200		176	
present work (P-W)	171		162	
μ_{exp} (μ_B)	[91]	0	[103] μ_{Fe1}=1.46	μ_{Fe}=2.14
μ_{theo} (μ_B)	[91][98][30]	0.69–0.71	[30] μ_{Fe1}=1.9	μ_{Fe}=2.25
present work (LDA)	0.64		μ_{Fe1}=1.63	μ_{Fe}=2.23
present work (P-W)	0.76		μ_{Fe1}=2.05	μ_{Fe}=2.48
$E^{cohe}{}_{exp}$ (eV)	[100]	-3.58		
$E^{cohe}{}_{theo}$(eV)	[91] [101]	7.66 -5.91		
present work	-5.90		-7.62	
$E^{f}{}_{exp}$ (eV)	[104] [105]	-0.26 -0.33		
$E^{f}{}_{theo}$ (eV)	[91][94][88]	-0.32 – -0.51	[27]	-0.22
present work	-0.46		-0.25	

Table II. Comparison of the obtained results with previous experimental and theoretical results.

Figure 17b shows the variation of the magnetic moment versus the lattice parameter in the B2 structure of $Fe_{50}Al_{50}$. In this structure 8 Al atoms surround the Fe atoms and there is a charge transfer between Al and Fe atoms, which makes the distribution of the density of states change with lattice parameter. Next to the equilibrium lattice-parameter there is a jump (from a low to a high moment state) of the mean magnetic moment per iron atom that goes from zero to around 0.6μ_B. This phenomenon can be explained by taking into account the DOS (see Fig. 18) and the difference between the majority- and minority-spin sub-bands.

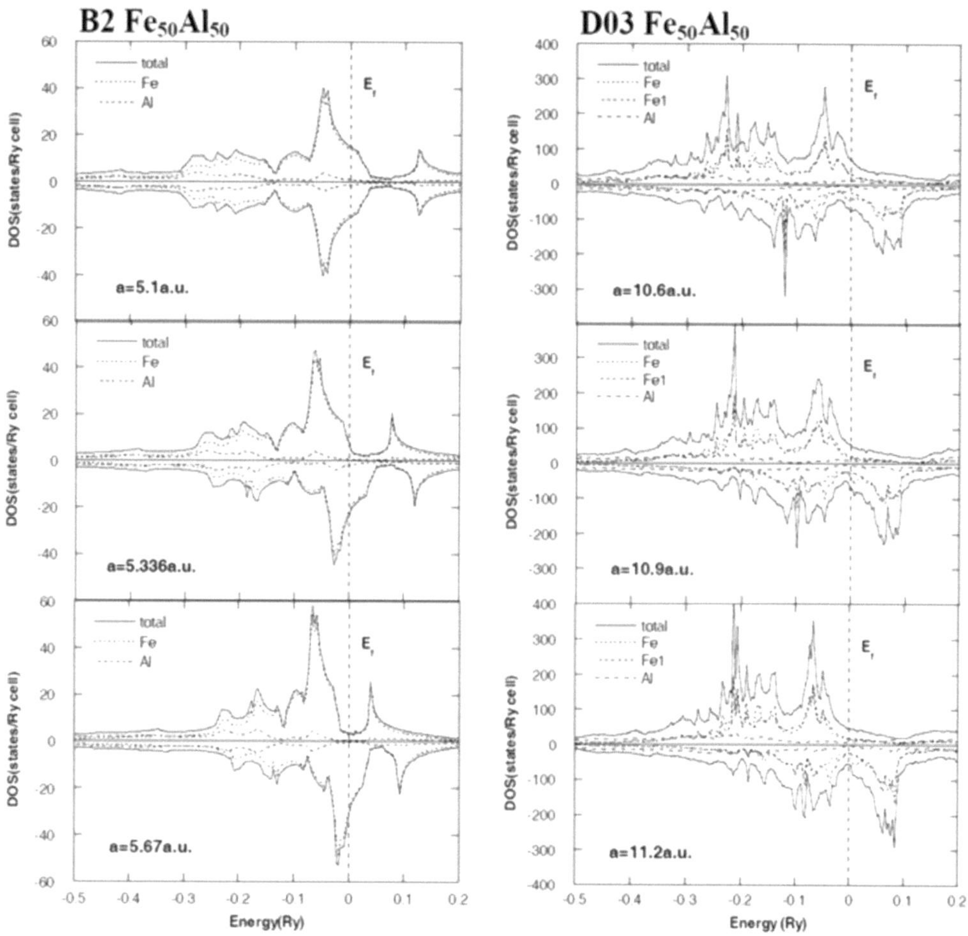

Fig. 18. Density of states calculated for three different lattice parameters of B2 and D03 structures corresponding to $Fe_{50}Al_{50}$ composition. Figure taken from reference [28].

The hybridization causes a charge transfer between the majority- and minority-spin sub-bands. This decreases the difference between the number of occupied states and consequently weakens the magnetism of the alloy.

If we take a closer look at the density of states versus energy (see Fig. 18), for this structure and this composition, the minority- and majority-spin sub-bands are very similar- they show two peaks separated by a large gap. For small lattice parameters both sub-bands are identical in shape and occupation and the main peak is not completely full. However, as the lattice parameter increases, owing to the decrease of hybridization, there is a charge transfer from the minority to the majority-spin sub-band, that makes the peak of this last band fill up and the one corresponding to the minority-spin sub-band empty. Therefore a magnetic moment appears.

For the D03 structure and $Fe_{50}Al_{50}$ composition a larger difference between the non-magnetic and magnetic states is obtained and we can state that for this structure and this composition the alloy is magnetic with an average magnetic moment of 1.71μ_B (see Table I). The minimum of the total energy is a little bit higher than the one obtained for the B2 structure, which is in agreement with the fact that B2 is the equilibrium phase for this composition.

The behavior of the average magnetic moment with varying the lattice parameter is completely different from the one shown previously (see Fig. 17, D03 structure). It increases monotonically from 0.5μ_B to 1.9μ_B, without showing any jump with the increase of the lattice parameter. There are two non-equivalent Fe atoms, which have the same next-nearest-neighborhood (4 Al and 4 Fe atoms) and show also a continuous increase of the magnetic moment with the lattice parameter. The magnetic moment of one of the Fe atoms in the equilibrium lattice-parameter is 2 % larger than for the other. This small difference could be attributed to the 2nd nearest neighbor environment; the one with the larger magnetic moment has 2 Fe and 4 Al atoms as 2nd nearest neighbors while the other has only Al atoms as 2nd nearest neighbors.

This behavior can be explained by taking into account the density of states, which is completely different from the one shown before. In this case, the majority-spin sub-band is almost full even for small lattice parameters and the Fermi energy is in the gap of the minority sub-band. As in the previous case, the lattice-parameter change causes a change in the hybridization of the sp-Al and d-Fe majority- and minority-spin sub-bands that induces a charge transfer between the minority-spin sub-band and the majority-spin sub-band that makes the magnetic moment increase with lattice parameter. However, the changes induced in the magnetic moment will be small because of the majority-spin sub-band is almost full even for small lattice parameters. (see Fig. 18).

For the B32 structure and $Fe_{50}Al_{50}$ the difference between the energy of the polarized and not polarized calculations tells us that this structure is magnetic for this composition (see Table I) with a mean magnetic moment that increases monotonically with the lattice parameter. The very low bulk modulus (44 Gpa) is related to the fact that this structure does not exist for the studied composition, and it indicates that the actual compositions with this structure are far from $Fe_{50}Al_{50}$. This is consistent with the phase diagram where this phase does not exist for this composition.

The results for $Fe_{75}Al_{25}$ are summarized in Table I. We can say that the D03 structure has the lowest energy in agreement with the phase diagram that shows this phase for this composition.

For higher Fe contents, two different concentrations in the Fe-richest side of the phase diagram were chosen and two cells that fulfill the DO3 and B32 structures conditions were built (see Table I). The calculations indicate that the stable structures are the stoichiometric ones, i.e. $Fe_{75}Al_{25}$ for the DO3 and $Fe_{50}Al_{50}$ for the B2 structure and this is in agreement with the phase diagram [106]. On the other hand, it has been shown that the bulk modulus for the B32 ($Fe_{50}Al_{50}$) structure is very low, but as the concentration of Fe grows, so does the the bulk modulus, and its values are closer to the values measured in the phases corresponding to this composition. This is a clear indication that the B32 structure can exist only for high Fe content compositions far from $Fe_{50}Al_{50}$ composition.

Summarizing, it can be concluded from the analyzed structures that in the previous composition the density of states has a really large importance for the behavior of the magnetic moment. When the majority subband is completely full the magnetic moment is high. This is due to the hybridization that induces intraband charge transfer in these alloys. This is also the reason why abrupt DOS changes with lattice parameter cause magnetic moment jump from low to high moment state.

It is interesting to notice that the behavior of the non-equivalent Fe positions with the lattice parameter is quite similar when they have similar environment, independently of the structure under which the calculations are performed.

The calculations indicate that the number of Fe atoms in the nearest neighborhood of one Fe atom, which are needed to cause an abrupt jump of its magnetic moment (with the variation of the lattice parameter) increases as the Fe-content in the alloy increases. In addition, table I indicates that the equilibrium lattice-parameter decreases with the increase of Fe content in the alloy. As expected, the magnetic moment increases with Fe content.

5.2 Disordered structure (A2)

As we have seen in the previous section most of the theoretical work done to study the reinforcement of magnetism in these alloys by disordering has been performed assuming point defects [90, 96, 107] and antiphase boundaries [27]. Nevertheless, several articles have studied, by self-consistent methods, the disordered structures for compositions near the equiatomic B2 one using the Coherent Potential Approximation (CPA) [29, 97]. Kulikov et al. [27] found magnetism in the range of disordered alloys studied, but contrary to the experimental results, they found a decrease of lattice parameter with disordering in all the studied composition range.

Taking into account that X-ray diffraction of severe cold deformed (mechanically milled) FeAl alloys shows diffraction peaks corresponding to the A2 structure [108, 109], and that this structure also appears in samples prepared by rapid quenching from the melt, we have simulated the disorder in FeAl alloys by means of the A2 structure. On the other hand, the magnetic properties depend strongly on the local environment, which at the same time depends on the chosen cell. Therefore, in order to make a good approximation, the average of seven different A2 supercells for $Fe_{50}Al_{50}$ composition and seven different A2 supercells for $Fe_{75}Al_{25}$ composition have been used to compare theoretical and experimental results.

The spin-polarized calculations have a lower energy than the non-polarized ones (not shown) and therefore, we can conclude that for these structures and compositions these alloys are magnetic in all the different cells built, independently of the Fe content ($Fe_{50}Al_{50}$ or $Fe_{75}Al_{25}$). Table III shows that the equilibrium lattice parameter decreases with the increasing Fe content in the alloy. The values of the lattice parameter obtained (see Tables II and III) underestimate by less than 3% the experimental value obtained by Frommeyer et al. [101] in deformed or ball milled samples of $Fe_{70}Al_{30}$ that possess the A2 structure. It is worth mentioning that this underestimate is of the same order as the one found between the $Fe_{75}Al_{25}$-D03 theoretical and experimental values.

Composition	Structure	a (a.u.)	$\mu_{Fe}(\mu_B)$	E^{pol}_{min}(Ry)
$Fe_{50}Al_{50}$	**B2**	10.74	0.64	-275.1255
	A2	10.96	1.75	-275.1210
$Fe_{75}Al_{25}$	**D03**	10.65	1.86	-300.3854
	A2	10.72	2.01	-300.3842

Table III. Comparison of the results between ordered and disordered structures.

Table III clearly shows that the mean magnetic moment corresponding to the structure with $Fe_{75}Al_{25}$ composition is larger than the one for $Fe_{50}Al_{50}$ composition. It also indicates that the calculated energy minimum in the disordered structures (A2) is above the ones corresponding to the ordered phases. This is in agreement with the fact that in the calculated ranges of compositions the stable structures are the ordered ones. For $Fe_{50}Al_{50}$ composition, the B2 structure has the lowest energy and this is the structure that appears in the phase diagram for this composition.

The equilibrium lattice-parameters for the calculated A2 structures are larger than the corresponding ordered (B2 and DO3) ones. This is again in good agreement with X-ray diffraction observations after severe deformation of alloys of similar composition, where the lattice parameter increases with deformation of the alloy. However, it disagrees with the calculations performed under the KKR-CPA approach [27]. Moreover, it must be mentioned that the theoretical increase of the lattice parameter between the A2 and the D03 structures of $Fe_{75}Al_{25}$ is 0.75%, and is in very good agreement with the experimental increase after deformation of about 0.7% found for $Fe_{70}Al_{30}$ [101; see section 3]. In the case of $Fe_{50}Al_{50}$ the lattice parameter increases by 2% due to disorder. That is to say, the results obtained are in good agreement with the experimental ones.

Table III shows that the magnetic moment per iron atom increases with increasing disorder. This is specially pronounced in the case of $Fe_{50}Al_{50}$ (B2), where the magnetic moment increases more than 1 μ_B after disordering. These theoretical results are in agreement with the experimentally observed increase of the magnetism in deformed or ball-milled alloys [13, 14, 84; see section 3], which, as previously cited, also show an increase in the lattice parameter with any type of deformation.

Figure 19 shows the density of states with respect to the energy for an ordered and a disordered structure of $Fe_{50}Al_{50}$ composition. Owing to the lower lattice parameter for the B2 structure, the hybridization is larger than for the disordered structure and therefore the difference between the two bands is larger in the disordered case, which causes the magnetic moment to be higher. In addition to this, the nearest neighbor configuration of each iron atom (in the disordered structure not every iron atom is surrounded by 8 Al atoms, as it happens in the B2 $Fe_{50}Al_{50}$) favors a larger magnetic moment.

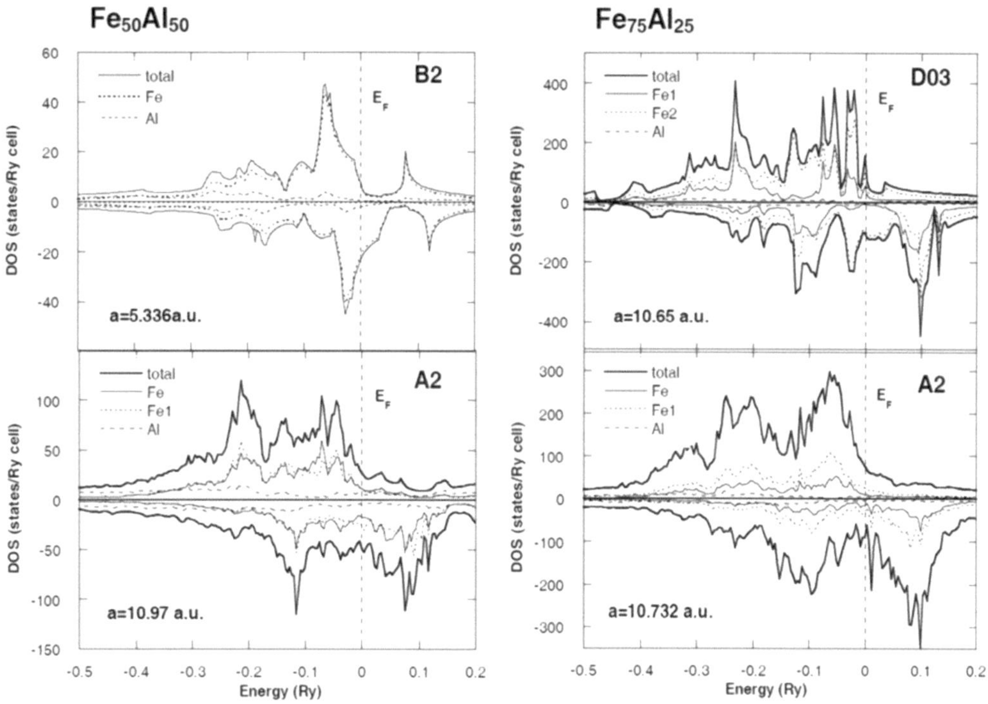

Fig. 19. Comparison of the DOS of ordered and disordered structures at the equilibrium volume for two different compositions.

Hernando et al. [14] found that an important contribution to the magnetism of these alloys comes from changes in the lattice parameter induced by the order-disorder transition. This contribution is linked to modifications in the electronic band structure induced by volume changes. In the present work, it was clearly shown, through the study of the lattice-parameter dependence of the magnetic moment, that this effect is important in a variety of structures and compositions that were studied. However, our results indicate that the disorder has also a large contribution. In the following table we have calculated the weight of each contribution for two compositions (see Table IV),

Composition	Structure	Weight (%)	
		Volume increase	Order-Disorder
$Fe_{50}Al_{50}$	B2	13	87
$Fe_{75}Al_{25}$	D03	45.5	54.5

Table IV. Average value (of the different ways to obtain theoretically the contributions, see [68]) of the weight of each of the contributions to the magnetic moment increase in the disordering process for two different concentrations.

It is interesting to notice that the values presented in table IV match the experimental values obtained by the XMCD experiments presented in section 3.1, where in the case of the disordered $Fe_{60}Al_{40}$ alloy the volume change contribution amounted to 35±5%.

Summarizing, the comparison between calculations performed in ordered and disordered structures of the same composition indicates that the disorder makes both the lattice parameter and the magnetism increase in comparison to the ordered structures. Indeed, the lattice parameter increase with disorder for $Fe_{50}Al_{50}$ and $Fe_{75}Al_{25}$ alloys is in good agreement with experimental results. The contribution of disorder to the magnetism of these alloys depends on the Fe content of the alloy. The disorder gives the largest contribution close to the equiatomic FeAl alloy, but in $Fe_{75}Al_{25}$ alloy its contribution is similar to the one given by the volume change.

6. Acknowledgments

We want to acknowledge the financial support from the UPV/EHU, Basque Government and the Spanish Government under Grants No. IT-443-10 and No. MAT2009-14398.

7. References

[1] J.H. Westbrook, R.L. Fleicher, Intermetallics compounds Vol. I: Principles, Vol. 2: Practice, New York, Wiley (1995) 4
[2] G. Sauthoff, Intermetallics, Werthaeim:VHC (1995)
[3] J.C. Wang, D.G. Liu, M.X. Chen, X.X. Cai, Scripta Metallurgica 25 (1991) 2581
[4] A. Taylor and R. M. Jones, J. Phys. Chem. Solids 6, 16-37 (1958)
[5] A. Arrot and H. Sato, Phys. Rev. 114, 1420 (1959).
[6] P. Huffman and R. M. Fisher, J. Appl. Phys. 38, 735 (1967)
[7] I. Vincze, phys. Stat. Sol. (a) 7 (1971) K43
[8] M.J. Besnus, A. Herr and A.J.P. Meyer, J. Phys. F: Metal Phys. 5 2138 (1975)
[9] D.A. Eelman, J.R. Dahn, G.R. Macklay, R.A. Dunlap, J. Alloys and Compounds 266 (1998) 1
[10] P.A. Beck, Metallugical Transactions (AIME) 2, 2015 (1971).
[11] L. Hedin, B.I. Lundqvist, J. Phys. C4 (1971) 2064
[12] R. Kuentzler, J. Physique 44 (1983)
[13] I. Turek, V. Drchal, J. Kudrnovský, M.Sov, P. Weinberger, Electronic structure of disordered alloys, surfaces and interfaces Kluwer Academic Publishers, Boston-London-Dordrecht (1997) p. 15

[14] A. Hernando, X. Amils, J. Nogués, S. Suriñach, M. D. Baró, and M. R. Ibarra, Phys. Rev. B 58, R11864 (1998)
[15] Iron binary phase diagrams. Berlin: Springer-Verlag, 1982.
[16] R. C. Hall, J. Appl. Phys. 30, 816 (1959)
[17] A.E. Clark, J.B. Restorff, M. Wun-Fogle, D. Wu and T.A. Lograsso, J. Appl. Phys., 103, 07B310 (2008)
[18] Z.H. Liu, G.D. Liu, M. Zhang, G.H. Wu, F.B. Meng, H.Y. Liu, L.Q. Yan, J.P. Qu and Y.X. Li. Appl. Phys. Lett., 85 , 1751 (2004)
[19] C. Jiang, X.X. Gao, J. Zhu and S.Z. Zhou. J. Appl. Phys., 99, 023903 (2006)
[20] R. Sato Turtelli, G. Vlásak, F. Kubel, N. Mehmood, M. Kriegisch, R. Grössinger, and H. Sassik, IEEE Trans. Magn., 46, 483 (2010)
[21] N. Mehmood, R. Sato Turtelli, R. Grössinger, M. Kriegisch, J. Magn. Magn. Mater. 322, 1609 (2010).
[22] P. Shukla, M. Wortis, Phys. Rev. B 21 (1980) 159
[23] B.D. Cullity, Introduction to Magnetic Materials, Addison-Wesley Publishing Company (1972) 134
[24] J.S. Kouvel, J. Appl. Phys, 30 313S (1959)
[25] J.S. Kouvel, Magnetism and Metallurgy, Vol. 2, p. 523. Academic Press. 1969
[26] Y. Yang, I. Baker and P. Martin, Phil. Mag. B 79, 449 (1999)
[27] N. I. Kulikov, A.V. Postkikov, G. Borstel and J. Braun, Phys. Rev. B 59, 6824 (1999)
[28] E. Apiñaniz, F. Plazaola, J.S. Garitaonandia, Eur. Phys. J. B 31 (2003) 167.
[29] S.K. Bose, V. Drchal, J. Kudrnovsky, O. Jepsen and O.K. Andersen, Phys. Rev. B 55, 8184 (1997).
[30] J. Bogner, W. Steiner, M. Rissner, P. Mohn, P. Blaha, K. Schwarz, R. Krachler and H. Ipser, B. Sepiol, Phys. Rew. B 58, 22 (1998)
[31] H. Gengnagel, M. J. Besnus, and H. Danan, Phys. Status Solidi A13, 499 (1972)
[32] D. Martin Rodriguez, F. Plazaola, J.S. Garitaonandia, J.A. Jimenez, E. Apiñaniz, Intermetallics 24, 38 (2012)
[33] D. S. Schmool, E. Araujo, M. M. Amado, M. Alegria Feio, D. Martin Rodriguez, J. S. Garitaonandia and F. Plazaola, J. Magn. Magn. Mater. 272-276, 1342 (2004).
[34] H. Sato and A. Arrot, Phys. Rev. 114, 1427 (1959).
[35] S.J. Pickart and R. Nathans, Phys. Rev. B 123, 1163 (1961).
[36] J.P. Perrier, B. Tissier and R. Tournier, Phys. Rev. Lett. 24 313 (1970).
[37] R.D. Shull, H. Okamoto and P.A. Bech, Solid State Commun. 20 863 (1976).
[38] K. Motoya, S.M. Shapiro and Y. Muraoka, Phys. Rev. B 28, 6183 (1983).
[39] G.P. Huffman, in Amorphous Magnetism, p. 283. Ed. H.O. Hooper and A.M. de Graaf, (Plenum New York, 1973).
[40] J. S. Garitaonandia, E. Apiñaniz, F. Plazaola, Bulletin of the APS 47, 1, p. 339 (2002)
[41] S. Mitsuda, H. Yoshizawa and Y. Endoh, Phys. Rev. B 45, 9788 (1992)
[42] P. Böni, S.M. Shapiro and K. Motoya, Phys. Rev. B 37, 243 (1988).
[43] Wei Bao, S. Raymond, S.M. Shapiro, K. Motoya, B. Fåk and R.W. Erwin, Phys. Rev. Lett. 82, 4711 (1999)

[44] J.W. Cable, L. David and R.Parra, Phys. Rev. B 16, 1132 (1977)
[45] T.M. Srinivasan, H. Claus, R. Viswanathan, P.A. Beck and D.I. Bardos in Phase Stability in Metals and Alloys, p.151. Ed. P.S. Rudman, J. Stringer and R.I. Jaffee (MacGraw-Hill New York, 1967)
[46] H. Maletta, G. Aeppli and S.M. Shapiro, Phys. Rev. Lett. 48, 1490 (1982)
[47] G. Aeppli, S.M. Shapiro, R.J. Birgeneau and H.S. Chen, Phys. Rev. B 25, 4882 (1982)
[48] E. P. Yelsukov, E. V. Voronina, and V. A. Barinov, J. Magn. Magn. Mater. 115, 271 (1992)
[49] S. Takahashi and Y. Umakoshi, J. Phys.: Condens. Matter 2, 4007(1990)
[50] X. Amils, J. Nogués, S. Suriñach, M. D. Baró, and J. S. Muñoz, IEEE Trans. Magn. 34, 1129 (1998)
[51] R. A. Varin, T. Czujko, J. Bystrzycki, and A. Calla, Mater. Sci. Eng., A 329-331, 213 (2002)
[52] M. Fujii, K. Katuyoshi, K. Wakayama, M. Kawasaki, T. Yoshioka, T. Ishiki, N. Nishio, and M. Shiojiri, Philos. Mag. A 79, 2013 (1999)
[53] W. Hu, T. Kato, and M. Fukumoto, Mater. Trans., JIM 44, 2678 (2003)
[54] X. Amils, J. Nogués, S. Suriñach et al. J. S. Muñoz, Phys. Rev. B 63, 52402 (2001)
[55] D. Negri, A. R. Yavari, and A. Deriu, Acta Mater. 47, 4545 (1999)
[56] L. F. Kiss, D. Kaptás, J. Balogh, L. Bujdosó, T. Kemény, I. Vincze, and J. Gubicza, Phys. Rev. B 70, 012408 (2004)
[57] G. K. Wertheim, V. Jaccarino, J. H. Wernick, and D. N. E. Buchanan, Phys. Rev. Lett. 12, 24 (1964)
[58] P. Shukla and M. Wortis, Phys. Rev. B 21, 159 (1980)
[59] J. A. Plascak, L. E. Zamora, and G. A. Pérez Alcazar, Phys. Rev. B 61, 3188 (2000)
[60] A. Arzhnikov, A. Bagrets, and D. Bagrets, J. Magn. Magn. Mater. 153, 195 (1996)
[61] M. Shiga, J. Phys. Soc. Jpn. 50, 2573 (1981)
[62] M. Shiga, IEEETransl. J. Magn. Jpn. 6, 1039 (1991)
[63] A. Hernando, J. M. Barandiarán, J. M. Rojo, and J. C. Gómez-Sal, J. Magn. Magn. Mater. 174, 181 (1997)
[64] B. V. Reddy, S. C. Deevi, F. A. Reuse, and S. N. Khanna, Phys. Rev. B 64, 132408 (2001)
[65] A. V. Smirnov, W. A. Shelton, and D. D. Johnson, Phys. Rev. B 71, 064408 (2005)
[66] V. L. Moruzzi, P. M. Marcus, Phys. Rev. B, 47, 7878 (1993)
[67] E. Apiñaniz, F. Plazaola, J. S. Garitaonandia, J. Non Cryst. Solids 287, 302 (2001)
[68] E. Apiñaniz, F. Plazaola et al. J. Magn. Magn. Mater. 272, 794 (2004)
[69] J. Deniszczyk, Acta Phys. Pol. A 97, 583 (2000)
[70] J. Nogués, E. Apiñaniz, J. Sort,1 M. Amboage, M. d'Astuto, O. Mathon, R. Puzniak, I. Fita, J. S. Garitaonandia, S. Suriñach, J. S. Muñoz, M. D. Baró, F. Plazaola, and F. Baudelet, Phys. Rev. B 74, 024407 (2006)
[71] E. Apiñaniz, F. Plazaola, J. S. Garitaonandia, D. Martin and J.A. Jimenez, J. Appl. Phys. 93, 7649 (2003)
[72] E. Apiñaniz, J.S. Garitaonandia, F. Plazaola, D. Martin, J.A. Jimenez, Sensors and Actuators A, 106 76 (2003)

[73] J. F. Humphreys, in *Materials Science and Technology: A Comprehensive Treatment*, edited by R. W. Cahn *et al.* (VCH, Weinheim 1992), Vol. 15.
[74] D. Wu, P. R. Munroe and I. Baker, Phil. Mag. 83, 295 (2003).
[75] D. Martin Rodriguez, E. Apinaniz, J. S. Garitaonandia, F. Plazaola, D. S. Schmool and G. Cuello, J. Magn. Magn. Mater. 272-276, 1510 (2004)
[76] D. Martin Rodriguez, E. Apinaniz, F. Plazaola, J. S. Garitaonandia, J. A. Jimenez, D. S. Schmool and G. J. Cuello, Phys. Rev. B 71, 212408 (2005).
[77] S. M. Allen and J. W. Cahn, Acta Metall. 24, 425 (1976)
[78] K. Yamashita, M. Imai, M. Matsuno and A. Sato, Phil. Mag. A 78, 285 (1998).
[79] A. E. Vidoz and L. M. Brown, Phil. Mag. 7, 1167 (1962).
[80] C. T. Chou and P. B. Hirsch, Phil. Mag. A 44, 1415 (1981).
[81] H. Y. Yasuda, R. Jimba and Y. Umakoshi, Scripta mater. 48, 589 (2003).
[82] M. van Schilfgaarde, I. A. Abrikosov and B. Johannsson, Nature 400, 46 (1999)
[83] Z. Arnold, J. Kamarad, P. A. Algarabel, B. Garcia-Landa and M. R. Ibarra, IEEE Trans. Magn. 30, 619-621 (1994)
[84] V. L. Moruzzi, A. R.Williams and J. F. Janak, Phys. Rev. B 10 (1974) 4856
[85] C. Müller, H. Wonn,W. Blau, P. Ziesche and V.P. Krivitskii, Phys. Status Solidi B 95 (1979) 215
[86] C. Müller and P. Ziesche, Phys. Status Solidi B 114 (1982) 523
[87] C. Müller, W. Blau and P. Ziesche, Phys. Status Solidi B 116 (1983) 561
[88] J. Zou and C. L. Fu, Phys. Rev. B 51 (1995) 4
[89] J. H. Westbrook and R. L. Fleischer, Intermetallic Compounds Wiley Chicester (1994), Vol. 1, p.127
[90] R. E. Watson and M. Weinert, Phys. Rev. B 58 (1998) 5981
[91] V. Sundararajan, B. R. Sahu, D. G. Kanhere, P. V. Panat and G. P. Das, J. Phys.: Condens. Matter 7 (1995) 6019
[92] B. I. Min, T. Oguchi, H. J. F. Jansen and A. J. Freeman, J. Magn. Mag. Mat. 54-57 (1986) 1091
[93] P. Mohn, C. Persson, P. Blaha, K. Schwarza, P.Novák, H. Eschirg, Phys. Rev. Letters 87 (2001) 6401 (edo 196401)
[94] Y. M. Gu and L. Fritsche, J. Phys.:Condens. Matter 4 (1992) 1905
[95] W. Lin, Jian-hua Xu and A. J. Freeman, J. Mater. Res. 7 (1992) 592
[96] C. L. Fu, Phys. Rev. B 52 (1995) 315
[97] J. Mayer, C. Elsässer and M. Fähnle, Phys. Stat. Sol.(b) 191, 283 (1995)
[98] V. L. Moruzzi and P. M. Marcus, Phys. Rev. B 46, 2864 (1993)
[99] S. Dorfman, V. Liubich, D. Fuks, Inter. J. Quantum Quem. 75, 4 (1999)
[100] F. Schmidt and K. Binder, J. Phys.: Condens. Matter 4, 3569-3588 (1992)
[101] G. Frommeyer, J.A. Jimenez, C. Derder, Z. Metallkd. 90, 43 (1999)
[102] R. Haydock, M.V. You, Solid State Comm. 33, 299 (1980)
[103] R. Nathans, H.T. Pigott, C.G. Shull, J. Phys. Solids 6, 38(1958)
[104] P. Villars, L.D. Calvert, *Pearson's Handbook of Crystallographic Data for Intermetallic Phases*, Vols. 1-3 (Metals Park, OH: American Society for Metals, 1985)
[105] F.R. de Boer, R. Boom, W.C. Mattens, A.R. Miedema, A.K. Niesen, *Cohesion Metals: Transition Metal Alloys* (North-Holland, Amsterdam, 1988)

[106] O. Kubaschewski, *Iron Binary Phase Diagrams* (Springer, Berlin, 1986)
[107] G. Bester, B. Meyer, M. Fähnle, Phys. Rev. B 60, No. 21, 14492 (1999)
[108] B. Fultz, Z. Gao, Phil. Mag. B 67, 6 (1993)
[109] D. Rafaja, Scripta Mat. 34, 9 (1996)

7

Magnetic Material Characterization Using an Inverse Problem Approach

Ahmed Abouelyazied Abdallh and Luc Dupré
Department of Electrical Energy, Systems and Automation, Ghent University
Belgium

1. Introduction

Soft magnetic materials are present in many electromagnetic devices and are widely used in industrial applications. In order to analyze magnetically such applications, e.g. by numerical techniques, the magnetic characteristics of the material inside the device have to be known.

There are several standard measurement techniques for characterizing magnetic materials, e.g. an Epstein frame is commonly used for identifying the magnetic properties of a material (Sievert, 2000). However, this requires extra samples of the magnetic material used in the electromagnetic device. These extra samples are often not available. Moreover, the characteristics of the magnetic material may be altered during the device manufacturing (Takahashi et al., 2008). Therefore, it is convenient to characterize the material on the specific geometry of the device itself.

In practice, the direct identification of the magnetic material exclusively based on magnetic measurements is quite difficult and sometimes impossible, due to the complexity of the electromagnetic device geometry. Therefore, an alternative method is needed. Recently, the coupled experimental-numerical inverse problem has gained a lot of interests for the identification problems (Tarantola, 2004).

The goal of this chapter is to design and develop a combined experimental-numerical algorithm for the magnetic characterization of the material, present in a geometrically complex structure, using an electromagnetic inverse problem approach. Specifically, we aim at identifying the single-valued *B-H* curve, hysteresis characteristics, and loss parameters of a magnetic material inside an electromagnetic device.

The proposed coupled experimental-numerical inverse algorithm mainly consists of two major parts: experimental measurements and numerical modeling of the device. The proposed algorithm in this chapter is robust because all uncertainties present in the two parts are taken into account in a stochastic framework.

Standard techniques for magnetic material characterization is presented below, followed by the state-of-the-art of the proposed scheme. The proposed algorithm is validated by applying it for the identification of the magnetic properties of the material in an electromagnetic device, an EI electromagnetic core inductor. However, the proposed algorithm can be applied for any other application.

2. Standard techniques for magnetic material characterization

Measurements are widely used when the measurement standards are fixed. Basic magnetic measurements in magnetic materials have been collected in a group of standards denoted by IEC 60404. The two main commonly used techniques to measure the global magnetic properties of sheets of magnetic materials are Epstein frame (IEC404-2, 1996) and single sheet tester (IEC404-3, 2000), which are considered as closed magnetic circuit and the magnetic path length inside the sample can be easily defined.

In section 2.1, we include the general principles for magnetic field and magnetic induction calculations in a closed magnetic circuit. The magnetic material characterization fully based on magnetic measurements are shown in section 2.2.

2.1 General principles for magnetic field and induction evaluation in a closed magnetic circuit

Consider a simple closed magnetic core, i.e. no air gap, consisting of a current carrying coil of N_1 turns and a magnetic core with a magnetic mean path length l_c and a cross sectional area A_c perpendicular to the field lines, as shown in Fig. 1(a). Assume that the relative magnetic permeability of the core material μ_r is very high so that all magnetic flux lines are confined within the core[1].

The magnetic field strength at the surface of the core H_s and the magnetic induction B_{av} inside the core material can be calculated using Ampere's law and Faraday's law, respectively.

Ampere's law in the integral form states that the total electric current $N_1 i$ through the surface S bounded by the specific closed contour C induces the magnetomotive force (m.m.f.) in this contour:

$$\oint_C \mathbf{H}.dl = \oint_S \mathbf{J}.ds \tag{1}$$

with $\mathbf{J}$ being the total current density confined by the closed contour C. Hence, as a first approximation:

$$H_s l_c = N_1 i \quad \therefore \quad H_s = \frac{N_1 i}{l_c} \tag{2}$$

where H_s is the magnetic field strength at the surface of the core, and $N_1 i$ the magnetomotive force.

Consider a ferromagnetic material uniformly magnetized along one of its symmetry lines using a excitation windings N_1, as shown in Fig. 1(a). The sample is subjected to a time dependent magnetic field H. The flux density can be derived from the induced voltage in the measurement winding 'pickup coil' by the time varying total magnetic flux, according to Faraday's law:

$$V_{sec}(t) = N_2 \frac{d\phi}{dt} = N_2 A_c \frac{dB_{av}}{dt} \tag{3}$$

$$B_{av}(t) = \frac{1}{N_2 A_c} \int_0^t V_{sec}(\tau) d\tau + B_{av}(0) = \frac{1}{N_2 A_c} \sum_{i=1}^{T/\Delta\tau} V_{sec_i} \Delta\tau + B_{av}(0) \tag{4}$$

[1] Indeed, in order to obtain homogenous magnetic field distributions in the magnetic sample, the excitation winding should be wound around the hole circumference of the magnetic sample.

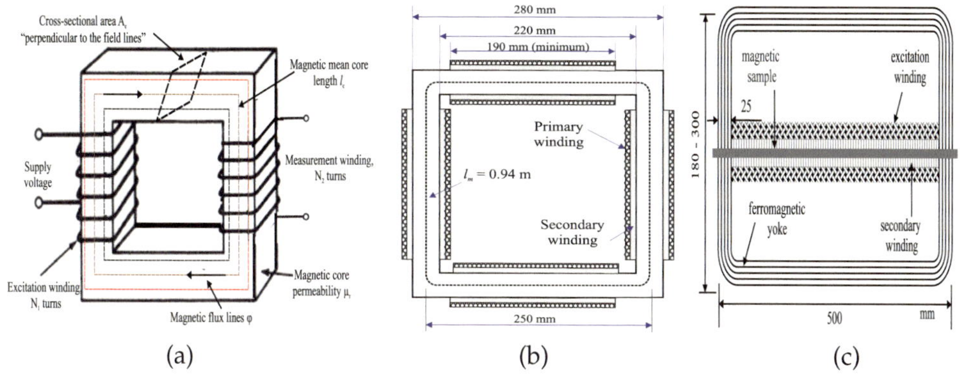

Fig. 1. (a) Schematic diagram of a simple magnetic circuit, (b) The standard Epstein frame according to IEC 60404-2 (IEC404-2, 1996), (c) The standard single sheet tester profile (IEC404-3, 2000).

where N_2 is the number of turns of the measurement winding, T is the period of the secondary voltage, $B_{av}(t)$ is the average flux density in the cross-section A_c and $\Delta\tau$ is the sampling period. Whether the magnetic flux density is uniformly distributed over the cross section of the sample or not, the sensing coil measures the average induction value over the cross section A_c[2].

2.2 Magnetic material characterization fully based on magnetic measurements

Figs. 1(b) and (c) show respectively the schematic diagram of the standard Epstein frame (IEC404-2, 1996) and single sheet tester (IEC404-3, 2000), which are considered as closed magnetic circuits. The *B*-*H* characterization of the material with these techniques are possible, only based on electromagnetic measurements as shown in section 2.1, under the conditions that the magnetic field pattern is homogeneous and known in advance. Moreover the measurements are standardized resulting in geometrical parameters that have a fixed known value. However, this requires extra samples of the magnetic material used in the electromagnetic device. These extra samples are often not available. Therefore, it is convenient to characterize the material on the specific geometry of the device itself.

Alternatively, the local magnetic measurements can be used for the identification of the magnetic material properties on the specific geometry of the device itself. The local magnetic measurements consist of measuring two magnetic quantities, i.e. H and B, locally at the same position. Several magnetic sensors are used for local magnetic measurements, such as the flat H-coil, fluxgate sensor, and the Hall-probe sensor 'for local magnetic field measurements', and the needle probes 'for local magnetic induction measurements', (Abdallh & Dupré, 2010c). In practice, the local magnetic measurements are useful and accurate enough in the case that the local place is accessible and the magnetic field pattern is sufficiently homogeneous at that

[2] In order to minimize the measurement of any flux in the air, the measurement winding should be wound as close as possible to the magnetic sample. In practice, since the diameter of the measurement winding coil is thinner than the diameter of the excitation winding coil, the measurement winding is wound closer to the magnetic sample beneath the excitation winding.

local position[3]. However, most of electromagnetic devices exhibit such non-uniform magnetic field patterns.

Therefore, we present in this chapter the inverse procedure for the identification of the magnetic properties of a magnetic material by interpreting well-defined measurements into a numerical model of the considered electromagnetic device.

3. State-of-the-art of the proposed methodology

The magnetic properties of a material can be recovered by the interpretation of well defined electromagnetic measurements with a numerical model of the device. The inverse problem minimizes iteratively the difference between the measurements and the simulated quantities, using a minimization algorithm, see section 5.

For the ideal case, i.e. measurements are noise-free and the mathematical model perfectly represents the reality, the actual magnetic properties of the material may be precisely retrieved. However, neither measurements nor modeling are perfect. The measurements contain noise, and the mathematical model simplifies the reality, i.e. some of the physical phenomena are not correctly modeled. Moreover, some of the model parameters, e.g. related to the geometry, are not known exactly. Their values are uncertain. Consequently, recovering the actual magnetic material properties is not guaranteed.

In the presented inverse problem approach, we propose the use of the Cramér-Rao lower bound technique for 'qualitatively' estimating the error in the recovered material properties, due to the measurement noise and uncertainties in the model parameters, see section 6. Based on the results obtained using this method, we are able to design *a priori* the inverse problem. For example, selecting the optimal condition for performing the measurements, and choosing the best measurement modalities which lead to the best recovery results.

Furthermore, we propose two efficient numerical techniques for reducing 'quantitatively' the modeling error in the inverse problem solution. In section 7.1, a deterministic technique is proposed for reducing the modeling error caused by the uncertain model parameter values, by modifying the minimization scheme of the inverse problem. On the other hand, in section 7.2, the stochastic Bayesian approach is utilized for reducing the modeling error originated from the modeling simplifications.

4. General overview of the forward and inverse problem

Generally, the problem of computing the model response of a physical system is called a forward problem. On the other hand, the inverse problem theory concerns the problem of making inferences about a physical system starting from 'indirect' noisy measurements (Scales & Snieder, 2000).

In electromagnetism, the accurate modeling of the electromagnetic phenomena is carried out by solving Maxwell's equations for certain given geometry, sources, and material characteristics. These so-called forward models can be used for analyzing the behavior of the system.

[3] The relative errors in the local H-measurements are higher than the relative errors in local B-measurements, especially for highly non-uniform magnetic field patterns. In practice, the performing accurate local H-measurements is not an easy task.

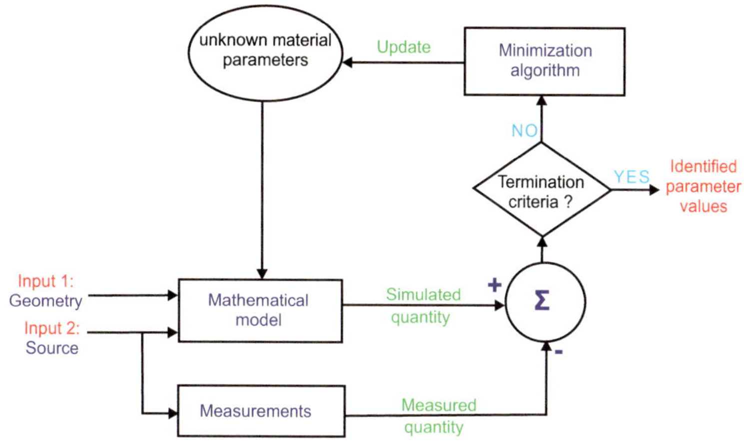

Fig. 2. The detailed schematic diagram of the 'ideal' inverse problem procedure. The termination criteria can be a predefined tolerance on the difference between the measured and simulated quantity.

Moreover, the forward problems also can be used for solving inverse problems. In fact, solving numerically inverse problems has been a subject of research for many years and arises when dealing with more complex systems. Notice that an inverse algorithm is using a certain number of forward model evaluations. The solution of the inverse problem is obtained by proposing an iterative scheme for the identification of the values of the unknown variables (geometrical, sources, or material characteristics) which need to be recovered.

In order to identify a certain unknown parameter of an electromagnetic device, as in an inverse problem, a predefined objective 'cost' function needs to be minimized. This objective function is the quadratic difference between the measured and simulated quantities.[4]

Due to the fact that the aim of this chapter is to characterize the magnetic material properties of an electromagnetic device on its complex geometry, we will restrict ourselves to this aim, i.e. the identification of the geometrical parameters or sources are not studied in this work. These issues have been discussed in (Crevecoeur, 2009), (Yitembe et al., 2011). The detailed schematic diagram of the 'ideal[5]' inverse problem[6] for magnetic material characterization, used in this chapter, is shown in Fig. 2.

---

[4] The well known least square non-linear algorithm, Levenberg-Marquardt method with line search (Marquardt, 1963), is often used for minimizing the cost function.

[5] Ideal inverse problem means that the mathematical model of the device represents perfectly the reality, i.e. no modeling error due to the simplification of reality and the model parameters are exactly known. Also, the measurements are assumed to be noise-free. For more detail concerning this issue, see sections 6 and 7.

[6] It is worth mentioning that electromagnetic optimization problems differ from electromagnetic inverse problems. The objective function in optimization problems is based on the minimization or maximization of the target, e.g. efficiency, maximum torque, power per unit volume of an electromagnetic device, coupled through simulated quantities in order to obtain the optimal values for the predefined parameters. The electromagnetic optimization problems are used for designing the electromagnetic devices. However, electromagnetic inverse problems are used for recovering properties of existing electromagnetic devices. In this chapter, we concern only with electromagnetic inverse problems.

5. Inverse problem formulation

The behavior of a magnetic system can be represented by a mathematical model with a set of partial differential equations. This model is parameterized by the following model parameters: the unknown parameters $\mathbf{u} \in \mathbb{R}^p$, the uncertain parameters $\mathbf{b} \in \mathbb{R}^q$, e.g. geometrical model parameters, and the precisely known parameters $\mathbf{d} \in \mathbb{R}^x$.

For solving inverse problems, the experimental observations $\mathbf{W} \in \mathbb{R}^K$ and the corresponding predicted model responses $\mathbf{\Phi} \in \mathbb{R}^K$ need to be obtained for the same conditions, e.g. same observation positions, same excitation source, etc. The experimental observations $\mathbf{W}$ of the state of a system are obtained for an excitation source I_k, and it can be represented by:

$$W_k = \{w(I_k)\} \quad k = 1, 2, \ldots, K \tag{5}$$

with K being the total number of discrete experimental observations.

On the other hand, the predicted response $\mathbf{\Phi}$ of a system based on the mathematical model, which involves the solution of the field equations, using the same excitation source I_k and the model parameters, can be represented by:

$$\Phi_k(\mathbf{u}) = \{\phi(\mathbf{u}, I_k)\} \quad k = 1, 2, \ldots, K \tag{6}$$

In order to estimate the unknown parameters $\mathbf{u}$, an inverse problem has to be solved by iteratively minimizing the quadratic *residuals* between the experimental observations of the magnetic system $\mathbf{W}$ and the modeled ones $\mathbf{\Phi}$. In other words, the functional $\Gamma(\mathbf{u})$

$$\Gamma(\mathbf{u}) = \sum_{k=1}^{K} \|W_k - \Phi_k(\mathbf{u})\|^2 \tag{7}$$

needs to be minimized:

$$\tilde{\mathbf{u}} = \arg\min_{\mathbf{u}} \Gamma(\mathbf{u}) \tag{8}$$

with $\tilde{\mathbf{u}}$ being the recovered unknown parameters using the inverse approach. The resolution of the inverse procedure highly depends on both measurements and modeling accuracy. Γ can depend on the definition of the inverse problem, i.e. the place where the measurements are conducted, the objective function to be minimized, etc. This resolution is due to measurement noise, available uncertainties in the forward mathematical model parameter values, or the simplifications in the used model.

Sources of errors

In practice, the actual measurements $\mathbf{W}$, can be expressed as:

$$\mathbf{W} = \mathbf{\Phi}(\mathbf{u}^*) + \mathbf{e} \tag{9}$$

with $\mathbf{e} \in \mathbb{R}^K$ being the error vector. The difference $\mathbf{e}$ between the simulated signals $\mathbf{\Phi}(\mathbf{u}^*)$ with the actual values for the model parameters $\mathbf{u}^*$ and the measured signals $\mathbf{W}$ may have two contributions:

$$\mathbf{e} = \mathbf{e}_n + \mathbf{e}_m \tag{10}$$

with $\mathbf{e}_n$ being the error due to measurement noise and $\mathbf{e}_m$ due to modeling uncertainties. In fact, the modeling error $\mathbf{e}_m$ can be divided into two main parts: the modeling error due to the uncertainty of some of the model parameters **b**, and the modeling error due to the simplification of the used mathematical model. Due to **e**, when minimizing the cost function (Equation 7), the values of the recovered parameters $\widetilde{\mathbf{u}}$ and the actual parameter values $\mathbf{u}^*$ are not necessary equal, i.e. $\widetilde{\mathbf{u}} \neq \mathbf{u}^*$.

6. Error estimation in the inverse problem solutions

As mentioned in section 5, a coupled experimental-numerical inverse approach is formulated in order to characterize the magnetic material of an electromagnetic device. The input of this inverse approach is a set of electromagnetic measurements, while the output is the magnetic material characteristic. It has been observed by numerical experiments "after solving the inverse problem" that the accuracy of the recovered material characteristics appreciably depends on the input nature of the inverse problem, e.g. the objective function to be minimized, the positions where the local magnetic measurements are carried out, etc, (Abdallh et al., 2009), (Abdallh et al., 2010d). However, it is desirable to define and estimate this accuracy *a priori* "before solving the inverse problem".

Generally, the inverse problem is an ill-posed problem, i.e. small deviation in the input data (measured quantities) leads to a considerable deviation in the output data (recovered material parameters). Consequently, the input of the inverse problem, i.e. 'measurements', and the mathematical model have to be perfectly accurate. However, neither measurements nor mathematical models are strictly accurate: measurements are distorted by measurement noise, and responses in the electromagnetic device model exhibit variations due to uncertainties of some parameter values used in these model calculations. Also, the model of an electromagnetic device does not simulate perfectly all phenomena as in the reality.

In this section, we aim at estimating *a priori* the error of the recovered magnetic material properties due to the measurement noise and the uncertainties in the electromagnetic device model parameter values.[7] State-of-the-art of the Monte Carlo simulations are able to achieve this goal but computations may become prohibitive, especially when dealing with time demanding numerical forward electromagnetic models, e.g. finite element models. Therefore, a mathematical technique based on the stochastic Cramér-Rao bound (sCRB) is presented for estimating the uncertainty in the inverse problem solutions. This sCRB offers the lower bound of the error within a rather small computational time compared to the well-know time-demanding techniques such as Monte Carlo simulations applied in e.g. (Leach et al., 2009), stochastic finite element method (Beck & Woodbury, 1998), polynomial chaos decomposition (Gaignaire et al., 2010).

In the stochastic Cramér-Rao bound method, it is assumed that the identification procedure is affected by the uncertainties in the measurements, and uncertainties in the model parameters. Several types of measurement uncertainties can be considered, in particular systematic and random uncertainty.

A systematic measurement uncertainty can be defined as a reproducible error that biases the measured value in a given direction (NIST, 2000), i.e. a systematic overestimation or underestimation of the true value. A systematic measurement uncertainty is by definition reproducible. Therefore, it cannot be reduced by averaging the values of a large number of

[7] In this section, we assume that the model is perfect, i.e. all physical phenomena are included in the model. The inaccuracy in the model structure is discussed in section 7.2.

measurements. However, the reproducible nature of the systematic measurement uncertainty makes it possible to estimate the bias on the measured value by means of a calibration procedure (Lauwagie, 2003). The results presented in this chapter restrict to cases where the measurement uncertainties contain only the random component 'noise'.

When carrying out magnetic measurements, noise can be caused by vibrations of steel sheets, noise originating from excitation current, noise due to the stray field, air flux noise, environmental noise, etc. (Fiorillo, 2004). Here, the measurement noise is assumed uncorrelated and Gaussian white distributed with zero mean and variance of σ_n^2.

Again, the least squares approach (Equation 7) and (Equation 8) may not yield to accurate identification results, because no information about $\mathbf{e}_n$ is utilized. So, when minimizing the cost function (Equation 7), the values of the recovered parameters $\tilde{\mathbf{u}}$ and the actual parameter values $\mathbf{u}^*$ are not necessary equal, i.e. $\tilde{\mathbf{u}} \neq \mathbf{u}^*$. Due to the fact of the poor posedness of the inverse problems, the search for the 'accurate' minimum value of $\Gamma(\mathbf{u})$ is difficult, and needs advanced mathematical techniques as regularization tools for determining the minimum and criteria for terminating the search (Engl et al., 1996), (Saitoh, 2007).

Due to this fact, it seems reasonable, when solving inverse problems, to use a criterion which reflects the statistical information available on the measurement noise. Therefore, the use of the Cramér-Rao bound method (CRB) is proposed for quantifying the possible uncertainties on the identified unknown parameter values $\mathbf{u}$. CRB is widely-used in many engineering applications; heat transfer applications (Fadale et al., 1995a), biomedical engineering applications (Radich & Buckley, 1995), and signal analysis applications (Stoica & Nehorai, 1989).

6.1 Traditional Cramér-Rao bound method (CRB)

In the traditional CRB method, it is assumed that the mathematical modeling is accurate, i.e. $\mathbf{e}_m = 0$. Hence, (Equation 9) can be rewritten as:

$$\mathbf{W} = \mathbf{\Phi}(\mathbf{u}^*) + \mathbf{e}_n \tag{11}$$

It is thus possible to represent the forward model as $\mathbf{\Phi}(\mathbf{u}) + \mathbf{e}_n$ with the incorporation of the measurement noise. Since the measurement noise is supposed to be known, the parameter vector, to be estimated, of this model is still $\mathbf{u}$.

We denote the unbiased estimation of this parameter vector 'after solving the inverse problem' by $\tilde{\mathbf{u}}$. The Cramér-Rao inequality theorem states that the covariance matrix of the deviation between the true and the estimated parameters is bounded from below by the inverse of the Fisher information matrix $\mathbf{F}$ (Goodwin & Payne, 1977), (Strang, 1986):

$$E\left\{(\tilde{\mathbf{u}} - \mathbf{u}^*)(\tilde{\mathbf{u}} - \mathbf{u}^*)^T\right\} \geq \mathbf{F}^{-1} \tag{12}$$

where E is the expectation. The Fisher information matrix $\mathbf{F}$ gives an indication about the probability of observing noisy measurements $\mathbf{W}$ given the model $\mathbf{\Phi}$ and the unknown model parameters $\mathbf{u}$. The probability distribution function of the random variable $\mathbf{W}$ given the value of $\mathbf{u}$, which is also the likelihood function of $\mathbf{u}$, is a function $L(\mathbf{W}|\mathbf{u})$. Indeed, the best estimates of the unknown parameter values $\mathbf{u}$ are those that maximizes the probability density function of the measurements $L(\mathbf{W}|\mathbf{u})$. This estimate of $\mathbf{u}$ makes the measurements most likely and therefore the probability density function $L(\mathbf{W}|\mathbf{u})$ is called the likelihood function.

The Fisher information matrix **F** is calculated based on the partial derivative of the log of the likelihood function with respect to the parameter vector **u**, which can be calculated by (Goodwin & Payne, 1977):

$$\mathbf{F} = E\left\{\left[\frac{\partial}{\partial \mathbf{u}} \ln L\left(\mathbf{W}|\mathbf{u}\right)\right]\left[\frac{\partial}{\partial \mathbf{u}} \ln L\left(\mathbf{W}|\mathbf{u}\right)\right]^T\right\} \tag{13}$$

with $\ln L(\mathbf{W}|\mathbf{u})$ being the log-likelihood of **W** given the parameter vector **u**. **F** is a matrix with $p \times p$ dimensions. The likelihood of the data is normally distributed and is given by $L(W_1, \ldots, W_k)$:

$$L(\mathbf{W}|\mathbf{u}) = \prod_{k=1}^{K} \frac{1}{(2\pi)^{1/2}\sigma_n} \cdot \exp\left(-\frac{1}{2\sigma_n^2}\left[W_k - \Phi_k\left(\mathbf{u}\right)\right]^2\right) \tag{14}$$

and can be rewritten as:

$$L(\mathbf{W}|\mathbf{u}) = \frac{1}{(2\pi)^{K/2}(\sigma_n)^K} \cdot \exp\left(-\frac{1}{2\sigma_n^2}\sum_{k=1}^{K}\left[W_k - \Phi_k\left(\mathbf{u}\right)\right]^2\right) \tag{15}$$

So that the log-likelihood function becomes:

$$\ln L(\mathbf{W}|\mathbf{u}) = \text{const} - K\ln(\sigma_n) - \frac{1}{2\sigma_n^2}\sum_{k=1}^{K}[W_k - \Phi_k(\mathbf{u})]^2 \tag{16}$$

So the derivative to **u** becomes:

$$\frac{\partial \ln L(\mathbf{W}|\mathbf{u})}{\partial \mathbf{u}} = \frac{1}{\sigma_n^2}\sum_{k=1}^{K}\left(W_k - \Phi_k\left(\mathbf{u}\right)\right)\frac{\partial \Phi_k(\mathbf{u})}{\partial \mathbf{u}} \tag{17}$$

Substituting (Equation 17) in (Equation 13), the Fisher information matrix can be written as (Goodwin et al., 1974):

$$\begin{aligned}(F)_{lm} = \sum_{k=1}^{K}&\left[\left\{\frac{\partial \Phi_k}{\partial u_m}\right\}^T S_k^{-1}\left\{\frac{\partial \Phi_k}{\partial u_l}\right\}\right.\\ &\left.+\frac{1}{2}Tr\left[S_k^{-1}\left\{\frac{\partial S_k}{\partial u_l}\right\}S_k^{-1}\left\{\frac{\partial S_k}{\partial u_m}\right\}\right]\right] \qquad l, m = 1, \ldots, p\end{aligned} \tag{18}$$

since $\sigma_n^2 = E\left\{(1/K)\sum_{k=1}^{K}[W_k - \Phi_k(\mathbf{u})]^2\right\}$, with $S_k = \sigma_n^2$ being the *measurement variance*, $k = 1, \ldots, K$.

Usually, the form of the Fisher information matrix (**F**) is simplified by assuming that the measurement noise is the same for all experimental observations K, uncorrelated, and its variance is independent on the unknown parameters **u**, i.e. $S_k = \sigma_n^2$ and $\frac{\partial S_k}{\partial u_p} = 0$ (Alifanov et al., 1995), (Fadale et al., 1995b). Thus, under these conditions, the Fisher information matrix (**F**) reduces to the classical form:

$$(F)_{lm} \cong \sum_{k=1}^{K}\left[\left\{\frac{\partial \Phi_k}{\partial u_m}\right\}^T S_k^{-1}\left\{\frac{\partial \Phi_k}{\partial u_l}\right\}\right] \qquad l, m = 1, \ldots, p \tag{19}$$

Using the inequality of Cramér-Rao bound (Equation 12), the lower bound for the p variances of the unknown parameters $\sigma^2_{\mathrm{u,F}}$ is given by (Stoica & Nehorai, 1989):

$$\sigma^2_{\mathrm{u,F}_{ii}} \geq \left(\mathbf{F}^{-1}\right)_{ii} \qquad i = 1, \ldots, p \tag{20}$$

$\mathbf{F}^{-1}$ expresses the lower bound for the covariance matrix of the unknown parameters where the variances of each unknown parameter can be deduced as the diagonal elements of $\mathbf{F}^{-1}$.

6.2 Stochastic Cramér-Rao bound method (sCRB)

Besides the recovery errors due to measurement uncertainties elaborated in section 6.1, errors are also introduced by the modeling uncertainty.

Specifically, the accuracy of the modeled response depends upon the numerical algorithm and the degree of approximation used, e.g. finite difference or finite element, coarse or fine discretization, etc. These errors can be reduced by using very fine discretizations, inclusion of more accurate material models, etc. This issue is discussed in more detail in section 7.2. In addition to these errors, the modeled response also exhibits variations which are due to the uncertainties in the uncertain parameters $\mathbf{b}$ used in these model calculations. We assume in this chapter that the mathematical forward model algorithm is exact by using a very fine mathematical mode so that the estimated values of the parameters are only influenced by the uncertain model parameters.

The traditional CRB method can be extended when dealing with stochastic uncertain model parameters $\mathbf{b}$, see (Fadale et al., 1995a), (Emery et al., 2000) with an unbiased estimator $\mathbf{u}$. The forward problem becomes now $\mathbf{\Phi}(\mathbf{u}, \mathbf{b}) + \mathbf{e}_n$. The predefined mean value of the uncertain parameters $\mathbf{b}$ defines the model $\mathbf{\Phi}$. We assume that the random $\mathbf{b}$ is a Gaussian prior with mean and variance values of μ_b and σ_b^2, respectively. In this case, the forward problem becomes $\mathbf{\Phi}(\mathbf{u}, \mu_b) + \mathbf{e}_n + \mathbf{e}_m$. Since $\mathbf{e}_n$ and $\mathbf{e}_m$ are normally distributed, it is possible to express the total uncertainty $\mathbf{e}_t$ in a normal distribution function as well, i.e. $\mathbf{e}_t = \mathbf{e}_n + \mathbf{e}_m$, with $\mathbf{e}_t \sim \mathcal{N}(\mu_t, \sigma_t)$, where $\mathcal{N}$ is the normal distribution function.

Similar to the derivation of the traditional CRB presented in previous section, the extended Fisher information matrix can be obtained. Here, the extended Fisher information matrix $\mathbf{M}$ is calculated based on the partial derivative of the log of the likelihood function with respect to the parameter vector $\mathbf{u}$, which can be calculated by (Goodwin & Payne, 1977):

$$\mathbf{M} = E\left\{\left[\frac{\partial}{\partial \mathbf{u}} \ln L(\mathbf{W}|\mathbf{u})\right]\left[\frac{\partial}{\partial \mathbf{u}} \ln L(\mathbf{W}|\mathbf{u})\right]^T\right\} \tag{21}$$

with $\ln L(\mathbf{W}|\mathbf{u})$ being the log-likelihood of the $\mathbf{W}$ given the parameter vector $\mathbf{u}$. $\mathbf{M}$ is a matrix with $p \times p$ dimensions. The likelihood of the data is normally distributed and is given by $L(W_1, \ldots, W_k)$:

$$L(\mathbf{W}|\mathbf{u}) = \prod_{k=1}^{K} \frac{1}{(2\pi)^{1/2}\sigma_t} \cdot \exp\left(-\frac{1}{2\sigma_t^2}\left[W_k - \Phi_k\left(\mathbf{u}, \mu_b\right)\right]^2\right) \tag{22}$$

and can be rewritten as:

$$L(\mathbf{W}|\mathbf{u}) = \frac{1}{(2\pi)^{K/2}(\sigma_t)^K} \cdot \exp\left(-\frac{1}{2\sigma_t^2}\sum_{k=1}^{K}\left[W_k - \Phi_k\left(\mathbf{u}, \mu_b\right)\right]^2\right) \tag{23}$$

So that the log-likelihood function becomes:

$$\ln L(\mathbf{W}|\mathbf{u}) = \text{const} - K\ln(\sigma_t) - \frac{1}{2\sigma_t^2}\sum_{k=1}^{K}[W_k - \Phi_k(\mathbf{u},\mu_b)]^2 \tag{24}$$

So the derivative to **u** becomes:

$$\frac{\partial \ln L(\mathbf{W}|\mathbf{u})}{\partial \mathbf{u}} = \frac{1}{\sigma_t^2}\sum_{k=1}^{K}(W_k - \Phi_k(\mathbf{u},\mu_b))\left(\frac{\partial \Phi_k(\mathbf{u},\mu_b)}{\partial \mathbf{u}}\right) \tag{25}$$

Substituting (Equation 25) in (Equation 21), the extended Fisher information matrix can be written as (Emery et al., 2000):

$$\begin{aligned}(M)_{lm} = \sum_{k=1}^{K}\Bigg[&\left\{\frac{\partial \Phi_k}{\partial u_m}\right\}^T V_k^{-1}\left\{\frac{\partial \Phi_k}{\partial u_l}\right\} \\ &+ \frac{1}{2}Tr\left[V_k^{-1}\left\{\frac{\partial V_k}{\partial u_l}\right\}V_k^{-1}\left\{\frac{\partial V_k}{\partial u_m}\right\}\right]\Bigg] \qquad l,m = 1,\ \ldots,\ p\end{aligned} \tag{26}$$

since $\sigma_t^2 = E\{(1/K)\sum_{k=1}^{K}[W_k - \Phi_k(\mathbf{u},\mu_b)]^2\}$, with $\mathbf{V}$ being the covariance matrix of the *total uncertainty*, and is given by:

$$\begin{aligned}V_k &= E\left[\{e_{t,k} - E\left[e_{t,k}\right]\}\{e_{t,k} - E\left[e_{t,k}\right]\}^T\right] \\ &= E\left[\delta\phi_k\delta\phi_k^T\right] + E\left[\delta W_k \delta W_k^T\right]\end{aligned} \tag{27}$$

The first-order estimate of the covariance matrix $E[\delta\Phi_k\delta\Phi_k^T]$ of the predictions is given by:

$$E[\delta\Phi_k\delta\Phi_k^T] = \Theta_k G \Theta_k^T \qquad k = 1,\ 2,\ \ldots,\ K \tag{28}$$

where Θ_k is the sensitivity matrix of the system prediction Φ with respect to the uncertain parameter $\mathbf{b}$, which can be calculated using any numerical differentiation techniques, e.g. finite difference technique. Θ_k is defined as:

$$\Theta_{k,q} = \partial\Phi_k/\partial b_q \tag{29}$$

while the covariance matrix $E[\delta W_k\delta W_k^T]$ is equal to the covariance of the measurement noise, i.e. $E[\delta W_k\delta W_k^T] = S_k$. Therefore (Equation 27) can be expressed as:

$$V_k = \Theta_k G \Theta_k^T + S_k \tag{30}$$

$$\frac{\partial V_k}{\partial u_l} = 2\Theta_k G\left(\frac{\partial \Theta_k}{\partial u_l}\right)^T + \frac{\partial S_k}{\partial u_l} \qquad l = 1,\ \ldots,\ p \tag{31}$$

According to (Emery et al., 2000), the effect of the trace term is very small, and can thus be neglected. The extended Fisher information matrix, **M**, can then be approximated by:

$$(M)_{lm} \cong \sum_{k=1}^{K}\left[\left\{\frac{\partial \Phi_k}{\partial u_m}\right\}^T V_k^{-1}\left\{\frac{\partial \Phi_k}{\partial u_l}\right\}\right] \qquad l,m = 1,\ \ldots,\ p \tag{32}$$

In this situation, a comparison of (Equation 19) and (Equation 32) shows that V_k can be considered as the *equivalent noise* of the experiment.

So, the lower bound for the p variances of the unknown parameter $\sigma^2_{\mathrm{u,M}}$ is (Radich & Buckley, 1995):

$$\sigma^2_{\mathrm{u,M}_{ii}} \geq \left(\mathbf{M}^{-1}\right)_{ii} \qquad i = 1, \ldots, p \tag{33}$$

In other words, $\mathbf{M}^{-1}$ expresses the lower bound for the covariance matrix of the unknown parameters where the variances of each unknown parameter can be deduced as the diagonal elements of $\mathbf{M}^{-1}$.

7. Error reduction in the inverse problem solutions

The material properties characterizing the magnetic circuit of an electromagnetic device can be identified by solving an inverse problem, where sets of measurements are properly interpreted using a forward numerical model of the device, as explained in section 5.

In practice, two major aspects can reduce the accuracy of the recovered solution when solving the inverse problem, specifically: measurement noise and inaccurate modeling. Measurement noise can be eliminated or reduced to some extend by accurately performing the measurements, see (Abdallh & Dupré, 2010c). On the other hand, modeling errors are basically originating from two main sources: the uncertain model parameter values, and the way of modeling of the physical phenomena of the electromagnetic device.

In section 6, a stochastic methodology for estimating *a priori* the uncertainty in the inverse problem solution is presented. Using this stochastic technique, one may be able to reduce the error in the solution of the inverse problem by incorporating, in the real inverse problem, the most accurate measurement modality, or performing the local measurements at the optimal positions. However, this is a qualitative error reduction. In order to reduce the error in the inverse problem solution quantitatively, we propose two approaches in this section.

First, we propose an effective technique, section 7.1, in which the influences of the uncertainties in the geometrical model parameters are minimized. In this proposed approach, the objective function, that needs to be minimized, is adapted 'iteratively' with respect to the uncertain geometrical model parameters.

On the other side, the second approach aims at reducing the modeling error due to the inaccurate modeling, i.e. not all physical phenomena are modeled, or in other words, the model does not exactly simulate the reality. The second approach uses the stochastic Bayesian technique, section 7.2.

7.1 Minimum path of the uncertainty method (MPU)

For the reconstruction of the magnetic material characteristics, the sources and the geometry of this electromagnetic device have to be exactly known. However, in practice, the data of the dimensions of the electromagnetic device, provided by the manufacture, may be uncertain. So, one may introduce errors in the modeling due to the geometrical inaccuracy. Consequently, the inverse results may be inaccurate, to some extend, see (Abdallh et al., 2011b).

Traditionally, the inverse problem is formulated by minimizing iteratively the residual between the measurements and simulated quantities. However, the uncertainty in the geometrical parameters in the forward model alters the shape of the "traditional" objective function and has thus an effect on the values of the recovered magnetic material properties. Therefore, a new formulation of the objective function is needed, in which the impact of the uncertain geometrical model parameters is reduced.

In case of noise-free measurements and correct modeling of the forward problem, the experimental observations of the magnetic system $\mathbf{W}$ can be expressed as

$$\mathbf{W} = \boldsymbol{\Phi}(\mathbf{u}^*, \mathbf{b}^*, \mathbf{d}) \tag{34}$$

where $\mathbf{u}^*$ and $\mathbf{b}^*$ are the actual values of the unknown and uncertain model parameters, respectively, while $\mathbf{d}$ are the precisely known model parameters.

Generally, in order to estimate the unknown parameters $\mathbf{u}$, an inverse problem has to be solved by iteratively minimizing the quadratic *residuals* between the experimental and modeled observations. In other words, the functional

$$\Gamma_{Trad}(\mathbf{u}) = \left\|\Phi(\mathbf{u}, \mathbf{b}, \mathbf{d}) - \mathbf{W}\right\|^2 \tag{35}$$

needs to be minimized:

$$\widetilde{\mathbf{u}} = \arg\min_{\mathbf{u}} \Gamma_{Trad}(\mathbf{u}) \tag{36}$$

with $\widetilde{\mathbf{u}}$ being the recovered values of the unknown model parameters. Particularly, when $\mathbf{b} = \mathbf{b}^*$ is satisfied, then the inverse problem is capable of recovering, in noise-free case and with a perfect forward model, the actual unknown parameters, i.e. $\widetilde{\mathbf{u}} \equiv \mathbf{u}^*$.

However, in practice, the knowledge of $\mathbf{b}$ is uncertain and the used value can differ from the actual value. For simplicity, assume only one uncertain model parameter, i.e. $q = 1$. If we consider an assumed value $b^\bullet$, i.e. $b^\bullet \neq b^*$, then

$$\mathbf{W} = \boldsymbol{\Phi}(\mathbf{u}^*, b^\bullet, \mathbf{d}) + \mathbf{e}_m \tag{37}$$

with $\mathbf{e}_m$ being the modeling error, which is the error due to the difference between the used $b^\bullet$ and the actual b^*, i.e. $\Delta b = b^* - b^\bullet$. Consequently, the inverse problem solution results in inaccurate recovered unknown model parameters ($\widetilde{\mathbf{u}} \neq \mathbf{u}^*$). Note that (Equation 34) can be rewritten as

$$\mathbf{W} = \boldsymbol{\Phi}(\mathbf{u}^*, b^\bullet + \Delta b, \mathbf{d}) \tag{38}$$

or its Taylor expansion to the first order:

$$\mathbf{W} \simeq \boldsymbol{\Phi}(\mathbf{u}^*, b^\bullet, \mathbf{d}) + \left.\frac{\partial \boldsymbol{\Phi}}{\partial b}\right|_{(b=b^\bullet)} .\Delta b \tag{39}$$

where we neglect the higher order terms $\partial^2\boldsymbol{\Phi}/\partial^2 b;\ \ldots$ etc. $\partial\boldsymbol{\Phi}/\partial b$ is the sensitivity of the modeled response $\boldsymbol{\Phi}$ with respect to the uncertain model parameter b, and is a measure for the "forward propagation" of the uncertainty to the forward solution.

Indeed, up to the first order, (Equation 39) reflects in a better way the actual forward solution. So that the proposed methodology, which is called minimum path of the uncertainty (MPU)

(Abdallh et al., 2011a), defines the following objective function $\Gamma_{MPU}(\mathbf{u})$:

$$\Gamma_{MPU}(\mathbf{u}) = \left\| \Phi(\mathbf{u}, b^{\bullet}, \mathbf{d}) + \left.\frac{\partial \boldsymbol{\Phi}}{\partial b}\right|_{(b=b^{\bullet})} .\Delta b - \mathbf{W} \right\|^2 \tag{40}$$

The proposed methodology is mainly based on adapting, at each iteration step, the objective function that needs to be minimized with respect to the sensitivity of the model responses to the uncertain model parameter b. In (Equation 40), since b^* and consequently Δb are unknown in practice, this can not be implemented unless an approximation can be made for the Taylor coefficient of Δb. As an approximation of Δb, (Equation 40) can be rewritten as:

$$\Gamma_{MPU}(\mathbf{u}) = \left\| \Phi(\mathbf{u}, b^{\bullet}, \mathbf{d}) + \alpha \left.\left(\frac{\partial \boldsymbol{\Phi}}{\partial b}\right)\right|_{(b=b^{\bullet})} - \mathbf{W} \right\|^2 \tag{41}$$

with α being a constant obtained from a linear fitting.

It is possible to make an approximation of the Taylor coefficient α if many measurement samples are available so as to make a linear fitting between the vector$(\mathbf{W} - \Phi(\mathbf{u}, b^{\bullet}, \mathbf{d}))$ and vector $\frac{\partial \boldsymbol{\Phi}}{\partial b} |_{b=b^{\bullet}}$, each vector consists of K points. The linear fitting is done at each iteration using the present value of $\mathbf{u}$ at this specific iteration (Abdallh et al., 2012d).

When minimizing the objective function (Equation 41), a path (parameter values $\mathbf{u}^{(j)}$ for the jth iteration) is followed that is minimally affected by the uncertainties. Indeed, a linear forward model, i.e. $\Phi(\mathbf{u}, b^{\bullet}, \mathbf{d}) + \alpha \left(\frac{\partial \boldsymbol{\Phi}}{\partial b}\right) |_{b=b^{\bullet}}$, is used with the incorporation of the dependence to the uncertain parameter values and with the calculation of α.

In the case of implementing more than one uncertain parameter, (Equation 41) can be rewritten as:

$$\begin{aligned}\Gamma_{MPU}(\mathbf{u}) = &\left\| \Phi(\mathbf{u}, \mathbf{b}^{\bullet}, \mathbf{d}) + \alpha \left.\left(\frac{\partial \boldsymbol{\Phi}}{\partial b_1}\right)\right|_{(b_1=b_1^{\bullet})} \right. \\ &\left. + \beta \left.\left(\frac{\partial \boldsymbol{\Phi}}{\partial b_2}\right)\right|_{(b_2=b_2^{\bullet})} + \cdots + \gamma \left.\left(\frac{\partial \boldsymbol{\Phi}}{\partial b_q}\right)\right|_{(b_q=b_q^{\bullet})} - \mathbf{W} \right\|^2\end{aligned} \tag{42}$$

where α, β, γ, ... are the fitting constants.

7.2 A Bayesian approach for modeling error reduction

Generally, there are several techniques to model an electromagnetic device, depending on the required accuracy. Analytical and numerical models are often used for modeling an electromagnetic device. Analytical models are computationally fast but less accurate. Numerical models can be divided into several accuracy levels depending on the degree of freedom and discretization. The two or three dimensional finite element (FE) models (2D-FE or 3D-FE) are commonly used with different mesh discretization levels. In general, numerical models are often more accurate than the analytical ones, but much more computational time consuming; the higher the degree of freedom, the more time consuming.

In order to solve an inverse problem with the highest solution accuracy, one may incorporate the measurements with the most accurate (fine) model, e.g. 3D numerical FE model, of an electromagnetic device. However, the computational time of this approach may be prohibitive. Alternatively, less accurate (coarse) models, e.g. analytical models, can be used with a demerit of an expected higher recovery error compared to the fine models.

In this section, we propose the use of the Bayesian approximation error approach (Nissinen et al., 2008) for improving the inverse problem solution when a coarse model is utilized. The proposed approach adapts the objective function to be minimized with the *a priori* misfit between fine and coarse forward models, in which the modeling error is represented in a stochastic way. The Bayesian approximation error approach is relatively fast and easy to implement compared to the two-level techniques, such as space mapping (Bandler et al., 2008), manifold mapping (Echeverría et al., 2006), two-level refined direct method (Crevecoeur et al., 2011a), etc.

Assume that we have two computer models of an electromagnetic device; a fine and a coarse model. The fine model is assumed to be close to the reality where we assume that the modeling error in the fine model is negligible[8]. However, the modeling error in the coarse model is refereed to the misfit between the fine and coarse forward model responses. We present the stochastic modeling error in section 7.2.1.

7.2.1 Stochastic representation of the modeling error

The fine and coarse forward model responses, $\mathbf{\Phi}_f \in \mathbb{R}^K$ and $\mathbf{\Phi}_c \in \mathbb{R}^K$ respectively, depend on $\mathbf{u}$. Since the exact value of $\mathbf{u}$ is not known, we assume Z hypothetical values of the unknown model parameters, with $\hat{\mathbf{u}}_z$, $(z = 1, \ldots, Z)$ being a hypothetical value. These Z hypothetical values are chosen in a such a way that they are random and cover the domain defined by the lower and upper bounds of these parameters.

The error between the fine and coarse forward models, at each model observation k and at each test value $\hat{\mathbf{u}}_z$, can be represented by:

$$e_{m,k}(\hat{\mathbf{u}}_z) = \Phi_{f,k}(\hat{\mathbf{u}}_z) - \Phi_{c,k}(\hat{\mathbf{u}}_z), \quad (k = 1, \ldots, K), \quad (z = 1, \ldots, Z) \tag{43}$$

By performing Z coarse and fine forward model computations, and assuming that the modeling error at each model observation k ($e_{m,k}$) follows the normal distribution, i.e. ($e_{m,k} \sim \mathcal{N}(\mu_{m,k}, \sigma^2_{m,k})$), one may calculate the mean modeling error and its covariance, $\mu_{m,k}$ and $\sigma^2_{m,k}$, respectively.

$$\mu_{m,k} = \frac{1}{Z} \sum_{z=1}^{Z} e_{m,k}(\hat{\mathbf{u}}_z) \tag{44}$$

$$\sigma^2_{m,k} = \frac{1}{Z} \sum_{z=1}^{Z} \left(e_{m,k}(\hat{\mathbf{u}}_z) - \mu_{m,k}\right)^2 \tag{45}$$

[8] In fact, it is impossible to construct an exact computer model combining the whole physical phenomena as in the reality. However, in order to solve an inverse problem, a computer model is needed with the highest possible accuracy: the most accurate model give rise to the most accurate recovery results.

Based on the calculated $\mu_{m,k}$ and $\sigma^2_{m,k}$, the approximate overall probability distribution function (PDF), at each model observation k, can be visualized as:

$$f\left(e_{m,k}\right) = \frac{1}{\sqrt{2\pi\sigma^2_{m,k}}} \cdot \exp^{\left(-\left(e_{m,k}-\mu_{m,k}\right)^2/\left(2\sigma^2_{m,k}\right)\right)} \tag{46}$$

The assumption of the modeling error to be Gaussian distributed is a possible means of statistically expressing the modeling error, see (Nissinen et al., 2008). The vector representations of the modeling mean error and its covariance at all model observations K are:

$$\overline{\boldsymbol{\mu}}_m = \left[\mu_{m,1}, \mu_{m,2}, \ldots, \mu_{m,K}\right]^T, \ \overline{\sigma}^2_m = \left[\sigma^2_{m,1}, \sigma^2_{m,2}, \ldots, \sigma^2_{m,K}\right]^T \tag{47}$$

Since the modeling error is assumed uncorrelated, i.e. $e_{m,i}$ does not depend on $e_{m,j}$ $(i, j = 1, \ldots, K, \ i \neq j)$, the covariance matrix of the modeling error ($\overline{\boldsymbol{\Sigma}}^2_m \in \mathbb{R}^{K\times K}$) can be written as: $\overline{\boldsymbol{\Sigma}}^2_m = \text{diag}(\sigma^2_{m,1}, \sigma^2_{m,2}, \ldots, \sigma^2_{m,K})$. The mean and the covariance of the modeling error are used in the following section for the modeling error compensation.

7.2.2 Bayesian approach: Traditional and approximation error

As mentioned before, the actual measurements $\mathbf{W}$ can be expressed as:

$$\mathbf{W} = \boldsymbol{\Phi}(\mathbf{u}^*) + \mathbf{e} \tag{48}$$

with $\mathbf{e}$ being the uncertainty "error" vector. A possible difference between the simulated signals $\boldsymbol{\Phi}(\mathbf{u}^*)$ with the actual model parameters $\mathbf{u}^*$, and the measured signals $\mathbf{W}$, can arise from two reasons, and is denoted by:

$$\mathbf{e} = \mathbf{e}_n + \mathbf{e}_m \tag{49}$$

with $\mathbf{e}_n$ being the uncertainty due to the measurement noise and $\mathbf{e}_m$ being the uncertainty due to modeling uncertainties. Due to the random nature of the measurement noise, it is assumed to be normally white distributed with zero mean ($\overline{\mu}_{n,k} = 0$) and a covariance of $\overline{\sigma}^2_{n,k}$, i.e. ($\mathbf{e}_{n,k} \sim \mathcal{N}(0, \overline{\sigma}^2_{n,k})$) (Abdallh et al., 2012c). Similarly,

$$\overline{\boldsymbol{\mu}}_n = \left[\mu_{n,1}, \mu_{n,2}, \ldots, \mu_{n,K}\right]^T = \mathbf{0}, \ \overline{\sigma}^2_n = \left[\sigma^2_{n,1}, \sigma^2_{n,2}, \ldots, \sigma^2_{n,K}\right]^T \tag{50}$$

Since the measurement noise is assumed uncorrelated, i.e. $e_{n,i}$ does not depend on $e_{n,j}$ $(i, j = 1, \ldots, K, \ i \neq j)$, the covariance matrix of the measurement noise ($\overline{\boldsymbol{\Sigma}}^2_n \in \mathbb{R}^{K\times K}$) can be written as: $\overline{\boldsymbol{\Sigma}}^2_n = \text{diag}(\sigma^2_{n,1}, \sigma^2_{n,2}, \ldots, \sigma^2_{n,K})$.

The modeling error is also represented in normal distribution, see section 7.2.1. Due to these uncertainties, when minimizing the cost function (Equation 35), the values of the recovered parameters $\tilde{\mathbf{u}}$ and $\mathbf{u}^*$ are not necessary equal, i.e. $\tilde{\mathbf{u}} \neq \mathbf{u}^*$. Therefore, we propose the use of the statistical Bayesian approach.

1. **Traditional Bayesian approach**:
   In the Bayesian framework, the identification problem is seen as a statistical inference problem, sometimes referred to as stochastic regularization (Emery et al., 2007), in which the measurements and the modeled response are assumed to be random (Nissinen et al.,

2008). In the traditional Bayesian approach, the modeling error is assumed to be negligible, i.e. $\mathbf{W} = \mathbf{\Phi}(\mathbf{u}^*) + \mathbf{e}_n$.

In the well known Bayes' formula, the posterior probability density function of the measurements $\mathbf{W}$ given the unknown model parameters $\mathbf{u}$ '$P(\mathbf{u}|\mathbf{W})$' is given by (Kaipio & Somersalo, 2005):

$$P(\mathbf{u}|\mathbf{W}) = \frac{P(\mathbf{u})P(\mathbf{W}|\mathbf{u})}{P(\mathbf{W})} \tag{51}$$

which can be written in a non-normalized form:

$$P(\mathbf{u}|\mathbf{W}) \propto P(\mathbf{u})P(\mathbf{W}|\mathbf{u}) \tag{52}$$

with $P(\mathbf{u})$ being the prior probability density function of the unknown model parameters. In our application, no information is given for the $P(\mathbf{u})$. So, we assume that the unknown model parameters follow the uniform distribution between lower and upper bounds:

$$P(\mathbf{u}) = \frac{1}{\mathbf{u}_{UB} - \mathbf{u}_{LB}}, \quad \mathbf{u} \in [\mathbf{u}_{LB}, \mathbf{u}_{UB}] \tag{53}$$

with $\mathbf{u}_{LB}$ and $\mathbf{u}_{UB}$ being the lower and upper bounds of the unknown model parameters, respectively, which can be known from the reasonable physical representation of $\mathbf{u}$.

Assuming that the measurement noise $\mathbf{e}_n$ does not depend on the unknown model parameters $\mathbf{u}$, the likelihood density function of the measurements $\mathbf{W}$ given the unknown model parameters $\mathbf{u}$ can be written as (Kaipio & Somersalo, 2005):

$$\begin{aligned} P(\mathbf{W}|\mathbf{u}) = & \frac{1}{(2\pi)^{K/2}\prod_{k=1}^{K}\sigma_{n,k}} \\ & \cdot \exp\left(-\frac{1}{2}\left[[\mathbf{W} - \mathbf{\Phi}(\mathbf{u}) - \overline{\mu}_n]^T \left(\overline{\mathbf{\Sigma}}_n^2\right)^{-1} [\mathbf{W} - \mathbf{\Phi}(\mathbf{u}) - \overline{\mu}_n]\right]\right) \end{aligned} \tag{54}$$

Therefore, in order to solve this inverse problem, the *maximum a posteriori* (MAP) estimates is used, in which the MAP of the unknown model parameters $\mathbf{u}$ is given by:

$$\mathbf{u}_{\mathrm{MAP}} = \arg\max_{\mathbf{u}} \; P(\mathbf{u}|\mathbf{W}) \tag{55}$$

Substituting (Equation 52) and (Equation 54) in (Equation 55), and ($\overline{\mu}_n = 0$):

$$\begin{aligned} \mathbf{u}_{\mathrm{MAP,\,Trad}} &= \arg\max_{\mathbf{u}} \; P(\mathbf{W}|\mathbf{u}) \\ &= \arg\max_{\mathbf{u}} \; \left\{ \exp\left(-\frac{1}{2}\left[[\mathbf{W} - \mathbf{\Phi}(\mathbf{u})]^T \left(\overline{\mathbf{\Sigma}}_n^2\right)^{-1} [\mathbf{W} - \mathbf{\Phi}(\mathbf{u})]\right]\right) \right\} \\ &= \arg\min_{\mathbf{u}} \; \left\{ [\mathbf{W} - \mathbf{\Phi}(\mathbf{u})]^T \left(\overline{\mathbf{\Sigma}}_n^2\right)^{-1} [\mathbf{W} - \mathbf{\Phi}(\mathbf{u})] \right\} \\ &= \arg\min_{\mathbf{u}} \; \left\| \mathbf{L}_n (\mathbf{W} - \mathbf{\Phi}(\mathbf{u})) \right\|^2 \end{aligned} \tag{56}$$

with $\mathbf{L}_n$ being the Cholesky factor of the covariance of the measurement noise, i.e. $\left(\overline{\mathbf{\Sigma}}_n^2\right)^{-1} = \mathbf{L}_n^T\mathbf{L}_n$. Solution of (Equation 56) is the recovered model parameter ($\mathbf{u}_{\mathrm{MAP,\,Trad}} \equiv \widetilde{\mathbf{u}}$) using the inverse problem in the traditional Bayesian framework.

2. **Bayesian approximation error approach**:

   In the traditional Bayesian approach, the modeling error $\mathbf{e}_m$ was assumed to be negligible. However, in the Bayesian approximation error approach, the $\mathbf{e}_m$ is taken into account. As discussed earlier in section 7.2.1, the modeling error exists only when the coarse model is incorporated in the inverse problem, so (Equation 48) can be rewritten as:

$$\mathbf{W} = \mathbf{\Phi}_c(\mathbf{u}^*) + \mathbf{e}_n + \mathbf{e}_m \tag{57}$$

   Due to the Gaussian distribution of both the measurement noise and modeling error, the overall error $\mathbf{e}$ is therefore also Gaussian distributed. Similarly, (Equation 56) can be reformulated as follows:

$$\begin{aligned} \mathbf{u}_{\text{MAP, Compensated}} &= \arg\max_{\mathbf{u}} \quad P(\mathbf{W}|\mathbf{u}) \\ &= \arg\min_{\mathbf{u}} \quad \left\| \mathbf{L}_{n+m}(\mathbf{W} - \mathbf{\Phi}_c(\mathbf{u}) - \overline{\boldsymbol{\mu}}_m) \right\|^2 \end{aligned} \tag{58}$$

   with $\mathbf{L}_{n+m}$ being the Cholesky factor of the covariance of the overall error, i.e. $\left(\overline{\mathbf{\Sigma}}_n^2 + \overline{\mathbf{\Sigma}}_m^2\right)^{-1} = \mathbf{L}_{n+m}^T \mathbf{L}_{n+m}$. Solution of (Equation 58) is the recovered model parameter ($\mathbf{u}_{\text{MAP, Compensated}} \equiv \tilde{\mathbf{u}}$) using the inverse problem in the Bayesian approximation error approach, in which the modeling error is compensated.

8. Magnetic material characteristics

There are several properties of the material to be recovered, however, we focus only on the magnetic properties. Specifically, the single-valued *B-H* curve, the hysteresis loops and loss parameters are being recovered here. In order to identify the magnetic characteristics of a material using the inverse problem approach, the magnetic parameters need to be formulated in mathematical formulas to be used in the model.

1. Single-valued *B-H* curve:

   The single-valued nonlinear constitutive characteristic of the *B-H* curve (normal magnetizing characteristic), to be reconstructed by the inverse problem, is modeled by means of three parameters $\mathbf{u}_1 = [H_0,\ B_0,\ \nu]$ of the (non-full) power-series formula (Abdallh et al., 2010a):

$$\frac{H}{H_0} = \left(\frac{B}{B_0}\right) + \left(\frac{B}{B_0}\right)^{\nu} \tag{59}$$

2. Hysteretic characteristic:

   The hysteretic characteristic of the magnetic material, is modeled by means of five parameters $\mathbf{u}_2 = [a,\ b,\ c,\ k_1,\ k_2]$ of the Lorentzian distribution of the scalar Preisach model (Abdallh et al., 2010d):

$$P(\alpha, \beta) = \frac{k_1}{\left(1 + \left(\frac{\alpha - a}{b}\right)^2\right)\left(1 + \left(\frac{\beta + a}{b}\right)^2\right)} + \delta_{\alpha,\beta} \frac{k_2}{1 + \left(\frac{\alpha}{c}\right)^2} \tag{60}$$

   where $P(\alpha, \beta)$ and $\delta_{\alpha,\beta}$ are the Preisach distribution function and the Kronecker delta symbol, respectively (Bertotti, 1998).

3. Loss parameters:

   The well-known Bertotti loss model, in its simplified form, is used for loss characterization (Bertotti, 1988):

$$P_{lp} = a_1 B_{lp}^{\alpha_1} f + b_1 B_{lp}^2 f^2 + c_1 B_{lp} f \left(\sqrt{\left(1 + d_1 B_{lp} f\right)} - 1 \right) \tag{61}$$

   with B_{lp} and f being the local peak magnetic induction in (T) and the frequency in (Hz), respectively. P_{lp} is the local iron loss. $\mathbf{u}_3 = [a_1, \alpha_1, b_1, c_1, d_1]$ is the vector of the loss model parameters. The values of these parameters are unknown and need to be identified using an inverse approach.

9. Application

The proposed inverse methodology is applied here into the following application, i.e. an EI electromagnetic core inductor, for the complete identification of the magnetic properties of the magnetic material. However, the proposed methodology can be applied for any other application, e.g. a permanent magnet synchronous machine (Sergeant et al., 2009), a switched reluctance motor (Abdallh et al., 2011c), (Abdallh et al., 2011a), an asynchronous motor (Abdallh et al., 2012b).

9.1 Problem definition

The magnetic material characteristics (the normal magnetizing curve, the hysteretic characteristics and the loss parameters) of the EI electromagnetic core inductor, shown in Fig. 3, are unknown and need to be recovered using the inverse approach.

In this application, we consider two magnetic measurements, i.e. the local and global measurements, as being the input of the inverse problem. The local magnetic induction measurements are carried out using the needle probe method as described in (Abdallh & Dupré, 2010c), at several positions on the EI profile. Global measurements use the current and voltage measurements, i.e. no local measurements are performed.

In practice, the uncertainties of all geometrical model parameters of the EI core are magnetically of second order, except the values of the two air gaps, g_1 and g_2. The mean value of the g_1 is given by the manufacturer, however the mean value of the g_2 is the thickness of the spacer inserted between the E and I yokes. In this study, we use g_2 as 0 or 0.25 mm. It is assumed that the local or global measurements are corrupted by Gaussian noise with zero mean and a standard deviation of $\sigma_{n,local}$ or $\sigma_{n,global}$, respectively. Moreover, the standard deviations of the uncertain geometrical model parameters are denoted by σ_{g_1} and σ_{g_2}.

9.2 Traditional inverse problem formulation

The three magnetic material parameters of the single-valued B-H characteristic $\mathbf{u}_1 = [H_0, B_0, \nu]$, the five parameters of the Lorentzian distribution of the scalar Preisach model $\mathbf{u}_2 = [a, b, c, k_1, k_2]$, and the five loss parameters $\mathbf{u}_3 = [a_1, \alpha_1, b_1, c_1, d_1]$ are recovered using the inverse problem approach. The unknown parameter values can be retrieved by iteratively minimizing predefined objective functions. The objective functions are formulated as the quadratic difference between the measured and the simulated local magnetic induction

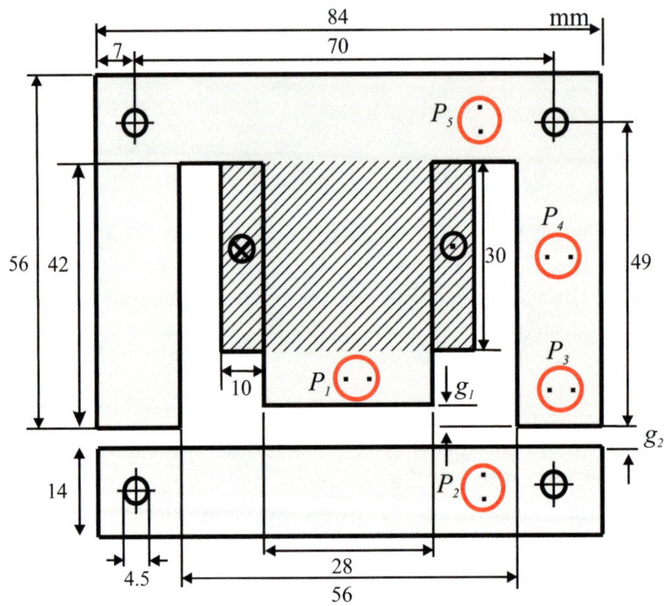

Fig. 3. Schematic diagram of the studied EI electromagnetic core inductor.

at a specific condition.

$$\widetilde{\mathbf{u}}_i = \arg\min_{\mathbf{u}_i} \; \Gamma(\mathbf{u}_i), \quad i = [1,\, 2,\, 3] \tag{62}$$

where $\widetilde{\mathbf{u}}_1$, $\widetilde{\mathbf{u}}_2$ and $\widetilde{\mathbf{u}}_3$ are the 'recovered' magnetic material parameters of the single-valued *B-H* curve, the hysteretic characteristics and loss parameters, respectively.

In the following, we formulate four objective functions. The first and second objective functions are devoted for recovering $\mathbf{u}_1$. The third and fourth objective functions are used for recovering $\mathbf{u}_2$ and $\mathbf{u}_3$, respectively.

1. First objective function:
   The first objective function Γ_1 is implemented using the amplitude (i_k) of the kth sinusoidal excitation current ($k = 1, \ldots, K = 40$) and the local magnetic induction measurements, at a fixed position as shown in Fig. 3:

$$\Gamma_1(\mathbf{u}_1) = \sum_{k=1}^{K} \left\| B_s(i_k, \mathbf{u}_1) - B_m(i_k) \right\|^2_{position\ j} \quad j = [1,\, 2,\, \ldots,\, 5] \tag{63}$$

   with $B_m(i_k)$ being the measured peak magnetic induction value of the kth excitation current and $B_s(i_k, \mathbf{u}_1)$ being the corresponding simulated local flux densities using the numerical model and the material parameter values $\mathbf{u}_1$.

2. Second objective function:
   The second objective function (Γ_2) is implemented using global measurements of the excitation current I and the voltage V of the excitation winding, where no local measurements are used. This global measurement gives rise to the flux density linked with the excitation winding:

$$\phi_m(t) = (1/N_1)\left[\int_0^t V(\tau)d\tau - R\int_0^t I(\tau)d\tau\right] \tag{64}$$

$$\Gamma_2(\mathbf{u}_1) = \sum_{k=1}^{K} \left\|\phi_s(i_k, \mathbf{u}_1) - \phi_m(i_k)\right\|^2 \tag{65}$$

with $\phi_m(i_k)$ being the measured peak magnetic flux value of the kth excitation current ($k = 1, \ldots, K = 40$), and $\phi_s(i_k, \mathbf{u})$ being the corresponding simulated value. N_1 and R are the number of turns of the excitation winding ($N_1 = 356$), and the resistance of the excitation coil ($R = 1.3$ Ohm), respectively.

3. Third objective function:

   The third objective function Γ_3 is constructed as follows:

$$\Gamma_3(\mathbf{u}_2) = \sum_{l=1}^{L}\sum_{n=1}^{T} \left\|b_{s,l}(t_n, \mathbf{u}_2) - b_{m,l}(t_n)\right\|^2_{position\ j} \quad j = [1, 2, \ldots, 5] \tag{66}$$

   where $b_{m,l}(t_n)$ is the measured magnetic induction value of the nth time step ($n = 1, ..., T = 1000$) for the lth magnetization loop and $b_{s,l}(t_n, \mathbf{u}_2)$ is the corresponding simulated local flux densities using the numerical model.

4. Fourth objective function:

   After recovering $\mathbf{u}_1$, the values of $\mathbf{u}_3$ are recovered *a posteriori*. Using a specific value of the frequency f considered in the iron loss measurements, which is not involved in the static computation mode, the iron loss P_{lp}, in W/kg, can be computed using (Equation 61). In order to calculate the overall iron loss of the electromagnetic device P_{total}, in (W), the local iron loss $P_{l,i}$ is multiplied by the mass of the EI inductor (m_{EI}), i.e. $P_{\text{total}} = P_{lp} \times m_{EI}$. In order to identify the five parameters of the loss model $\mathbf{u}_3$, the following objective function is formulated:

$$\Gamma_4(\mathbf{u}_3) = \left\|P_{\text{total}}(\mathbf{u}_3, \tilde{\mathbf{u}}_1, f) - P_m(f)\right\|^2 \tag{67}$$

   with P_m being the measured iron loss of the EI core inductor.

9.3 *A priori* error estimation

Depending on the nature of the measurements that are used as input for the inverse problem, a certain resolution or accuracy of the recovered magnetic material parameter values is achieved (Abdallh et al., 2011b). Possible measurements are local (on a specific part of the geometry) and global (whole considered geometry) magnetic measurements. These measurement modalities contain measurement noise which decreases the accuracy of the inverse problem solution. Additionally, the uncertainties of important model parameters, i.e. air gaps, influence the resolution. We assume here $\sigma_{n,local} = \sigma_{n,global} = 0.025$ and $\sigma_{g_1} = \sigma_{g_2} = 0.025$. Therefore, the sCRB method, presented in section 6.2, is used for selecting *a priori* the best measurement modality that results in the highest accuracy, taking into account the measurement noise and the geometrical uncertainties.

The optimum measurement modality follows the criterion that this modality has the minimum estimated uncertainty when using the sCRB technique. The prior uncertainty estimation can be implemented before carrying out the real experimental measurements.

Therefore, numerical experiments are carried out in the sense that the 'numerical measurements' quantities (B_m or ϕ_m) are 'modeled' as the output of the numerical direct model that has as input the following fictitious single-valued characteristics $\mathbf{u}_{\text{fic}} = [100, 1.1, 8]$. The output is furthermore corrupted by Gaussian noise with zero mean and a standard deviation of σ_n, assumed as $\sigma_{n,local} = \sigma_{n,global} = 0.025$. The 'numerical' local magnetic induction measurements are carried out at the five different positions P_1 - P_5, shown in Fig. 3.

In fact, and for clarity, the selection process is done in two stages. The first stage aims at determining the optimum placement for carrying out the local measurements among the five different positions P_1 - P_5. The second stage aims at selecting the type of the measurement, i.e. global or local measurements. In the second stage, we compare the global measurements with the local measurements at the optimum position.

For the comparative issue, the estimated error (EE) is defined as the percentage estimated error in the recovered material characteristics [9] and is formulated as:

$$EE = \left| \frac{RMS_{BH}(\mathbf{u}_{\text{fic}} + \sigma_{u,M})}{RMS_{BH}(\mathbf{u}_{\text{fic}})} - 1 \right| \times 100\% \tag{68}$$

with $\sigma_{u,M}$ being the lower bound of the standard deviation of the unknown parameters obtained from the sCRB analysis, see (Equation 33). RMS_{BH} is the root mean square of the B-H curve, given by $RMS_{BH} = \sqrt{\sum_{k=1}^{K} \frac{B^2(H_k)}{K}}$.

9.3.1 Optimal needle placement

Table 1 shows the "theoretical" EE values based on the sCRB due to the measurement noise and the uncertainty in g_1 and g_2. It is clear from this table that the local measurements carried out at position 2 (P_2) results in the best inverse problem results, and the local measurements carried out at positions 1 (P_1) and 5 (P_5) result in the worst inverse problem results. Also, Table 1 depicts that the g_2 is a more critical parameter than g_1. Similar results are obtained using a deterministic methodology based on first order sensitivity analysis, see (Abdallh et al., 2012a).

Position	1	2	3	4	5
(a) The EE value due to σ_n and σ_{g_1}	44.68	34.60	39.72	36.17	36.28
(b) The EE value due to σ_n and σ_{g_2}	50.94	43.91	51.62	44.90	44.72

Table 1. The EE values based on the sCRB due to (a) the measurement noise and the uncertainty in g_1 ($\sigma_n = 0.025, \sigma_{g_1} = 0.025$), (b) the measurement noise and the uncertainty in g_2 ($\sigma_n = 0.025, \sigma_{g_2} = 0.025$).

9.3.2 Selection measurement modality

In this section, the sCRB is used for selecting the best measurement modality, i.e. global or local, that results in the best solution. We compare the inverse problem based on global

[9] It is worth mentioning that EE can not be calculated in practice, because the knowledge of the sought-after parameters ($\mathbf{u}$) is unknown. However, we use EE in order to estimate the uncertainty in the inverse problem solution in a 'qualitative' way rather than a 'quantitative' way.

measurements, i.e. Γ_2, with the inverse problem based on the local magnetic induction, i.e. Γ_1, at position 2 (P_2) because at this position the best results are observed. Table 2 shows the "theoretical" *EE* values based on the sCRB due to the measurement noise and the uncertainty in g_2. It is clear from this table that the inverse problem based on Γ_1 gives better results than the one based on Γ_2.

Objective function	Γ_1	Γ_2
The *EE* value	43.91	53.24

Table 2. The *EE* values based on the sCRB due to the measurement noise and the uncertainty in g_2 ($\sigma_n = 0.025, \sigma_{g_2} = 0.025$) for Γ_1 and Γ_2.

9.4 Traditional inverse problem

The magnetic properties of the magnetic material are recovered using the traditional formulation of the inverse approach. Traditional inverse problem means that no quantitative error reduction techniques are used, e.g. MPU or Bayesian techniques. Therefore, the value of g_2 is kept zero in order to eliminate the error initiates from the uncertainty in its value.

9.4.1 Identification of the single-valued *B-H* curve

In order to validate experimentally the results obtained from the sCRB analysis, three 'traditional' inverse problems are solved. The local magnetic induction measurements at position 1 (P_1) and position 2 (P_2) are used for the first two inverse problems. The third inverse problem is solved starting from the global measurements.

Fig. 4 shows the recovered *B-H* curve, using the measured signals at P_1 and P_2 and the global measurements compared to the actual *B-H* characteristic[10]. It is clear that the recovered characteristic based on measurements at P_2 is much closer to the actual *B-H* characteristic than the recovered characteristic based on measurements at P_1. Moreover, it is also clear that the recovered characteristic based on the local measurements at P_2 is better than the one based on the global measurements. The values of the recovered parameters compared to the actual values are also shown in Fig. 4. These results validate the sCRB theoretical results.

9.4.2 Identification of the hysteretic characteristics

In order to recover the hysteretic characteristics, the 'traditional' inverse problem based on (Equation 62) and (Equation 66) is solved. For simplicity, and the best results, we solve the inverse problem based on the local measurements carried out at position 2 (P_2). Fig. 5 shows the recovered hysteresis loops of the EI core inductor material using the proposed inverse problem method, compared to the actual magnetic hysteresis loops. A good correspondence between the measured hysteresis loops on the fully wound magnetic ring core, and the recovered one is observed.

[10] The actual magnetic characteristics were measured on the fully wound magnetic ring core made from the same material as the material of the EI core inductor, $\mathbf{u}_1^* = [292.03, 1.35, 11.99]$, $\mathbf{u}_2^* = [124.99, 38.01, 504.08, 1.89, 5.15]$, $\mathbf{u}_3^* = [0.0351, 1.671, 7.2557e^{-4}, 0, 0]$.

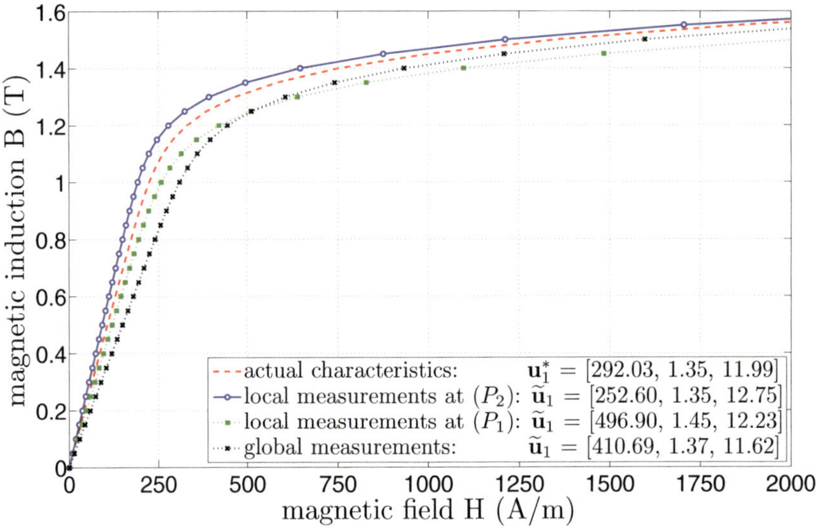

Fig. 4. Recovered *B*-*H* characteristics based on real B_m at P_2 and P_1 and ϕ_m compared to the actual characteristic, ($g_2 = 0$).

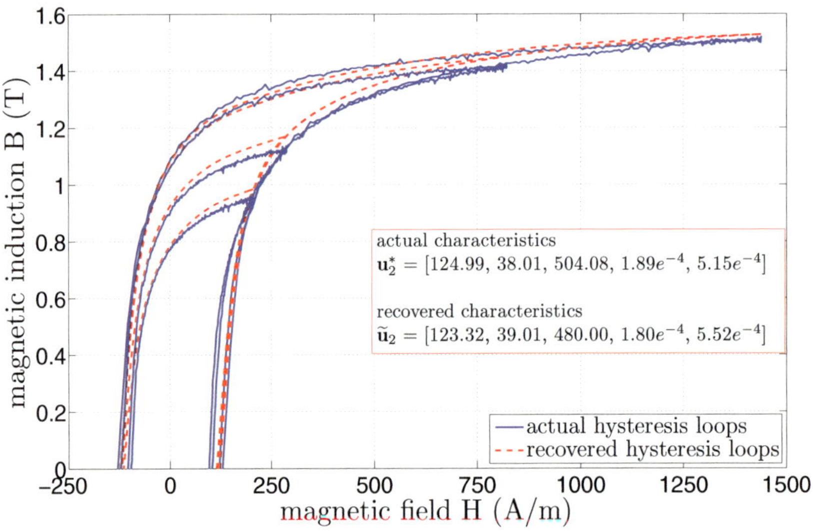

Fig. 5. The half of the hysteresis loops using Γ_2 compared to the actual hysteretic characteristic at ($f = 1$ Hz.).

9.4.3 Identification of the loss parameters

After retrieving the values of $\tilde{\mathbf{u}}_1$, an *a posteriori* inverse problem, based on (Equation 62) and (Equation 67), is solved for the identification of the loss parameters of the EI core material.

For simplicity, and due to the fact that the excess loss is absent in the studied material c_1, $d_1 = 0$ in (Equation 61), we restrict ourselves only to the identification of the first three parameters in $\mathbf{u}_2$. Fig. 6 shows the reconstructed loss characteristics of the EI core material, for different

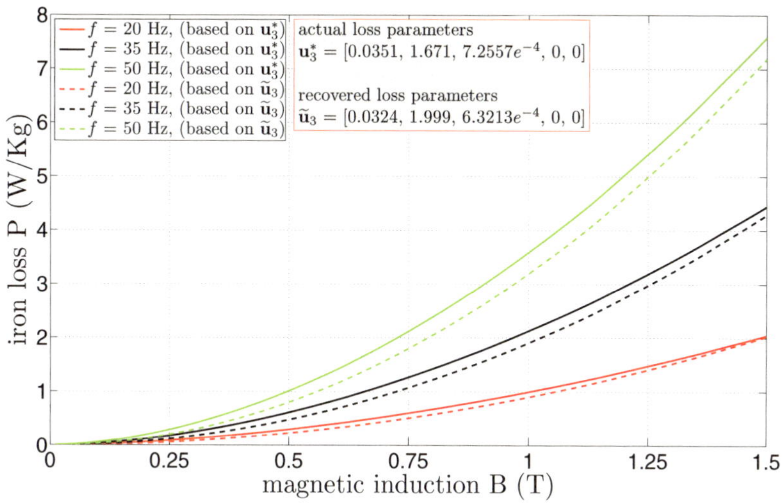

Fig. 6. The loss characteristics of the EI core material based on the reconstructed loss parameters compared to loss characteristics based on the original loss parameters.

frequency and magnetic induction levels, compared to the actual loss characteristics. Again, a good correspondence is observed. It is clear from Fig. 6 that the original grade of the EI core material at 1.5 T is 7.5 W/kg, however, the estimated one at 1.5 T is 7.15 W/kg.

9.5 Error reduction

In this section, the error originating from the uncertain geometrical model parameters is decreased by implementing the proposed MPU technique presented in section 7.1. Furthermore, a Bayesian approximation error approach, presented in section 7.2, is utilized for the error reduction when a coarse model is used in the inverse problem.

9.5.1 MPU inverse problem formulation

Due to the fact that g_2 is the most critical geometrical parameter, we use the MPU technique for reducing the error originating only form the uncertainty in the value of g_2. The traditional objective function Γ_1 is reformulated as follows:

$$\Gamma_{1,MPU}(\mathbf{u}) = \sum_{k=1}^{K} \left\| \mathbf{B}_s(\mathbf{u}, g_2) + \alpha \left(\frac{\partial \mathbf{B}_s(\mathbf{u}, g_2)}{\partial g_2} \right) - \mathbf{B}_m \right\|^2_{(g_2 = g_2^\bullet,\ position\ 2)} \tag{69}$$

with constant α. $g_2^\bullet$ is the value of g_2 used in the forward model. It is possible to make an approximation of the Taylor coefficient if many measurement samples are available so as to make a linear fitting between the vector $(\mathbf{B}_m - \mathbf{B}_s(\mathbf{u}, g_2))$ and vector $\left(\frac{\partial \mathbf{B}_s(\mathbf{u}, g_2)}{\partial g_2}\right)$. When minimizing the objective function (Equation 69), a path (parameter values $\mathbf{u}^{(j)}$ for the jth iteration) is followed that is minimally affected by the uncertainties. Indeed, a linear forward model, i.e. $B_s(I_k, \mathbf{u}, g_2) + \alpha \left(\frac{\partial B_s(I_k, \mathbf{u}, g_2)}{\partial g_2}\right)$, is used with the incorporation of the dependence to the uncertain parameter values and with the estimate of the uncertainty (g_2).

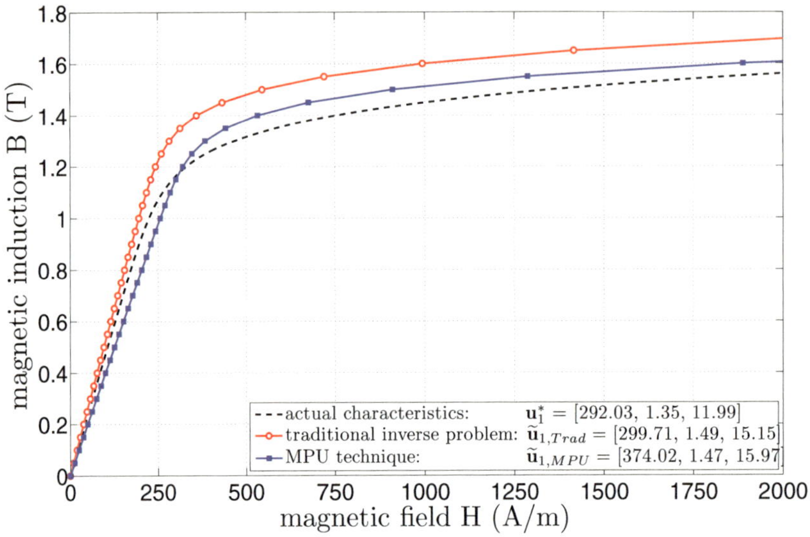

Fig. 7. The recovered *B-H* curves based on the traditional and the MPU techniques compared to the original characteristics, for the experimental validation of the proposed technique on the EI core inductor application, ($g_2 = 0.25$ mm).

In order to validate 'experimentally' the proposed MPU technique, two inverse problems (traditional and MPU) are solved starting from real measurements. The local magnetic induction is measured, at P_2, for $K = 16$ sinusoidal excitation current values $I = [0.5, 1, 1.5, \ldots, 8]$ A, and for $g_2 = 0.25$ mm. The value of g_2 is taken from the thickness of the inserted solid spacer between the E- and I-yokes. The inverse problems in the traditional and MPU formulation are solved yielding two recovered *B-H* curves. These *B-H* curve results are compared with the original *B-H* curve. Fig. 7 depicts the recovered *B-H* curves based on the traditional and the MPU techniques compared to the original characteristics, for $g_2 = 0.25$ mm. It is clear from Fig. 7 that the MPU technique results in a more accurate recovered characteristics compared to the traditional inverse problem.

9.5.2 Bayesian approximation error approach

In all results presented above, we use a high fidelity model of the EI core, i.e. 3D finite element model with very fine mesh discretizations. However, this model is very time consuming. One may reduce the burden of the inverse problem by incorporating a relatively faster but not accurate coarse model. Using the Bayesian approximation error approach, presented in section 7.2, it is possible to reduce the error initiated from incorporating a coarse model instead of a fine model in the inverse problem.

For the sake of the comparison, we build three computer models: a very fine model based on 3D-FE with fine mesh discretizations 'model-a', a moderate fine model based on 3D-FE with coarse mesh discretizations 'model-b', and a coarse model based on a magnetic reluctance network 'model-c' in a similar way as presented in (Cale et al., 2006). Here, we consider 'model-a' as the fine model, however, the other two models, i.e. 'model-b' and 'model-c', are considered as 'relatively' coarse models.

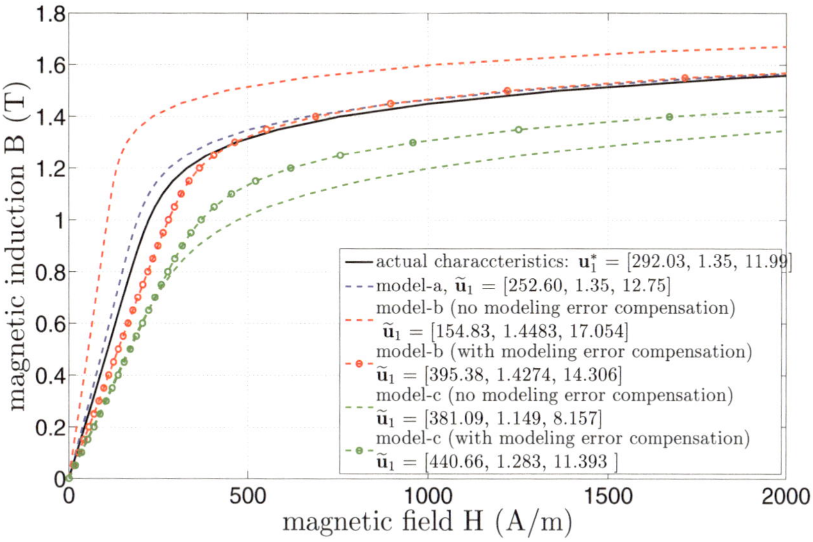

Fig. 8. The recovered *B-H* curve using the two inverse problems based on the traditional Bayesian approach 'no modeling error compensation' and the Bayesian approximation error approach 'with modeling error compensation' for the models-b and c, and the recovered characteristics for model-a, compared to the actual characteristics, ($g_2 = 0$).

The value of g_2 is kept zero in order to eliminate the modeling uncertainty caused by the uncertain value of g_2, as previously explained. The local magnetic induction measurements at position 2 (P_2) are used as being the input of the inverse problem.

In order to represent the modeling error between the fine and the coarse model stochastically, the procedure presented earlier in section 7.2.1 is used. Z hypothetical values of $\mathbf{u}$ are generated using the Latin hypercube sampling technique (Viana et al., 2010). Z forward fine and coarse model computations are solved, and their responses 'local magnetic induction measurements' (i.e. $\mathbf{B}_s$) are compared.

Different inverse problems are solved, with the assumption[11] of $\mathbf{e}_n = 0$, for each computer model. Then, the identified magnetic characteristics (single-valued *B-H* curve) are compared. For model-a, only one inverse problem is solved based on the traditional Bayesian approach, (Equation 56). However, for models-b and c, four inverse problems are solved, two for each computer model. The first inverse problem is based on the traditional Bayesian approach, (Equation 56), in which the modeling error is not compensated. While, the other inverse problem is based on the Bayesian approximation error approach, (Equation 58), in which the modeling error is compensated.

Fig. 8 illustrates the solution of the five inverse problems compared to the original characteristics. It is clear from Fig. 8 that the inverse problem based on (Equation 58) gives better results compared to the one based on (Equation 56) for both relatively coarse models. Of course, model-b results in a better solution compared to model-c due to the fact that model-c

[11] This assumption is reasonable because the local magnetic induction measurements contain only a very limited amount of noise. In addition, this small measurement noise is difficult to quantify precisely in practice.

is coarser than model-b. The results presented in this section validate 'experimentally' the proposed approach.

10. General discussion

In the previous sections, we applied the proposed inverse procedure for identifying the properties of the magnetic material inside an electromagnetic device (EI core electromagnetic inductor). The recovery error is 'qualitatively' estimated using the Cramér-Rao lower bound technique. Moreover, two different techniques are shown for a 'quantitative' reduction in the recovery error. The MPU technique 'iteratively' adapts the objective function to be minimized with respect to the uncertain model parameter. The Bayesian approximation error approach is used for compensating the modeling error originated from simplification of the mathematical model.

We are convinced that the stochastic Cramér-Rao bound method is the crucial part in the proposed methodology; it is fast, accurate and gives a better 'qualitative' view for the inverse problem results. Indeed, it neither depends on the fictitious parameter values, nor the accuracy of the used mathematical model. Moreover, it can be applied *a priori* (before solving the real inverse problem). The two other 'relatively time consuming' techniques, i.e. MPU and the Bayesian approximation error, are important for a 'quantitative' reduction of the recovery error. In fact, the three techniques, i.e. CRB, MPU and Bayesian, are not competitive techniques. Rather, they are integrative techniques. Also, it is worth mentioning that the 'true' values of the magnetic material properties are not needed in the identification procedure, we use it here only for validating the inverse procedure. All results presented using the three techniques are consistent.

In general, we think that the inverse problems based on local magnetic measurements are better than that based on global magnetic measurements or mechanical measurements, i.e. torque measurements in rotating electrical machines (Abdallh et al., 2011b). In practice, the local measurement should be carried out in regions with less stray fields, e.g. far from the excitation sources, sample edges and air gap (Abdallh & Dupré, 2010b). Also, the sensors should be precisely positioned and calibrated (Abdallh & Dupré, 2010c). The optimal number of sensors can be determined using the CRB technique, or using a special selection procedure, based on the presented MPU technique, such as the one presented in (Yitembe et al., 2011).

The most accurate knowledge about the model parameters increases the accuracy of the inverse problem. However, it is not essential to know precisely the values of the model parameters. MPU technique can be used for reducing the error of the uncertainties in these model parameters provided that the sensitivity analysis with respect to these model parameters can be calculated. Furthermore, coupling the measurements with the most accurate fine 'time demanding' model results in the best recovery solution. Again, the Bayesian approximation error approach can be used for reducing the modeling error when a 'time efficient' coarse model is incorporated in the inverse procedure, provided that a stochastic modeling error is existed.

Although we applied, in this chapter, the proposed methodology for the identification of the magnetic material inside relatively large scale industrial applications, such as the considered EI electromagnetic core inductor, the inverse procedure can be used also in small scale applications, such as the reconstruction of magnetic nanoparticles (Crevecoeur et al., 2012).

Also, the inverse procedure can be used for nondestructive magnetic evaluation of defects in magnetic materials (Durand et al., 2006).

11. Conclusions

An efficient coupled experimental-numerical inverse problem is proposed for characterizing the properties of a magnetic material. The proposed methodology is capable of estimating *a priori* the error in the inverse problem solution. This *a priori* 'qualitative' error estimation takes into account the measurement noise and the uncertainties in the model parameters in a stochastic framework based on the Cramér-Rao lower bound technique. Moreover, a deterministic method based on the sensitivity analysis is presented for reducing 'quantitatively' the error due to the uncertainties in the model parameters. Furthermore, the stochastic Bayesian approach is adapted in order to reduce the error in the inverse problem solution present when a coarse model is used. The complete procedure of the presented inverse problem methodology ensures not only the optimum experimental design but also the best numerical scheme. The proposed methodology is applied for identifying the magnetic properties of a magnetic material inside an EI electromagnetic core. The methodology is tested numerically and validated experimentally. The obtained results reveal the success and the reliability of the proposed scheme, which can be used for any other application. Finally, the main advantage of the proposed procedure is that it offers an efficient and consistent framework for the magnetic material characterization using the coupled experimental-numerical inverse problem approach. To our knowledge, it is the first time to propose a complete inverse problem procedure for identifying the properties of magnetic materials. On the other hand, it is the case that the proposed methodology is much more time consuming compared to the direct identification techniques based only on measurement results. However, this is the 'cost' that is needed for the 'non-destructive' identification of a magnetic material inside a complex geometry. Further research will focus on solving stochastic inverse problems in order to identify more magnetic material properties.

12. Acknowledgment

The authors gratefully acknowledge the financial support of the projects FWO-G.0082.06 of the Fund of Scientific Research-Flanders, and GOA07/GOA/006 of the "Bijzonder Onderzoeksfonds" of Ghent University.

13. References

Abdallh, A.; Sergeant, P.; Crevecoeur, G.; Vandenbossche, L.; Dupré, L. & Sablik, M. (2009). Magnetic material identification in geometries with non-uniform electromagnetic fields using global and local magnetic measurements, *IEEE Transactions on Magnetics*, Vol. 45, pp. 4157-4160.

Abdallh, A.; Crevecoeur, G. & Dupré, L. (2010a). Optimal needle placement for the accurate magnetic material quantification based on uncertainty analysis in the inverse approach, *Measurement Science and Technology*, Vol. 21, pp. 115703(16pp).

Abdallh, A. & Dupré, L. (2010b). A Rogowski-Chattock coil for local magnetic field measurements: sources of error, *Measurement Science and Technology*, Vol. 21, pp. 107003(5pp).

Abdallh, A. & Dupré, L. (2010c). Local magnetic measurements in magnetic circuits with highly non-uniform electromagnetic fields, *Measurement Science and Technology*, Vol. 21, pp. 045109(10pp).

Abdallh, A.; Sergeant, P.; Crevecoeur, G. & Dupré, L. (2010d). An inverse approach for magnetic material characterization of an EI core electromagnetic inductor, *IEEE Transactions on Magnetics*, Vol. 46, pp. 622-625.

Abdallh, A.; Crevecoeur, G. & Dupré, L. (2011a). A robust inverse approach for magnetic material characterization in electromagnetic devices with minimum influence of the air gap uncertainty, *IEEE Transactions on Magnetics*, Vol. 47, pp. 4364-4367.

Abdallh, A.; Crevecoeur, G. & Dupré, L. (2011b). Selection of measurement modality for magnetic material characterization of an electromagnetic device using stochastic uncertainty analysis, *IEEE Transactions on Magnetics*, Vol. 47, pp. 4564-4573.

Abdallh, A.; Sergeant, P.; Crevecoeur, G. & Dupré, L. (2011c). Magnetic material identification of a switched reluctance motor, *International Journal of Applied Electromagnetics and Mechanics*, Vol. 37, pp. 35-49.

Abdallh, A.; Crevecoeur, G. & Dupré, L. (2012a). A *priori* experimental design for inverse identification of magnetic material properties of an electromagnetic device using uncertainty analysis, *International Journal for Computation and Mathematics in Electrical and Electronic Engineering (COMPEL)*, Vol. 31, pp. 972-984.

Abdallh, A.; Sergeant, P. & Dupré, L. (2012b). A non-destructive methodology for estimating the magnetic material properties of an asynchronous motor, *IEEE Transactions on Magnetics*, Vol. 48, No. 4, in press.

Abdallh, A.; Crevecoeur, G. & Dupré, L. (2012c). A Bayesian approach for stochastic modeling error reduction of magnetic material identification of an electromagnetic device, *Measurement Science and Technology*, Vol. 23, pp. 035601(12pp).

Abdallh, A.; Crevecoeur, G. & Dupré, L. (2012d). Impact reduction of the uncertain geometrical parameters on magnetic material identification of an EI electromagnetic inductor using an adaptive inverse algorithm, *Journal of Magnetism and Magnetic Materials*, Vol. 324, pp. 1353-1359.

Alifanov, O.; Artyukhin, E. & Rumyantsev, S. (1995). *Extreme methods for solving ill-posed problems with applications to inverse problems*, Begell House, ISBN: 978-1567000382, Wallingford, UK.

Bandler, J.; Cheng, Q.; Dakroury, S.; Mohamed, A.; Bakr, M.; Madsen, K. & Sφndergaard, J. (2008). Space mapping: state of the art, *IEEE Transactions Microwave Theory and Techniques*, Vol. 52, pp. 337-361.

Beck, J. & Woodbury, K. (1998). Inverse problems and parameter estimation: integration of measurements and analysis, *Measurement Science and Technology*, Vol. 9, pp. 839-847.

Bertotti, G. (1988). General properties of power losses in soft ferromagnetic materials, *IEEE Transactions on Magnetics*, Vol. 24, pp. 621-630.

Bertotti, G. (1998). *Hysteresis in magnetism: for physicists, materials scientists, and engineers*, Academic Press, ISBN: 978-012093270, San Diego, California, USA.

Cale, J.; Sudhoff, S. & Turner, J. (2006). An improved magnetic characterization method for highly permeable materials, *IEEE Transactions on Magnetics*, Vol. 42, pp. 1974-1981.

Crevecoeur, G. (2009). Numerical methods for low frequency electromagnetic optimization and inverse problems using multi-level techniques, *Ph.D. Thesis, Ghent University, Belgium*.

Crevecoeur, G.; Abdallh, A.; Couckuyt, I.; Dupré, L. & Dhaene, T. (2011a). Two-level refined direct optimization scheme using intermediate surrogate models for electromagnetic optimization of a switched reluctance motor, *Engineering with Computers*, in press.

Crevecoeur, G.; Baumgarten,D.; Steinhoff, U.; Haueisen, J.; Trahms, L. & Dupré, L. (2012). Advancements in magnetic nanoparticle reconstruction using sequential activation of excitation coil arrays using magnetorelaxometry, *IEEE Transactions on Magnetics*, in press.

Durand, S.; Cimrak, I.; Sergeant, P. & Abdallh, A. (2010). Analysis of a non-destructive evaluation technique for defect characterization in magnetic materials using local magnetic measurements, *Mathematical Problems in Engineering*, article ID 574153 (19 pages).

Echeverría, D.; Lahaye, D.; Encica, L.; Lomonova, E.; Hemker, P. & Vandenput, A. (2006). Manifold-Mapping optimization applied to linear actuator design, *IEEE Transactions on Magnetics*, Vol. 42, pp. 1183-1186.

Emery, A.; Nenarokomov, A. & Fadale, T. (2000). Uncertainties in parameter estimation: The optimal experiment, *International Journal of Heat and Mass Transfer*, Vol. 43, pp. 3331-3339.

Emery, A.; Valenti, E. & Bardot, D. (2007). Using Bayesian inference for parameter estimation when the system response and experimental conditions are measured with error and some variables are considered as nuisance variables, *Measurement Science and Technology*, Vol. 18, pp. 19-29.

Engl, H.; Hanke, M. & Neubauer, A. (1996). *Regularization of Inverse Problems (Mathematics and its Applications)*, Kluwer Academic Publishers, ISBN: 978-0792361404, Nederlands.

Fadale, T.; Nenarokomov, A. & Emery, A. (1995a). Uncertainties in parameter estimation: The inverse problem, *International Journal of Heat and Mass Transfer*, Vol. 38, pp. 511-518.

Fadale, T.; Nenarokomov, A. & Emery, A. (1995b). Two approaches to optimal sensor locations, *Journal of Heat Transfer*, Vol. 117, pp. 373-379.

Fiorillo, F. (2004). *Measurement and characterization of magnetic materials*, Academic Press, ISBN: 978-0122572517, San Diego, California, USA.

Gaignaire, R.; Crevecoeur, G.; Dupré, L.; Sabariego, R.; Dular, P. & Geuzaine, C. (2010). Stochastic uncertainty quantification of the conductivity in EEG source analysis by using polynomial chaos decomposition, *IEEE Transactions on Magnetics*, Vol. 46, pp. 3457-3460.

Goodwin, G.; Zarrop, M. & Payne, R. (1974). Coupled design of test signals, sampling intervals, and filters for system identification, *IEEE Transactions on Automatic Control*, Vol. 19, pp. 748-752.

Goodwin, G. & Payne, R. (1977). *Dynamic system identification. Experiment design and data analysis*, Academic Press, ISBN: 978-0122897504, New York, USA.

International Electrotechnical Commission 'IEC' International Standard 404-2. (1996). Methods of measurement of the magnetic properties of electrical steel sheet and strip by means of an Epstein frame. Geneva, Switserland.

International Electrotechnical Commission 'IEC' International Standard 404-3. (2000). Methods of measurement of the magnetic properties of electrical steel sheet and strip by means of a single sheet tester. Geneva, Switserland.

Kaipio, J. & Somersalo, E. (2005). *Statistical and computational inverse problems (Applied Mathematical Sciences Vol. 160)*, Springer, ISBN: 978-0387220734 , New York, USA.

Lauwagie, T. (2003). Vibration-based methods for the identification of the elastic properties of layered materials , *Ph.D. Thesis, Katholieke Universiteit Leuven, Belgium.* http://www.mech.kuleuven.be/dept/resources/docs/lauwagie.pdf.

Leach, R.; Giusca, C. & Naoi, K. (2009). Development and characterization of a new instrument for the traceable measurement of areal surface texture, *Measurement Science and Technology*, Vol. 20, pp. 125102(8pp).

Marquardt, D. (1963). An algorithm for least-squares estimation of nonlinear parameters, *SIAM Journal on applied Mathematics*, Vol. 11, pp. 431-441.

National Institute of Standards and Technology 'NIST' (2000). Essentials of expressing measurement uncertainty. http://physics.nist.gov/cuu/Uncertainty/index.html.

Nissinen, A; Heikkinen, L. & Kaipio, J. (2008). The Bayesian approximation error approach for electrical impedance tomography – experimental results, *Measurement Science and Technology*, Vol. 19, pp. 015501(9pp).

Radich, B. & Buckley, K. (1995). EEG dipole localization bounds and MAP algorithms for head models with parameter uncertainties, *IEEE Transactions Biomedical Engineering*, Vol. 42, pp. 233-241.

Saitoh, S. (2007). Applications of Tikhonov regularization to inverse problems using reproducing kernels, *Journal of Physics: Conference Series*, Vol. 73, ID. 012019.

Scales, J. & Snieder, R. (2000). The anatomy of inverse problems, *Geophysics*, Vol. 65, pp. 1708-1710.

Sergeant, P.; Crevecoeur, G.; Dupré, L. & Van den Bossche, A. (2009). Characterization and optimization of a permanent magnet synchronous machine, *COMPEL: The International Journal for Computation and Mathematics in Electrical and Electronic Engineering*, Vol. 28, pp. 272-285.

Sievert, J. (2000). The measurement of magnetic properties of electrical sheet steel – survey on methods and situation of standards, *Journal of Magnetism and Magnetic Materials*, Vol. 215, pp. 647-651.

Stoica, P. & Nehorai, A. (1989). MUSIC, Maximum-likelihood, and Cramér-Rao bound, *IEEE Transactions on Acoustics, Speech and Signal Processing*, Vol. 37, pp. 720-774.

Strang, G. (1986). *Introduction to applied mathematics*, Wellesley, ISBN: 978-0961408800, Cambridge, UK.

Takahashi, N.; Morimoto, H.; Yunoki, Y. & Miyagi, D. (2008). Effect of shrink fitting and cutting on iron loss of permanent magnet motor, *Journal of Magnetism and Magnetic Materials*, Vol. 320, pp. E925-E928.

Tarantola, A. (2004). *Inverse problem theory and methods for model parameter estimation*, Society for industrial and applied mathematics (SIAM), ISBN: 978-0898715729, Philadelphia, USA.

Viana, F.; Venter, G. & Balabanov, V. (2004). An algorithm for fast optimal Latin hypercube design of experiments, *International Journal for Numerical Methods in Engineering*, Vol. 82, pp. 135-156.

Yitembe, B.; Crevecoeur, G.; Van Keer, R. & Dupré, L. (2011) Reduced conductivity dependence method for increase of dipole localization accuracy in the EEG inverse problem, *IEEE Transactions Biomedical Engineering*, Vol. 58, pp. 1430-1440.

8

The Everett Integral and Its Analytical Approximation

Jenõ Takács
Department of Engineering Science, University of Oxford, Oxford
UK

1.Introduction

The advancement in modern technology revolutionized our approach to the design of magnetic devices. The design requires concentrated effort from the designer to make the device more efficient. As a result, accurate modeling of magnetic materials in industrial application becomes a vital part of the design procedure. Amongst the number of hysteretic models for magnetic substances available for the user, the best known and commonly used is still the geometrical model suggested by Ferenc Preisach in 1935 (Preisach, 1935). Although various authors suggested a number of modifications for different applications, (Della Torre, 1999; Mayergoyz, 2003) somehow in various forms it still survives as the dominating hysteresis model of our time. The modified versions include the classical, static, dynamic, state-independent, state dependent, reversible, and vector, etc. Detailed discussion of the model can be found in the literature, for instance by Mayergoyz, Bertotti and Della Torre (Bertotti, 1998; Della Torre, 1999; Mayergoyz, 2003).

The principal assumption of this model, is that any magnetic material is a composition of basic particles called hysterons with the characteristic switching up U and switching down V magnetic field values, with a characteristic square like hysteresis loop. Hysterons are elementary hysteresis loops or hysteresis operators characterized by the switching fields.

Fig.1 Shows two hysterons displaced by H_s in the magnetic domain, schematically.

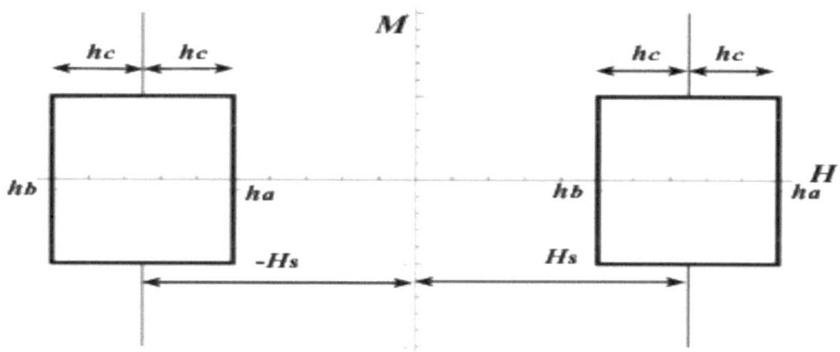

Fig. 1. Schematic representation of two elementary hysteresis operators or hysterons, displaced by H_s to the negative and the positive direction.

These hysterons, in Fig 1, populate the hysteresis loop in a statistical manner, whose distribution density is described mathematically by the Preisach function (Ivanyi, 1997; Della Torre, 1999; Mayergoyz, 2003). In the classical Preisach scalar model (CPSM) this leads to the magnetisation expression versus time in the following form , known as Everett integral (Mayergoyz, 2003).

$$M(t) = \iint_{T} P(U,V)dUdV - \iint_{-T} P(U,V)dUdV \tag{1}$$

Here, *M(t)* is the sum of the magnetisation of the hysterons and T and $-T$ are the positive and negative domains of the Preisach triangle, where hysterons contribute positively or negatively to the overall time dependent magnetisation (Kadar et al, 1989). Fig. 2 shows the Preisach triangle in the plane of the switching field.

In Preisach based models the fundamental difficulty is the determination of the *P(U,V)* distribution function, which is called in general terms as "the identification problem" (Mayergoyz, 2003; Ivanyi, 1997) determined by the set of the symmetrical first order reversal loops. To solve the problem of so called "identification", some authors suggested substituting the Preisach distribution function with suitable approximations. Mayergoyz suggested power series (Mayegoys et al., 1990), while Kadar and Della Torre used the bilinear product of Gaussian functions (Kadar & Della Torre, 1988) for the solution of the Everett integral (1).

In spite of the large variation of modified Preisach models all variants have failed to model the complicated variety of hysteresis loops, such as the spin valve and anti-spin valve configurations (see Section 9). The only successful approach so far was, to our knowledge, the hyperbolic analytical approximation, in that field.

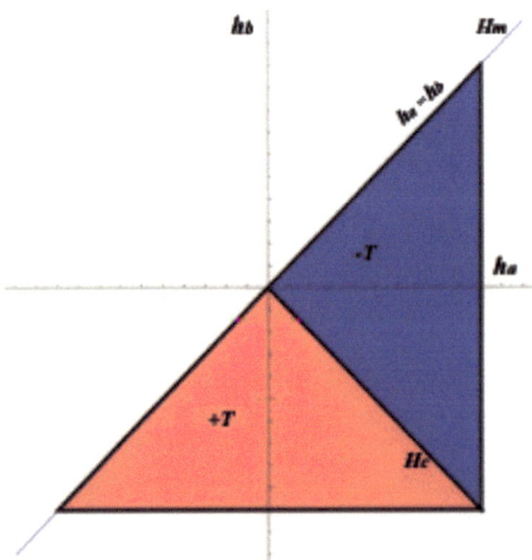

Fig. 2. The Preisach triangle in the plane of switching field.

Here h_a and h_b are the normalised switching fields (for normalization see Section 10) and *Hc* is the coercivity.

For soft materials, where the Gaussian approach failed to provide acceptable results, Finocchio and his co-workers suggested Lorentzian distribution (Finocchio et al., 2006). All approximations have their positive and negative sides and can only be used for certain type of materials.

The two latest approximations will allow analytical evaluation of the Everett integral in closed mathematical terms.

Takacs in his paper published in 2000 introduced the hyperbolic distribution (Takacs, 2000) to approximate the sigmoid- like hysteresis loops and to mathematically describe the different properties of various magnetic substances. At the time this model, based on Langevin's theory (Bertotti, 1998), was purely phenomenological and no connection was assumed to the Preisach classical approach.

The hyperbolic distribution *T(x)* can be described mathematically in its simplest form as

$$T(x) = A\ \mathbf{sech}^2\alpha(x-a) \tag{2}$$

Here A represents the amplitude, a is the shift in position of the function and α controls the steepness of the curve. A typical hyperbolic and a Gaussian distribution are shown in Fig 3. As we can see from the graphs, the hyperbolic distribution is running very close to the Gaussian line in spite of having quite different mathematical properties. The close relation between the two models (classical Preisach and hyperbolic) was only discovered recently and published here for the first time.

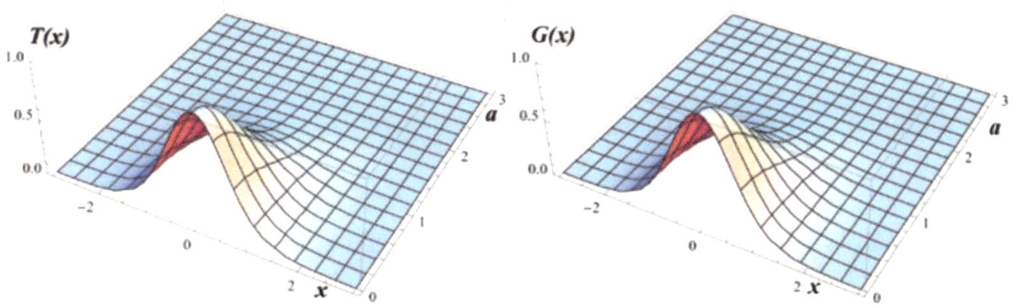

(a) (b)

Fig. 3. Distribution functions: (a) hysteretic, (b) Gaussian.

Bertotti pointed out (for details see (Bertotti, 1998)), that in most cases, the Everett integral in (1), particularly for symmetrical loops, can be replaced finally with a function of a single running variable, representing the excitation field H. In this argument we always assumed, point symmetry in the system, therefore

$$T(H-H_c) = T(H+H_c) \tag{3}$$

There is particular interest in functions, where the distribution can be formed as the product of two terms in the form of

$$T(H,H_c)=T(h_a)T(h_b)=T(H+H_c)T(H-H_c) \tag{4}$$

where h_a and h_b are the switching fields (see Fig. 2)

Using the hyperbolic distribution, shown in (2) substituted into (1), the Everett double integral will take the following form:

$$M(t)=\iint_{T} T(H,H_c)\gamma(H,H_c)dHdH_c-\iint_{-T} T(H,H_c)\gamma(H,H_c)dHdH_c \tag{5}$$

where
$$\gamma(H,H_c)=\begin{cases}+1 \text{ if } (H,H_c)\in +T\\ -1 \text{ if } (H,H_c)\in -T\end{cases}. \tag{6}$$

When
$$T(H,H_c)=A\mathbf{sech}^2[\alpha(H-H_c)]\mathbf{sech}^2[\alpha(H+H_c)] \tag{7}$$

By substituting (7) and (6) into (5) we obtain (Kadar et al., 1987 and 1988)

$$M(t)=\iint_{T} T(H,H_c)dHdH_c-\iint_{-T} T(H,-H_c)dHdH_c \tag{8}$$

A Preisach elementary operator, populating the Preisach plane with the hyperbolic character is shown in Fig. 4. We can now solve Everett's integral in (8), by using hyperbolic distribution in (2), leading to the following expressions for the ascending and descending magnetization correspondingly:

$$M_u=A\ tanh[\alpha(H-H_c)]+F(H_\mathrm{m}) \tag{9}$$

$$M_d=A\ tanh[\alpha(H+H_c)]-F(H_\mathrm{m}) \tag{10}$$

Here A represents the maximum amplitude of the magnetization, H_c is the coercivity, H_m is the maximum excitation, α is the differential permeability at $H = H_c$ or the angle of the tangent of the hysteresis loop at the point where it crosses the field axis. The F integration constant can be calculated from the condition, that at the first return point, where M_u and M_d must be equal, for all minor and major loops (Della Torre, 1999; Mayergoyz at al.,1990), therefore per definition:

$$F(H_\mathrm{m})=\frac{A}{2}[\tanh\alpha(H_m+H_c)-\tanh\alpha(H_\mathrm{m}-H_c)] \tag{11}$$

The capital letters represent the physical quantities and the corresponding lower case letters will refer to the normalized units. The normalization used in this chapter is not related necessarily to the maximum value. It is the free choice of the user; the base can be any convenient number, which helps to carry out the mathematical operations. A long detailed description of the free normalization process can be found in Section 10.

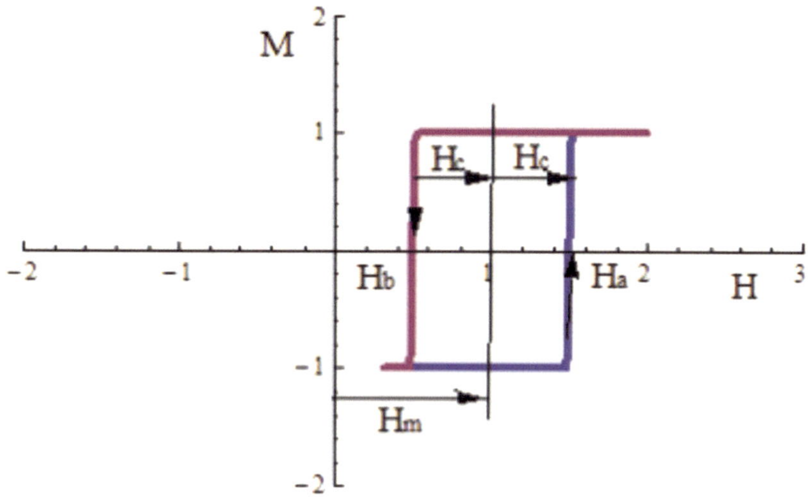

Fig. 4. Preisach elementary operator (hysteron) with hyperbolic character.

After normalization the ascending m_u and the descending m_d branches of the hysteresis loop can be written as:

$$m_u = a\ tanh[\alpha(h - h_c)] + f_0 \tag{12}$$

$$m_d = a\ tanh[\alpha(h + h_c)] - f_0 \tag{13}$$

respectively, where

$$f_0 = \frac{a}{2}((\tanh[\alpha(h_m + h_c)] - (\tanh[\alpha(h_m - h_c)]) \text{ is the normalized form of } F(H_m) \tag{14}$$

Here, all lower case letters represent the appropriate normalized quantities. In this study, where-ever possible, the normalized form is going to be used.

2. Fundamentals of the hyperbolic model

The hyperbolic model analytical approach is based on the practical assumption, that in general, there are at least, three parallel processes, including very soft irons, dominating the overall magnetization process; i.e. the reversible and irreversible domain wall movement (DWM), the reversible and irreversible domain rotation (DR) and the domain wall annihilation and nucleation (DWAN) processes (Varga at al., 2008). Although these processes are interlinked, they can be mathematically formulated separately and combined, by using Maxwell's superposition principle. They individually dominate the low, middle and near saturated region of magnetization and are supposed to have also have sigmoid shapes. This model is already used successfully in number of applications (Takacs at al, 2008; Nemcsics at al., 2011; Jedlicska at al., 2010) and some aspects are already well documented in the literature (Takacs, 2003).

When numerous processes are running simultaneously, it is difficult to describe the overall magnetization with one function, resulting from the integration of one Everett integral, but the combination of all the concurrent processes brings the model nearer to the experimental results (Takacs, 2006; 2008). The magnetization processes of magnetic materials, particularly for ultrasoft substances and the role of the domain rotation (DR) and domain wall movement (DWM) in the process has been the subject of recent experimental and theoretical studies. Fitting the experimentally obtained curves with analytical expressions of the hyperbolic model, provided a useful tool for obtaining the mixed second derivative of the curves, eliminating the deviation, between the model and the experimental data. The diagram of mixed second derivatives, as conceived by Pike (Pike et al., 1999) can be obtained for both the set of first order reversal curves (FORC diagram) and for the set of biased first magnetization curves (BFMC diagram), respectively. These FORC and BFMC diagrams of any magnetic system can be measured and modelled irrespective of their magnetic softness/hardness. It was assumed that all processes have hyperbolic (sigmoid) character, but they differ in the model parameters including the (near) reversible part as suggested by Della Torre (Della Torre, 1999; Jiles et al. 1983). All these components can be formulated mathematically as:

$$m_u = \sum_{k=1}^{n} [a_k f_{uk} + f_{0k}(h_{\mathrm{m}})] \tag{15}$$

$$m_d = \sum_{k=1}^{n} [a_k f_{dk} - f_{0k}(h_{\mathrm{m}})] \tag{16}$$

$$f_{uk,dk} = \tanh[\alpha_k (h \mp h_{ck}) \tag{17}$$

Where n is the total number of processes present and k is the running variable. Normally n is running between 1 and 3.

Fig. 5 demonstrates a typical application of the hysteresis model. Fig. 5 a. shows the major hysteresis loop of a toroid made of NO Fe-Si and Fig.5 b depicts the three constituent components. For fitting the measured loop (broken line), in Fig. 5 a, we used equations (15), (16) and (17), with the following normalised and physical parameter values.

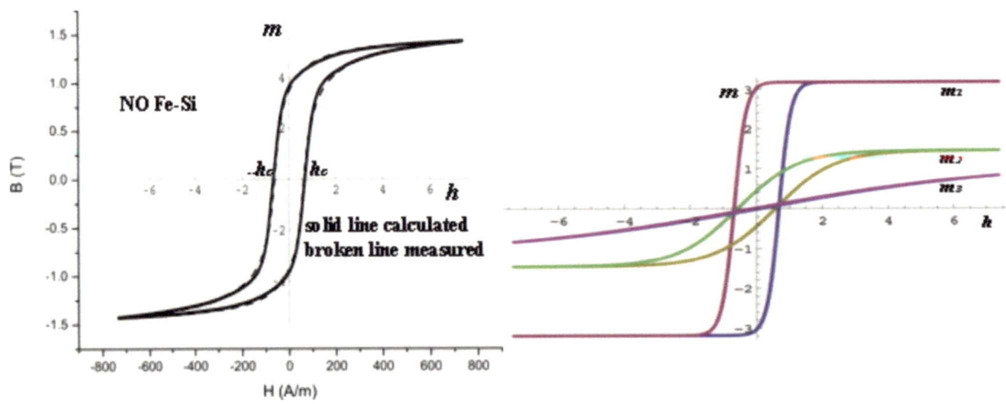

(a) (b)

Fig. 5. Measured and calculated major hysteresis loop of NO Fe-Si toroid, measured broken line, calculated solid line. The components are drawn in different colours.

$a_1 = 3.18$, $a_2 = 1.45$, $a_3 = 1.09$,
$\alpha_1 = 2.75$, $\alpha_2 = 0.56$, $\alpha_3 = 0.134$
$h_{c1} = 0.675$, $h_{c2} = 0.57$, $h_{c3} = 0.2$, $h_m = 7.35$
Corresponding to: $A_1 = 0{,}842$ T, $A_2 = 0.384$ T, $A_3 = 0.288$ T
$H_{c1} = 67.5$ A/m, $H_{c2} = 57$ A/m, $H_{c3} = 20$ A/m, $H_m = 57$ A/m

with free (arbitrary) normalization (see Section 10) of 1 h = 100 A/m for the field and 1 m = 0.53 T for the magnetization.

The normalization factors can be read from the two coordinate systems (normalized in the middle and outside one measured). For detailed explanation for the normalisation process see Section 10.

The other well-known hysteresis model, based on Lagevin function is the Jiles-Atherton model (Jiles, 1994). This model fundamentally considers, as its principle, the interaction between domains and already includes the internal demagnetisation in the form of the Weiss field (see Section 4); therefore it falls into the category of dynamic models. Its five parameters however are often difficult to associate with parameters used in practical magnetism. While in the hyperbolic model the number of parameters is the choice of the user and can vary between three and nine (depending on the accuracy required), in the J-A model this number is fixed. In complex cases the set parameters are not enough to model the phenomenon, therefore for a number of applications the model had to be modified for the specific purpose (Carpenter, 1991; Carpenter et al. 1992; Korman et al. 1994). In the iteration process, to obtain the model parameters, there is very little difference between the models of hysteresis but the hyperbolic model with its utmost simplicity provides a clear and practical method for the practical user. In the following a number of useful examples will illustrate possible applications of the model for the potential user.

3. Symmetrical and biased major and minor loops

Modeling of a regular shape hysteresis loop is based on the functions defined in (9) and (10). By using functions, mimicking the sigmoid shape of the hysteresis loop, one can formulate both phenomena of saturation and hysteresis. Changing h_m between zero and saturation field values, with a slow changing ac field we can obtain a set of minor and major loops for a given set of a, α, h_c and h_m (amplitude, differential susceptibility, coercivity, maximum magnetization) values. The substitution of these values into (15), (16) and (17) will give the solution of the Everett integral for every loop, minor or major in the set. At steady state, the first return points will sit on the theoretical or intrinsic loci, the only theoretical line, which belongs to both the ascending and the descending set of branches. It is un-hysteretic, but carries all the properties of the hysteresis loop. Due to the ever present internal demagnetization, it is an entirely theoretical concept. It cannot be realized experimentally, (internal demagnetization cannot be reduced to zero), nevertheless both f_0 and m_0 secondary functions will help in later calculations (see Part 6). By definition m_0 can be formulated as:

$$m_0 = (m_u + m_d)/2 = \frac{a}{2}(\tanh[\alpha(h+h_c)] + \tanh[\alpha(h-h_c)]) \tag{18}$$

For the definition of f_0 see (14).

Fig. 6a shows a set of symmetrical major and minor loops, calculated from (15), (16) and (17).

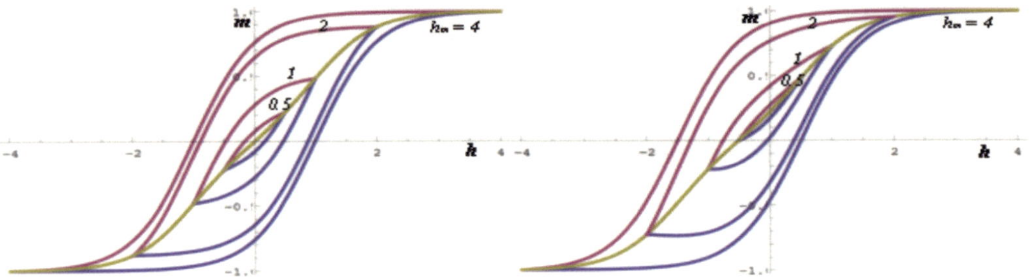

(a) (b)

Fig. 6. (a) and (b) show typical symmetric and biased major and minor loops.

In practice very often, like in power transformers, dc current is passing through the windings creating a constant dc field with the ac superimposed on it. This represents a shift in the first reverse points in the direction of the bias. The solution of the integral for the biased hysteresis loops is shown in (19), (20) and (21). For the detailed explanation the reader is referred to Ref. (Takacs, 2003).

Let us assume the dc field represents a normalize b bias on the magnetic object. This increases the effect of the magnetizing field in the following way:

$$m_u = a\ \tanh[\alpha(h - h_c + b)] + f_{00} \tag{19}$$

$$m_d = a\ \tanh[\alpha(h + h_c + b)] - f_{00} \tag{20}$$

$$m_0 = a\ \{\tanh[\alpha(h + h_c + b)] + \tanh[\alpha(h - h_c + b)]\} / 2 \tag{21}$$

where by using laws of a line going through two point in the hyperbolic domain.

$$f_{00} = f_1 \frac{m_0(-x_m) - m_0(x)}{m_0(-x_m) - m_0(x_m)} + f_2 \frac{m_0(x_m) - m_0(x)}{m_0(x_m) - m_0(-x_m)} \tag{22}$$

$$f_1 = \{tanh[\alpha(-h_m + h_c + b)] - tanh[\alpha(-h_m - h_c + b)]\} a / 2 \tag{23}$$

$$f_2 = \{tanh[\alpha(h_m + h_c + b)] - tanh[\alpha(h_m - h_c + b)]\} a / 2 \tag{24}$$

where b represents the bias applied, in normalized form.

4. Experimental loci of the first return points

The prime aim of most magnetic measurement is to find the intrinsic magnetic properties of the tested material. Due to the ever presence of the demagnetization field, (Fiorillo, 2004) a number of measuring methods have been developed to minimize its effect. The most commonly accepted way is to make the sample into a closed magnetic circuit, like a toroid or an Epstein square (Korman et al. 1994; Fiorillo, 2004). Although these two are not completely free from the presence of the internal demagnetization, they suffer the least from it. Authors went into great length to include the internal demagnetization force into the known models like Preisach, Stoner-Wohlfarth, Jiles (Preisach, 1935; Stoner et al. 1991; Jiles, 1996) etc. leading to so called, dynamic versions.

Within the saturation loop lie the loci of the vertices (first return points) of the symmetrical minor loops, which is when measured, differs from that of the theoretical intrinsic line calculated from saturated hyteresis loop data, free from internal demagnetization (i.e. Classical Preisach, hyperbolic model, etc). The relationship between the two curves can be formulated in closed mathematical form, independent of any models. To describe the internal demagnetising field, we can use the concept of the effective field, analogous to the Weiss mean field, used by other authors also (Jiles 1998; Barkhausen, 1919; Alessandro et al. 1990). Generally the H_d demagnetizing field vector is antiparallel to M vector and proportionality is assumed between H_D interaction field and M magnetization. This accepted linear approximation is

$$H_{eff} = H + N_d M \tag{25}$$

where, by using Jiles notations N_d is the internal demagnetization factor.

The internal demagnetization factor N_d usually given in tables with unity dimension. We must remember that its numerical value and dimension depends on the units used. This dimension is only unity, when both M and H_D measured in ampere per meter. When other unit convention is used, then N_d -s dimension is not unity and must be normalised like other magnetic quantities.

In normalised form with the normalised magnetization with effective field is

$$f(h_{eff}) = f(h + n_d m) \tag{26}$$

where f function is the analytical approximation of the integrated Preisach function. After substitution of (26) into the expression of the intrinsic loci we arrive to a model free expression of m_0 of the following form:

$$m_0 = f(h + n_d m_0) \tag{27}$$

The first derivative of m_0 by h of equation in (27) leads to an expression, which shows a character, similar to the feedback in an electrical circuit (Fiorillo, 2004; Jiles, et al.1986).

$$\frac{dm_0}{dh} = \frac{\dfrac{dm_0}{dh_{eff}}}{1 - n_d \dfrac{dm_0}{dh_{eff}}} \tag{28}$$

Expression (28) describes the relationship between the inherent (χ_{in}) and the effective (χ_{eff}) susceptibility (Fiorillo, 2004; Jiles, 1998). Here the inherent susceptibility is a material property, while the effective one is measured and depends on the geometry of the sample and the air gap included in the magnetic circuit, used in the measuring setup.

For most magnetic substances the value of N_d is small in the order of -10^{-5} when given in unity dimension (Jiles, 1998). Expansion of (28) into its geometric progression (Tranter, 1971), truncated at three terms ($n = 3$) will yield the following expression:

$$\frac{dm_0}{dh} = \frac{dm_0}{dh_{eff}}(1 + n_d \frac{dm_0}{dh_{eff}} + (n_d \frac{dm_0}{dh_{eff}})^2 + ...) \tag{29}$$

The first term in equation (29) depends on the intrinsic (N_d = *0,* see Part 6) material parameters only.

Following the integration of expression in (29) by h_{eff} will lead us to the measured loci m_0. (We assume n_d is small and $dh \approx dh_{eff}$). When the integration, carried out by using generalized integration by parts (Tranter, 1971) we arrive at the following expression.

$$m_0(h) \approx m_{01}(1 + n_d \frac{dm_{01}}{dh} + (n_d \frac{dm_{01}}{dh})^2 + ...) \tag{30}$$

Here $m_{01,}$ for $n_d = 0$ is representing the intrinsic or theoretical loci of the vertices (the same theoretical concept is used at the free energy calculation in Section 6). With the appropriately selected n_d in most practical cases the first two or three terms gives good enough accuracy in calculation. Fig. 7 depicts a measured, intrinsic and a calculated curve using (30). Fig.8 shows one measured minor loop and its equivalent, modelled with expression in (30).

We can conclude from (30) that the losses due to the internal demagnetization are proportional to the Preisach function in first approximation.

Based on experimental evidence, N_D (not normalised) can be approximated as

$$N_D \approx \mu_0 \frac{H_c}{B_s} \tag{31}$$

where B_s is the saturation induction.

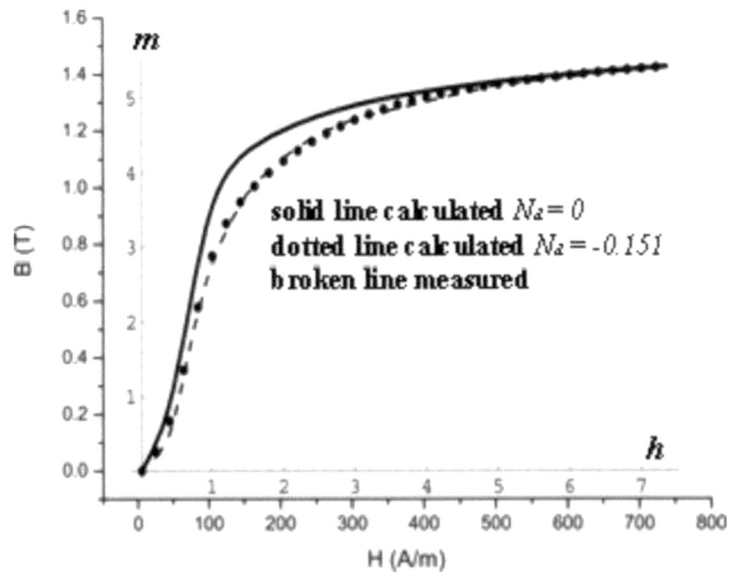

Fig. 7. Measured (broken line), modelled (dotted line) and intrinsic (solid line) loci for NO Fe-Si

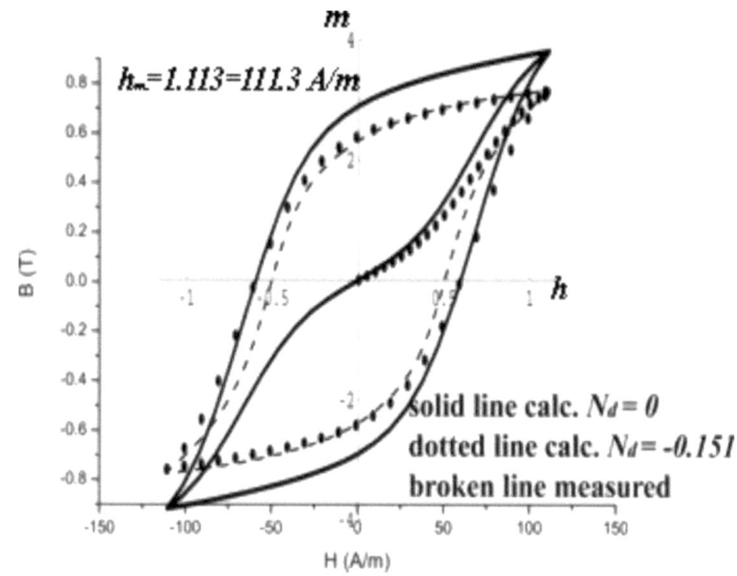

Fig. 8. Calculated NO Fe-Si minor hysteresis loop for $N_d = 0$ (solid line) and $N_d = -0.151$ (dotted line), fitted to the measured loop (broken line) with internal demagnetization.

5. Hysteresis loss and stored energy

In ac applications one of the vital properties of the magnetic material is its electrical loss. This represents the energy dissipated by the device as heat. The hysteresis loss is shown to be proportional to the area enclosed inside the hysteresis loop (Steinmetz, 1891). As an indicator, it is the measure of the loss per unit volume over one cycle of the periodic excitation (Steinmetz, 1892). While the other losses (eddy current, excess etc.) depend on the geometry of the sample, frequency, conductivity and other parameters, (Bertotti, 1998) the hysteresis loss is related primarily to the area enclosed by the hysteresis loop. We intend to formulate this enclosed area, therefore show an analytical way to calculate the hysteresis loss in magnetic substances.

The total area T (not to be taken as the Preisach triangle or Tesla) inside the hysteresis loop is represented by the difference between the integrals of the m_u and m_d functions in (12) and (13) by h, between the limits of $\pm h_m$ maximum field excitations, in the following way:

$$T = \int_{-hm}^{hm} a\,(\tanh[\alpha(h + h_c)] - f_0)dh - \int_{-hm}^{hm} a\,(\tanh[\alpha(h - h_c)] + f_0)dh \tag{32}$$

The final result of this integration is shown in equation (33) for n number of components.

$$T = \sum_{k=0}^{n} \{2\frac{a_k}{\alpha_k}[\ln\cosh\alpha_k(h_m - h_{0k}) - \ln\cosh\alpha_k(h_m + h_{0k})] - 2f_k h_m\} \tag{33}$$

It is customary after Steinmetz, to plot the losses proportional to the area as the function of the maximum magnetization. Fig.9 shows the losses versus h_m. The graph depicts also the

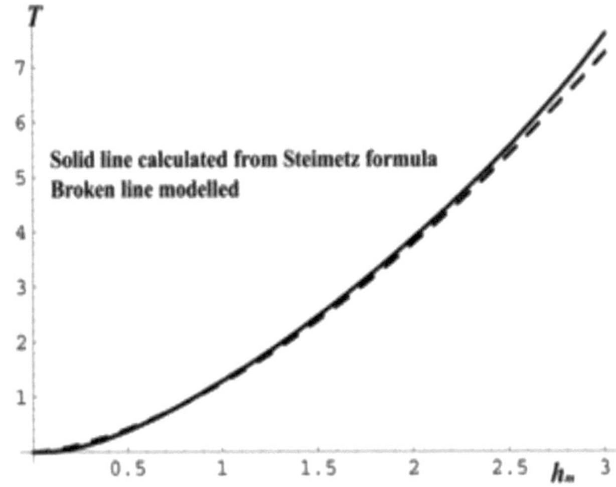

Fig. 9. The T area, (proportional to the hysteresis loss) versus h_m and its Steinmetz' approximation.

$Kh_m^{1.6}$ curve, or Steinmetz's approximation. It shows that for soft steel, with coercivity around 70 A/m, the Steinmetz approximation is within a few per cent to the theoretical value, up to near saturation for the materials constant K = *1.25*, which is very close to *1.1* value, predicted by the model.

For permanent magnet manufacturers one of the most important parameters is the maximum energy product. It is a measure of the total energy that can be produced by the magnet, or the maximum amount of work, that the magnet can do outside the volume of the magnet. When it is related to its volume, we come to the energy product density, which by nature is independent of the geometry of the magnet; therefore it is entirely a material parameter.

For a permanent magnet the integral of the field - induction product for the whole space must be zero as there is no external energy introduced to the closed system (Jiles, 1998). The system is in an equilibrium state. The total energy can be divided in to the energy inside the volume of the magnet v and the energy outside. It is self-explanatory therefore, from (34) that the two energies inside and outside of the magnet must be equal.

$$\int^{space} HB\, dv = \int^{inside} HB\, dv + \int^{outside} HB\, dv = 0 \qquad (34)$$

Outside of the magnet $B = \mu_0 H$ therefore we can write the following equation

$$\int^{inside} \mu_0 H^2\, dv = \int^{outside} HB\, dv \qquad (35)$$

We can say that the energy of the field H outside the magnet is equal to the HB product inside the magnet integrated over the whole volume of the magnet. The geometry of a magnet is normally well defined, therefore it is enough to calculate the energy stored in a unity volume called the "energy density". The energy density w stored in a medium permeated by a magnetic field can be expressed as

$$w = \int_0^B H\, dB = HB - \int_0^H B\, dH \tag{36}$$

Let us use Preisach function and substitute it into (34). This substitution will yield the following:

$$w = \int h\, [a\alpha\, \mathbf{sech}\alpha(h + h_0)]^2\, dh \tag{37}$$

A simple but representative quantity for the energy stored, often used in practice (Jiles, 1998), is the product of the magnetic induction and the magnetic field per unit volume. The larger the $(HB)_{max}$ product for a magnetic material the better are the magnetic properties of a permanent magnet. Expressing h from (13) the energy product hb (the normalised HB product) can be described as

$$hb = \mu_0 m_d h = \frac{m_d}{\alpha} \mathbf{arctanh}\, \left(\frac{\mu_0 m_d + f_0}{a}\right) - h_c m_d \tag{38}$$

According to an accepted custom, it is usually shown as a function of the magnetic induction. Fig. 10 shows a typical schematic hb curve as a function of m_d with its maximum around m_d = 0.4327 in normalized units. In the case of more than one magnetic component present in the core (usual case), the sum of the hb curves will apply.

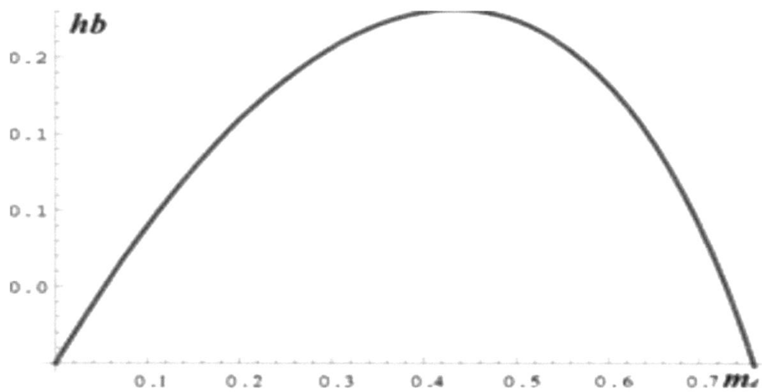

Fig. 10. Typical energy product as a function of induction.

6. Barkhausen jump and instability

Weiss postulated the domain structure for magnetic materials in 1907 but not until only 1919 was the existence of the ferromagnetic domains experimentally (Barkhausen, 1919) verified. In that year Barkhausen put a secondary coil around the specimen, under investigation and connected it to an amplifier with a loudspeaker connected to its output. By Faraday's law the induced voltage in the coil is proportional to the rate of change of the induced flux. As the magnetic field was smoothly changed up or down at a constant rate, a series of pulses was heard on the loudspeaker, indicating, that a series of jumps were taking place in the magnetization process. That was the first verification that a magnetic substance is not composed of molecular size components but larger regions, called domains. The series of clicks indicated the movement, the size change or rotation of these domains. The continuous looking magnetization curve obtained experimentally is composed of a number of small jumps, or many small discontinuous flux changes. These jumps have an unpredictable random nature in space and time. The jumps represent an instant change in the microscopic magnetic state of the material. The system abruptly leaves a higher energy state for a lower one. Although it is random, its statistical average is characteristic to the material and is correlated with the previous magnetization or demagnetization period. These clicks are called Barkhausen jumps and the phenomenon is the Barkhausen effect or Barkhausen instability.

The mathematical description of the phenomenon proved extremely difficult. After several unsuccessful attempts recently Bertotti developed a comprehensive model based on stochastic principles. Although with one assumed statistical state, a large number of real domain structures can be associated, the calculated results of this model are so far in good agreement with experimental observations. For further details and wider discussion on the Barkhausen effect, the reader is referred to the literature (Bertotti, 1998).

Until now we have concentrated on the microscopic nature of the Barkhausen jump but the question arises whether similar jumps could occur at macroscopic level.

Let us look at equation (28), which shows the formula with a character of a feedback amplifier. It is obvious that, when the Weiss coefficient is a certain value the denominator goes to zero and the equation shows, that instability sets in the circuit. Expression (28) is independently applicable to any functions; therefore it is valid for (12) and (13) as well.

By introducing Weiss effective field, as defined in (25), into (12) and (13), they become:

$$m_u = a\ tanh[\alpha(h - h_c) + \beta m_u] + f_0(h_m) \tag{39}$$

$$m_d = a\ tanh[\alpha(h + h_c) + \beta m_d] - f_0(h_m) \tag{40}$$

$$f_0(h_m) = (m_u + m_d)/2 \text{ for h = h}_\text{m} \tag{41}$$

here β is the Weiss coefficient. From the equations above m_u or m_d cannot be expressed in closed mathematical form. Instead we express h as the function of $m_{u,d}$ and we obtain the following:

$$h_u = \frac{\textbf{arctanh}(\frac{m_u - f_0(h_m)}{a}) - \beta m_u}{\alpha} + h_c \tag{42}$$

$$h_d = \frac{\textbf{arctanh}(\frac{m_d - f_0(h_m)}{a}) - \beta m_d}{\alpha} - h_c \tag{43}$$

We can parametric plot these two expressions m_u and m_d as a function of h and from this graphical solution (Takacs, 2003), by using available computer technology, we can investigate graphically the behavior of m_u and m_d due to the variation in Weiss coefficient.

At values of $\beta < 1$ The angle of intersection between the major loop and the horizontal axis is less than a right angle. At $\beta = 1$ the intersection is 90^0. At $\beta > 1$, Barkhausen jump occurs in the hysteresis curve, between the two magnetization values, which belong to the same excitation field values. Fig. 11 shows the coexistence of the states simultaneously with two free energy values. Here the system can freely move between the two energy states (see later). The system jumps into a lower energy state (Bertotti, 1998) and this energy jump results in a loop similar to a hysteresis loop but the loop intersection with the horizontal axis is governed by the magnitude of β as shown in Fig. 11. On the example the jump occurs at the values of $m = \mp 0.70613$ and $h = \pm 1.0308$ as marked on Fig. 11. Starting with a demagnetized sample, beyond the critical value of β, when the first jump occurs, the sample cannot be demagnetized by any ac signal. For detailed explanation of the macroscopic Barkhausen jump and its effect on magnetization, the reader is referred to the literature (Della Torre, 1999). The shape of the loop is representative of the energy state of the system.

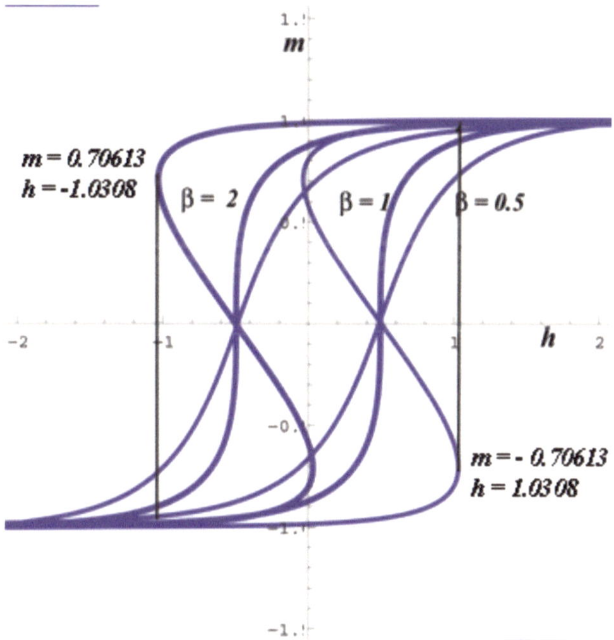

Fig. 11. Schematic hysteresis loops calculated from (42) and (43) showing the Barkhausen jump on macroscopic scales, β as parameter. Curves are for $\beta = 0.5$, 1 and 2 in arbitrary units.

In the following we are going to discuss the important relationship between system stability and the bistable system represented by the system with positive feedback.

Let us consider a system, where the m moment can freely move around under the influence of H external excitation (for the theoretical or intrinsic concept see Part 4) field. Then the energy E of m_i moment without any hindrance from the interaction between the moments can be written as:

$$E = -\mu_0 H \sum_i m_i \tag{44}$$

According to Weiss' theory however the interaction between domains introduces an additional field as we see in Section 4 equ. (25), opposing the external magnetizing field. This internal field has an additional energy contribution in the following form:

$$E_{add} = -\frac{N_D}{2}\mu_0 \sum_{i,j} m_i m_j \;=\; -\frac{N_D}{2}\mu_0 M^2 \tag{45}$$

The energy resulting from effective field (see Section 4) of a magnetic moment can be written as:

$$E_i = -\mu_0 m_i H_{eff} + \frac{N_D}{2}\mu_0 M^2 \tag{46}$$

Based on the law of statistical mechanics and resulting from averaging process, the partition function of the system defined as:

$$Q_i = \exp(-\frac{1}{k_b T}E_i(m)) + \exp(-\frac{1}{k_b T}E_i(-m)) \tag{47}$$

By substituting Equ. 46 into (47) and using known mathematical relations we come to the following formulation:

$$Q_i = 2\exp(\frac{N_D \mu_0}{2k_b T}M^2)\cosh\frac{\mu_0 m_i}{k_b T}H_{eff} \tag{48}$$

Per definition the total free energy of N number of moments in a unity volume can be written as:

$$A = -k_b \sum_i^N \ln Q_i = \sum_i^N -\frac{N_D \mu_0}{2} m_i\, M + k_b T \ln(2\cosh\frac{\mu_0 m_i}{k_b T}H_{eff}) \tag{49}$$

After summation and some simplification we come to the following form

$$A = -\frac{N_D \mu_0}{2} m_i M + k_b T N \ln(2\cosh\frac{\mu_0 m_i}{k_b T}(H + \frac{N_D}{2}M) \tag{50}$$

Dividing (50) by $k_b T_c N_D$ where T_c is the Curie temperature defined as:

$$T_c = \frac{\mu_0 m M_m N_D}{2k_b} \tag{51}$$

and M_m is the maximum magnetization and with the material constant of b_0 representative of the magnetic properties of the sample under test

$$b_0 = \frac{N_D \mu_0 M_m^2}{2k_b T_c N} \tag{52}$$

The normalized form of the Gibbs free energy for the up and down going branch of the hysteresis loop will be given in form of Equ.53 and 54 respectively

$$a_{u,free} = \ln[2\cos(h - n_d m_u)] + b_0 m_u \tag{53}$$

$$a_{d,free} = \ln[2\cos(h + n_d m_d)] - b_0 m_d \tag{54}$$

In Fig. 12 the free energy changes are shown for the up and down going branch of the hysteresis loop as the function of maximum magnetization, the parameter is the coercivity in arbitrary units.

Figs. 12 a and b clearly show the identical energy states, which belong to two different magnetization conditions, where the system can move freely between the two energy states. The system then moves to the lower energy state satisfying the laws of thermodynamics.

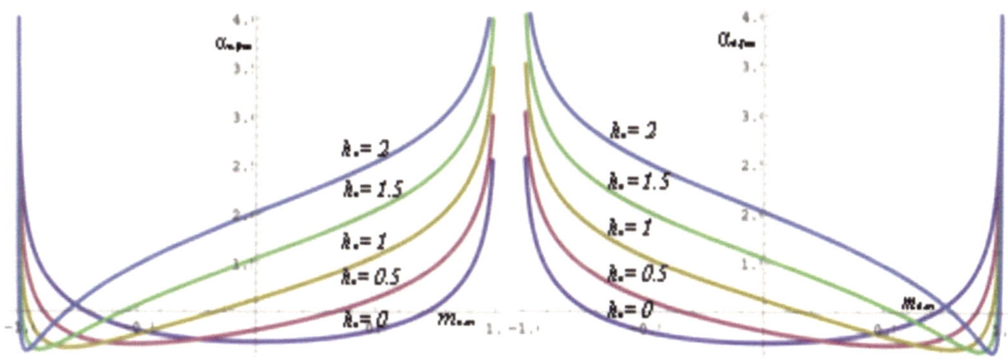

(a) (b)

Fig. 12. Gibbs free energy as the function of maximum magnetization. The parameter is the coercivity, in arbitrary units.

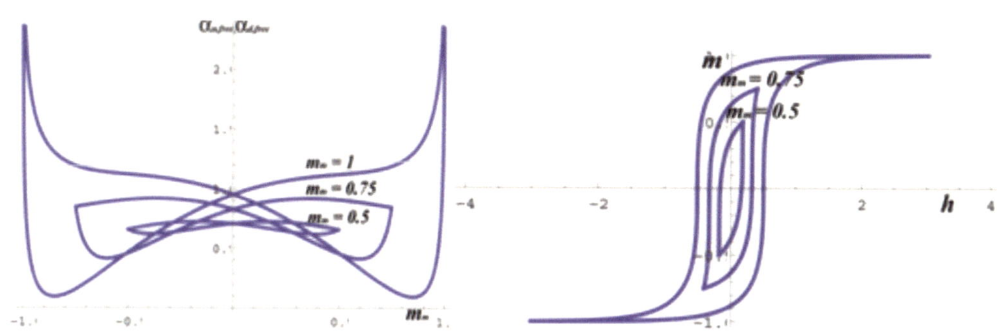

(a) (b)

Fig. 13. The free energy flow for major and minor hysteresis loops (a) Figure (b) shows the corresponding major and minor loops.

These are representing the points where the Barkhausen macroscopic instability can occur, when a positive $\beta > 1$ value creates a positive feedback.

We have to note, that during this calculation the effect of the temperature was neglected. We have assumed that the sample, under investigation, was kept around normal room temperature, well below the critical Curie point (see above). For further studies of the subject and for the impact of the temperature on the magnetization process, we refer the reader to the literature (Takacs, 2005; Bertotti, 1998). We did not feel that the temperature dependence fits into the framework of this study.

7. Eddy current loss

The hysteretic loss is by no means the only one that a magnetic substance suffers from, during magnetization. Most magnetic materials, in practical applications, (power transformers, relays etc.), are subjected to ac magnetization or the combination of both ac and dc fields. Fields, if they are not changing extremely slowly in time (quasi-static or rate independent) will alter the shape of the hysteresis loop due to the countering effect of the induced field in the magnetic medium, reducing the effectiveness of the applied external excitation. This phenomenon separates the rate independent or static hysteresis from the rate dependent hysteresis behavior. The magnetization process becomes a function of time and space. The mathematical treatment of the phenomenon is not easy, due to the fact that this phenomenon is strongly dependent on the shape of the sample and more other factors contribute to the loss. Eddy current losses depend on the conductivity of the magnetic substance and the size and shape of the magnetic circuit i.e. transformer. The analysis is further complicated by the inhomogeneity of the magnetic material used in practical applications. The theoretical treatment of these losses is difficult primarily because at no time can the sample be regarded as an infinite continuum, with homogeneous magnetic properties even when we neglect of the effect of the shape and the un-reproducible conductivity, which depends also on the previous treatment of the magnetic substance. The application of Maxwell's equations, for calculating the losses, therefore is difficult. Shockley, Williams and Kittel (Williams et al 1950; Cullity, 1972) were the first to look at this rather complex problem theoretically. The best approach to losses was made recently by Bertotti (Bertotti, 1998) on statistical principles. He introduced the concept of separation of losses and gave a theoretical verification to the dependence of the shape of the hysteresis loop on the various loss contributory factors. He expressed the separated losses in the following form.

$$W = W_h + W_{ed} + W_{ex} \tag{55}$$

Here W, the energy loss per cycle is given by the sum of W_h hysteresis, W_{ed} eddy current and the W_{ex} excess losses. Here the W_h hysteresis loss is proportional to the area enclosed by the static hysteresis loop (Steinmetz law, see section 5), the W_{ed} is the function of the first time derivative of the magnetization and W_{ex} covers all the other losses not included in the first two categories. Our intention is not to cover the whole problem of losses, only to apply a hyperbolic solution to the Everett integral to demonstrate the effect of W_{ed} on the shape of the hysteresis loop. This will give an approximate idea for the practical user on the relative magnitude of the eddy current effect. The following calculation relates to one unit cube of the magnetic sample and other effects will be disregarded. We assume that the eddy current effect is the dominating factor in the magnetic losses. All other losses are included in W_h and W_{ex} category, which are not regarded as part of this Section.

Let us use the Weiss effective field, to express the effect of the eddy current as specified in (27). We can write the magnetization as

$$\begin{aligned} m_u &= a\, tanh[\alpha(h-h_c) - \beta\frac{dm_u}{dt}] + \\ &\frac{a}{2}\{\tanh[\alpha(h_{\mathrm{m}}+h_c)+\beta\frac{dm_u(h_m)}{dt}] - \tanh[\alpha(h_{\mathrm{m}}-h_c)+\beta\frac{dm_u(h_m)}{dt}]\} \end{aligned} \tag{56}$$

$$\begin{aligned} m_d &= a\, tanh[\alpha(h+h_c) + \beta\frac{dm_d}{dt}] - \\ &\frac{a}{2}\{\tanh[\alpha(h_{\mathrm{m}}+h_c)+\beta\frac{dm_d(h_m)}{dt}] - \tanh[\alpha(h_{\mathrm{m}}-h_c)+\beta\frac{dm_d(h_m)}{dt}]\} \end{aligned} \tag{57}$$

where β represents the eddy current loss factor and the additional field is proportional to the first time derivative of the magnetization.

The time derivatives of the magnetizations after using well known functional relations are

$$\frac{dm_u}{dt} = \frac{dm_u}{dh}\frac{dh}{dt} = a\alpha(1-\frac{1}{a^2}m_u^2)\frac{dh}{dt} \tag{58}$$

$$\frac{dm_d}{dt} = \frac{dm_d}{dh}\frac{dh}{dt} = a\alpha(1-\frac{1}{a^2}m_d^2)\frac{dh}{dt} \tag{59}$$

For triangular excitation, when T (not to be mistaken to the Preisach triangle or Tesla) is the duration of one period and ω is the repetition frequency

$$\frac{dm_u}{dt} = \frac{2h_m\omega}{\pi} a\alpha(1-\frac{1}{a^2}m_u^2) \quad \text{for } -\frac{\pi}{2} < \mathrm{t} < \frac{\pi}{2} \tag{60}$$

$$\frac{dm_d}{dt} = -\frac{2h_m\omega}{\pi} a\alpha(1-\frac{1}{a^2}m_d^2) \quad \text{for } \frac{\pi}{2} > \mathrm{t} > -\frac{\pi}{2} \tag{61}$$

Substitution of (58) and (59) into (56) and (57) respectively will give us the working expressions for the magnetization process, when the effect of eddy current is included in the calculation. In (56) and (57) we left out the already negligible second order terms in the calculations.

$$\begin{aligned} m_u &= a\, tanh[\alpha(h-h_c) + \beta\frac{2h_m\omega}{\pi} a\alpha(1-\frac{1}{a^2}m_u^2)] \\ &+\frac{a}{2}\{\tanh[\alpha(h_m+h_c)+\beta a\alpha\frac{2h_m\omega}{\pi}(1-\frac{1}{a^2}m_u^2(h_m)] - \\ &-\tanh[\alpha(h_{\mathrm{m}}-h_c)+\beta a\alpha\frac{2h_m\omega}{\pi}(1-\frac{1}{a^2}m_u^2(h_m)]\} \end{aligned} \tag{62}$$

$$
\begin{aligned}
m_d &= a\,tanh[\alpha(h-h_c)+\beta a\alpha\frac{2h_m\omega}{\pi}(1-\frac{1}{a^2}m_d^2)] \\
&+\frac{a}{2}\{\tanh[\alpha(h_m+h_c)+\beta a\alpha\frac{2h_m\omega}{\pi}(1-\frac{1}{a^2}m_d^2(h_m)]- \\
&-\tanh[\alpha(h_m-h_c)+\beta a\alpha\frac{2h_m\omega}{\pi}(1-\frac{1}{a^2}m_d^2(h_m)]\}
\end{aligned}
\tag{63}
$$

Fig. 14 shows the change in the shape of the hysteresis loop due to eddy current for $\beta = 0$, 0.1, 0.2, 0.3 and 0.4 loss factor, for an arbitrary hysteresis loop in arbitrary units. The losses are always proportional with the area enclosed by the hysteresis loop (Steinmetz law). In practical case therefore, the user need to know the static (dc) and the dynamic (ac) hysteresis loop area. The difference between the two is related to the total losses due to the ac magnetization. The eddy current loss is regarded as the dominant loss factor after the hysteretic loss.

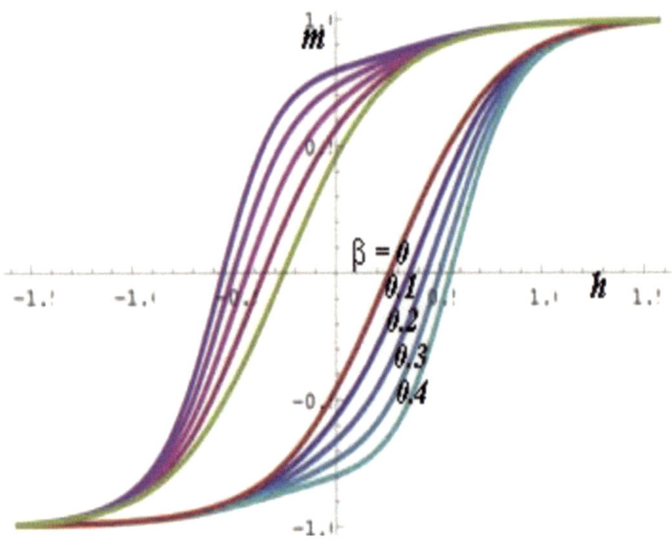

Fig. 14. Showing the effect of eddy current for β = 0, 0.1, 0.2, 0.3 and 0.4 in arbitrary units.

For sinusoidal excitation the substitution is

$$h = h_m \sin\frac{T}{2\pi}\omega t \tag{64}$$

The excess losses are taken as the square root of the time derivative of the magnetization function. Its mathematical treatment is only possible in special cases. In most cases can only be handled by computerized numerical iteration.

The increase in area of the loop is proportional to energy loss in one period due to eddy current in an infinitely large sample of the modeled material. The frequency dependency of the eddy current loss can also be calculated from (62) and (63) without taking into account the frequency dependent skin effect. The penetration depth depends primarily on the conductivity of the material under test and the frequency used.

8. Coercivity and remanent magnetism

The coercive force is the magnetic field needed to reduce the magnetization level to zero. Generally, it is referred to as coercivity and it is one of the representative values of a ferromagnetic material. Its numerical value strongly depends of the pre-history of the sample, like mechanical treatment, temperature etc. It has a strong relationship with another characteristic parameter, i.e. the remanent magnetism at maximum magnetization, called remanence. As the field gradually reduced to zero, from positive saturation, the process, following the descending branch of the hysteresis loop, the magnetization recedes to this remanent magnetization value usually marked as M_r.

When the field gradually is reduced to zero from the positively saturated state $(h = h_m)$, where $f_0 \sim 0$, the magnetization remains at a constant value, called remanence.

Fig. 15 depicts the measured and modelled remanence B_r as the function of the maximum excitation H_m for NO Fe-Si.

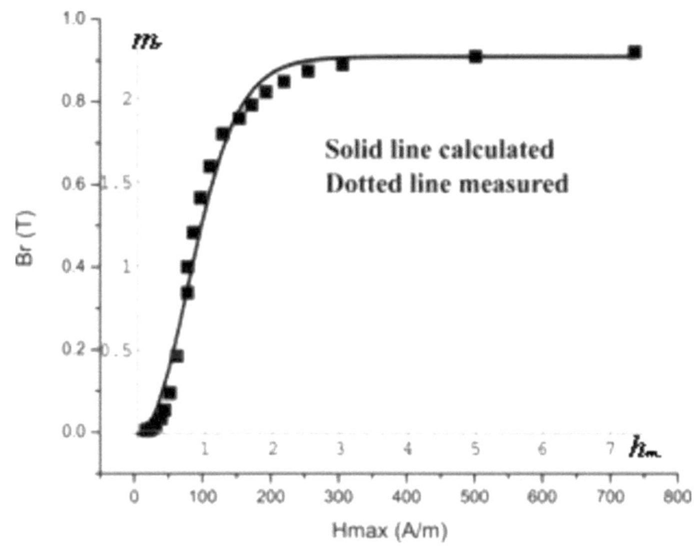

Fig. 15. Measured and calculated B_r remanence versus H_m maximum field excitation.

Take expression (13) and substitute $h = 0$. Expressing h_c we come to the mathematical relationship between coercivity (h_c) and remanence (m_r) in this form, for a single component.

$$h_c = \frac{1}{\alpha} \arctan \frac{m_r}{a} \tag{65}$$

This mathematical expression is valid for all major and minor loops.

By equating m with 0 in (16), the remanence can be calculated as the function of the maximum magnetizing field in the following form:

$$m_r = \sum_{k=1}^{n} [a_k \tanh\ \alpha_k h_{0k} - a_k \frac{\tanh\ \alpha_k h_{0k} - \tanh\ \alpha_k h_{0k} (\tanh\ \alpha_k h_m)^2}{1 - (\tanh\ \alpha_k h_{0k} \tanh\ \alpha_k h_m)^2}] \tag{66}$$

The total value of remanent magnetism is the sum of the remanence of the constituent components.

In Fig.16 a measured remanence curve for NO Fe-Si is shown with the modeled equivalent, calculated from (66).

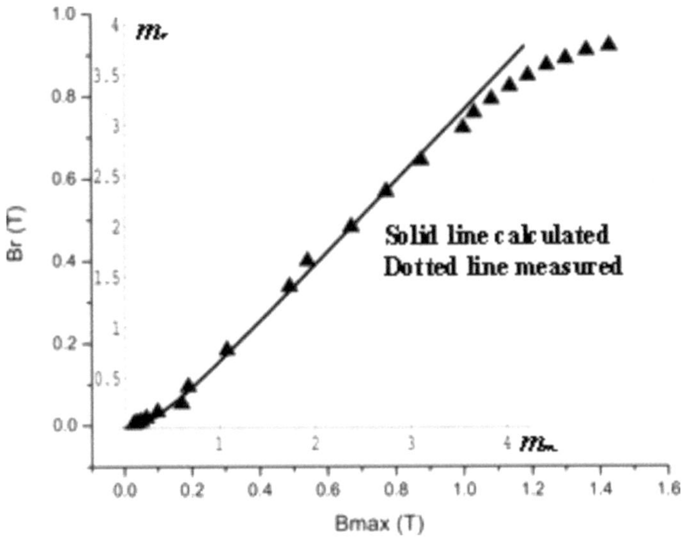

Fig. 16. Remanence versus maximum induction.

9. Spin valve and inverse spin valve character, negative coercivity

Magnetoresistive (MR) and Giant Magnetoresistive effects (GMR), the principles of spin valves, have been discovered in 1988. For many years it was only a scientific curiosity confined to research laboratories. The inverse spin valve, inverted variety of spin valve, based on similarity and not always associated with magnetism (Nemcsics et al. 2011), was discovered many years later. Today a large variety of sensors are based on these principles in vital industrial applications (Mallison, 2002).

Spin valve devices fundamentally consist of two ferromagnetic layers separated by a thin diamagnetic metal layer. One of the ferromagnetic layers, the so called reference layer is usually configured as an artificial anti-ferromagnet and pinned by exchange biasing to an anti-ferromagnetic layer. The other ferromagnetic layers (free layers) can freely change their magnetization direction under the influence of an external magnetic field (Jedlicska et al. 2010). Their $M = f(H)$ response characterized by the so called wasp-waisted loop shape. Its character show two interlinked hysteresis loops representing the switching of individual layers constituting the spin valve structure.

Fig. 17 depicts a typical characteristic spin valve, coupled double hysteresis loop. To our knowledge none of the hysteresis models can simulate any of these complicated structures. The hyperbolic model, with its flexibility and adaptability can model intricate and sophisticated structures, which are getting more popular in industrial applications. The model applications are not restricted to simple investigations of material parameters, but

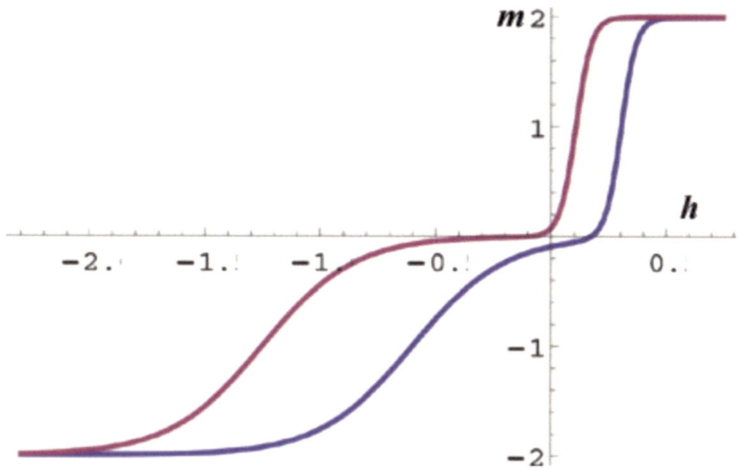

Fig. 17. Typical spin valve coupled hysteresis loops in arbitrary units.

can be applied to modeling instrument characteristic behavior and for the calculation of characteristic parameters of instruments.

Associated with the Magnetoresisrtive (MR) and Giant Magnetoresistive effects (GMR) is the phenomenon of negative coercivity in the hysteresis loop, which leads to three- looped hysteretic processes. This is due to the presence of the anti-ferromagnetic layer, forming part of the composite loop. By using (15), (16) and (17) for $n = 2$ and substituting negative numerical values for h_{c2} (anti-ferrous) we can model the overall character, where h_{cm} final coercivity changes from h_{cm} positive to $h_{cm} = 0$ and h_{cm} negative as shown in Fig. 18.

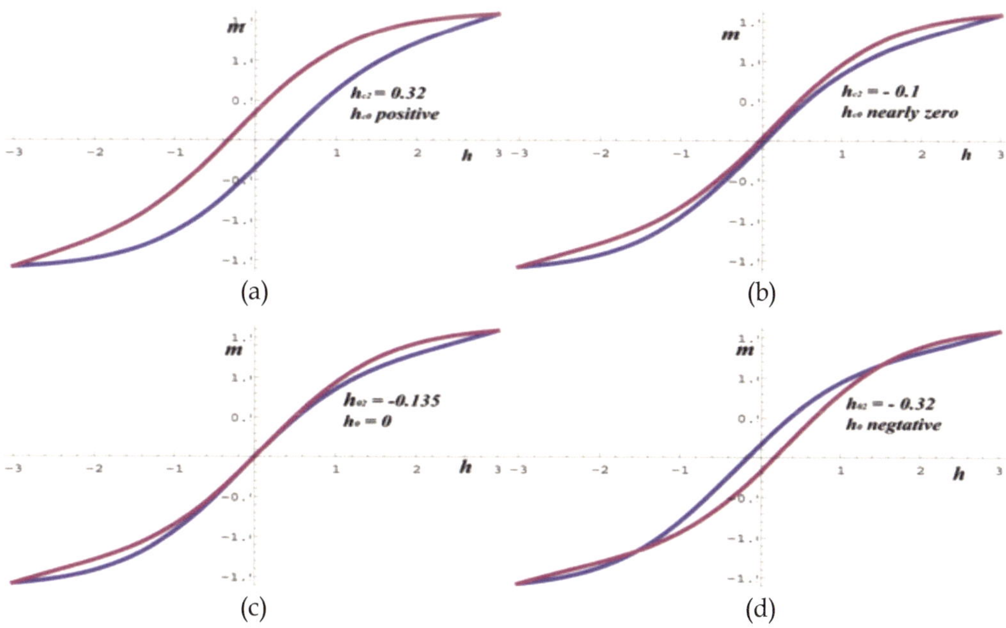

Fig. 18. h_{c0} at various values of h_{c2} a. Positive h_{20} b. Small negative h_{20} c. Medium negative h_{20} d. Larger negative h_{20}

The phenomena of hysteresis are not confined to magnetism. In other branches of the sciences like biology and semiconductor physics it is very common. Here, however we will demonstrate, what the model is capable of, not on a simple hysteresis loop, but on an inverse spin valve like loop, picked from the field of semiconductor surface physics. In Fig. 19 a RHEED specular spot intensity versus temperature diagram and its modelled curve is shown, which manifests itself in this complicated inverse spin like manner.

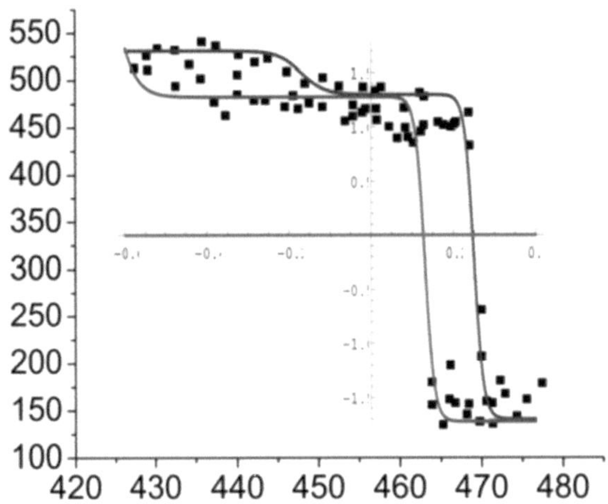

Fig. 19. An inverse spin valve character example from surface physics in arbitrary units.

10. Conclusion, process of identification, iteration, normalization and other practicalities

The hysteretic approach described here, based on Preisach model has a number of advantages. The most relevant once are:

a. The defining parameters can be the usual quantities used in practice: The maximum magnetization, coercivity, initial differential permeability (inclination). There are three in the simplest case (A, H_c, α). All tabulated in catalogues (Jiles, 1998)
b. The analytical relation between magnetization and field is a "soft" function, easy for mathematical operations (like integration). This is opposite of the Gaussian and Lorentzian approximations, which result in the error function and the inverse tangent functions, both are "hard " functions (Della Torre, 1999)
c. The hyperbolic approach does not suffer from the congruency restriction and correctly describes the minor loops between identical limits (Takacs, 2003)
d. The hysteron in the hyperbolic system retains its square like character, although it is hyperbolic in mathematical terms
e. All parameters are analytically identifiable from the major loop (Varga et al. 2008).

The hyperbolic model has its easy applicability in practical cases and has high accuracy as demonstrated on the practical examples. It has been shown, starting on biased minor loops, plus followed by hysteretic losses and other examples. The Barkhausen macroscopic jump analytical approach has been done here for the first time, to our knowledge. An extra section is devoted to the internal demagnetization, which forced researchers to modify all other simple models, including that of CPSM. Here a model-independent mathematical approach is given to resolve this problem in a form, which is applicable to all simple models.

At the start the iteration in the curve fitting exercise to model the hysteresis curve from the experimental data the reader needs a set of initial or starting parameters. This is the first step in the identification process. In the simplest case this parameter number is three, which grows to nine for the most complicated composite material. a_0 is the amplitude of the maximum measured magnetization. When, more than one process is present, then $\sum_n a_k = a_0$. The ratio between the h_m and the starting value of coercitivity h_c is the same as that of the experimental data. The relation between coercivity and remanence is set by (63). The starting value for α can easily read from the measured data. With the appropriately selected initial set of data the iteration for fitting the experimental curve is converging whether the iteration is assisted by the computer software or purely carried out by manual process.

Most researchers normalize to the maximum values of the parameters. For instance, the magnetization is divided with M_m to get unity value for the normalized maximum magnetization. This however is not always an advantage.

Normalization is to help the user with the very often complicated mathematics and numerical calculations and must not be restricted to a single value like the maximum amplitude. Often is more convenient to choose other values or use free normalization, where the normalization base is only determined after the successful iteration, giving free hand to the user. Iteration can be based on the free transform facility, available in most graphic and Photo oriented software. The iteration can be carried out by using Mathematica, Mathlab, Origin or a number of other mathematical packages and it can be computer assisted or entirely manually interactive. The measured data and the calculated results should be saved in identical formatting such as PDF, EPS, JPEG, TIFF or others. When the calculated hysteresis loop or curve is reached a certain stage in similarity, it should be compared with the measured one, which can be stored in an appropriate formatting in transparent form (see Fig.20). The modeled loop should be copied then onto the measured data and stretched over the measured loop (conform transformation), as it is depicted in Fig. 21. Then one can see how much the parameter values need to be changed to make a closer fit to the experimental data. This iteration process will be repeated several times, until the experimenter is satisfied with the fit between the modeled and the experimental data either visually or numerically. When the iteration is finished and the experimenter is satisfied with the fit, the normalization for the final parameter values can be read from the two coordinate (measured and modeled) systems with the equivalent values on the scales (see Fig 21). Often the best approach is to start with manual iteration (human mind can make shortcuts). The final, tedious fine tuning can be left to the computer for time saving. Optimization packages used for optimization processes, like Genetic algorithm (GA) (Leite et al. 2004), Direct search (DS) (Kolda et al. 2003), Particle Swarm Optimization (PSO) (Marion et al. 2008), Differential Evolution (DE) (Toman et al. 2008) and others, have already been applied in similar iterative applications.

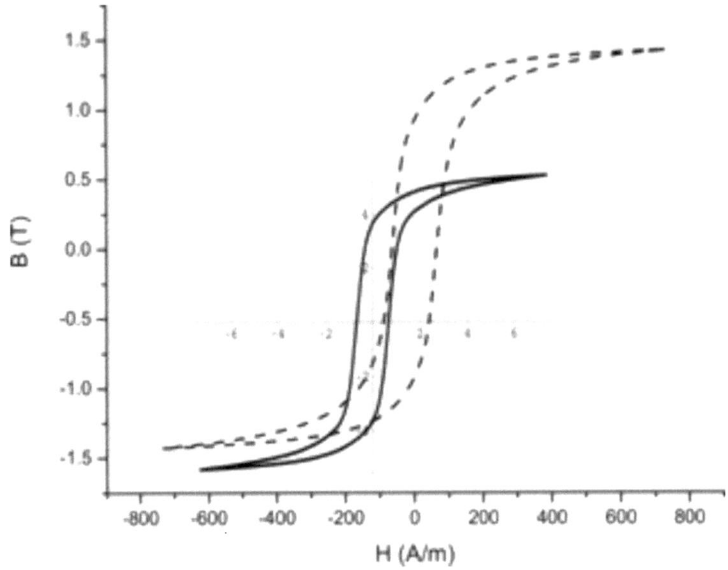

Fig. 20. Calculated (solid) loop is copied over the measured (dotted) curve in transparent mode.

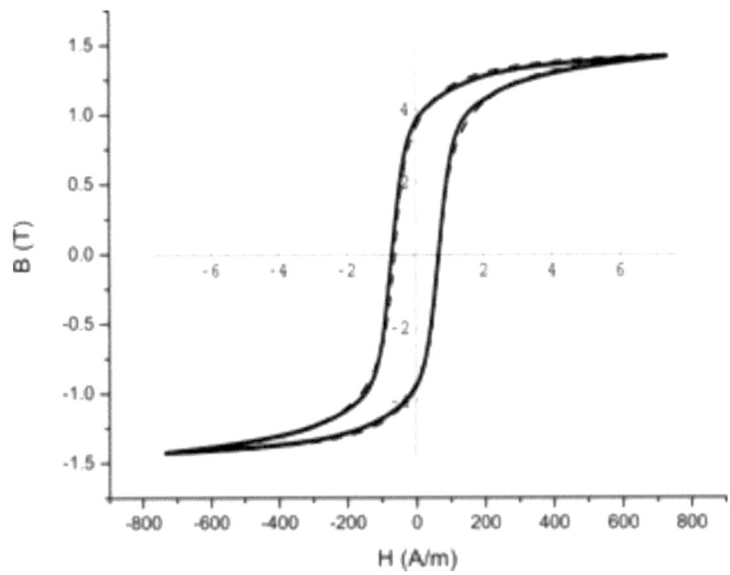

Fig. 21. The calculated loop is stretched over the experimental one

The number of calculations in the proposal is directed to the practitioner, who is interested in quick, easily achievable accurate results. Practical suggestions are included to help the potential users and encourage the model penetration into general engineering and other practices for fast and accurate modeling of large variation of magnetic materials used in industry and other fields of science.

11. Acknowledgement

I am grateful to I. Jedlicska, Gy Kovacs, L.K. Varga and R. Weiss for providing some of the experimental data for this contribution.

12. References

Alessandro B., Beatrice C. Bertotti G. and Montorsi A. (1990), pp. 2901 *J. Appl. Phys.* Vol. 68.
Barkhausen H. (1919), pp.(401) *Phyz. Z.* Vol. 29.,
Bertotti G. (1998) *Hysteresis in Magnetism*, Academic Press, London,.
Carpenter K. H. (1991) pp. 4404 *IEEE, Trans. on Magn.*, Vol. 27.
Carpenter K. H. and Warren S. (1992), pp. 2037, *IEEE Trans. on Magn.*, Vol. 28.
Cullity B. D. (1972), *Introduction to Magnetic Materials*, Addison - Wesley, Reading Mass.
Della Torre E. (1999), *Magnetic hysteresis*. IEEE Press, New York.
Finocchio G., Carpentieri M, Cardelli E. and Azzerboni B., (2006), pp. 451 *JMMM*, 300.
Fiorillo F. (2004), *Measurement and characterization of magnetic materials*, Elsevier Academic Press, Oxford.
Harison R. G. (2009), pp. 1922 *IEEE Trans. Magn.* 45., 4.
Ivanyi A. (1997), *Hysteresis Models in electromagnetic Computation*. Akademiai Kiado, Budapest.
Jiles D. C. and Atherton D. I. (1983), pp. 2183 IEEE *Trans.on Magn.*,Vol. 19.
Jiles D. C. and Atherton D. I. (1986), pp. 48 , *JMMM*, Vol 61.
Jiles D. C. IEEE Trans. on Magn. (1994), pp. 4326, Vol. 30.
Jiles D. C. (1998), *Magnetism and Magnetic Materials*, Chapman and Hall, London.
Jedlicska I., Weiss R. and Weigel R. (2008), pp. 884, *IEEE Trans. on Ind. El.*
Jedlicska, I.; Weiss, R.; Weigel, R. (2010), pp. 1728, *IEEE Trans. on Ind. Electronics* 57, 5.
Kadar Gy., (1987), pp. 4013, *J. of Appl. Phys.* 61.
Kadar Gy. and Della Torre E. (1988), pp. 3001, *J. of Applied Phys.*, 63.
Kadar Gy., Kisdi-Koszo, Kiss E. L. Potocky L., Zatroch M.and Della Torre E. (1989), pp. 3931 *IEEE Trans. on Magn.* 25.,.
Kolda T. G., Lewis M. R., and Torczon V. (2003), pp. 385, *SIAM Review*, Vol. 45., No. 3.
Korman C. E. and Mayergoyz I. D. (1994), pp. 4368, *IEEE Trans. on Magn.*, Vol. 30.
Leite J. V., Avila S. L., Batistela N. J., Carpes W. P., N. Sadowski N., Kuo-Peng P. and Bastos J. P. A. (2004), pp. 888, *IEEE Trans. on Magn.* Vol. 40., No. 2.
Mallinson J. C. (2002) *Magneto-Resistive and Spin valve Heads.*, A.P. N.Y
Marion R., Scorretti R., N. Siauve N., M. A. Raulett M. A. and L. Krahenbuhl L. (2008) pp. 894, *IEEE Trans. on Magn.* Vol. 44. No. 6.
Mayergoyz I. D., Adly A A. and Friedmam G., (1990), pp. 5373, *J. of Appl. Phys.* 67.
Mayergoyz I. D., (2003), *Mathematical Models of Hysteresis and their Applications*, Academic Press, Elsevier, New York.
Nemcsics A. and Takacs J., (2011), pp. 91, *Semiconductors* 45.
Stoner E. C. and Wohlfarth E. P., (1991), pp. 3475, *IEEE Trans. on Magn.* 27.
Pike C. R., Roberts A. P. and K. L. Verosub K. L., (1999), pp. 6660, *J. Appl. Phys.* 85.
Preisach F., (1935), pp. 227, *Phys. Z.* 94 .
Steinmetz C., (1891), pp. 261, *The Electrician*, Jan 2.
Steimetz C., (1892), pp. 3, *IEEE Transactions*, 9.
Takacs J., (2000), pp. 1002 *COMPEL*, 20, 4,.

Takacs J., (2003), *Mathematics of Hysteretic Phenomena*, Wiley-VCH, Wenheim.
Takacs J., (2005), pp. 220, *COMPEL*, 24, 1.
Takacs J., (2006), pp. 57, *Physica B.*, 372.
Takacs J. and Meszaros I., (2008), pp. 3137, *Physica B*. 403.
Takacs J., Kovacs Gy. and Varga L. K., (2008), pp. e1016, *JMMM*, 320.
Toman M., Stumberger G. and Dolinar D., (2008), pp. 1098, *IEEE Trans. on Magn.*, Vol. 44, No. 6.
Tranter C. J., (1971), *Advanced Level of Pure Mathematics*, E.U.P., London.
Varga L. K., Kovacs Gy. and Takacs J., (2008), pp. L26, *JMMM*, 320.
Varga L. K., Kovacs Gy. and Takacs J., (2008), pp. e814, *JMMM* 320.
Williams H. J., Shockley W. and Kittel C., (1950), pp. 1090, *Phys. Rev*. Vol. 80.